Adipose Tissue Protocols

Second Edition

METHODS IN MOLECULAR BIOLOGY™

John M. Walker, SERIES EDITOR

Adipose Tissue Protocols

Second Edition

Editor
Kaiping Yang

The University of Western Ontario, London, Ontario, Canada

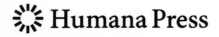 Humana Press

Editor
Kaiping Yang
The University of Western Ontario
London, Ontario
Canada

Series Editor
John M. Walker
University of Hertfordshire
Hatfield, Hertfordshire
UK

ISBN 978-1-58829-916-1   e-ISBN 978-1-59745-245-8
ISSN 1064-3745   e-ISSN 1940-6029
DOI: 10.1007/ 978-1-59745-245-8

1006531410
Library of Congress Control Number: 2008922757

Cover illustration: Fig. 2A–C, Chapter 11, "Application of DNA Microarray to the Study of Human Adipose Tissue/Cells," by Paska A. Permana, Saraswathy Nair, and Yong-Ho Lee.

Printed on acid-free paper

9 8 7 6 5 4 3 2

springer.com

Preface

In the past decade, adipose tissue has emerged as a dynamic organ, producing a wide spectrum of adipokines (e.g., hormones, growth factors, cytokines) and playing an important role in the pathogenesis of obesity and its associated metabolic disorders. Consequently, research into the biology and pathophysiology of adipose tissue has attained an unparalleled stature. Since the publication of the first edition of *Adipose Tissue Protocols*, the field has grown enormously, and the techniques and experimental approaches used have undergone radical expansion and revision, warranting a new edition of this text. The second edition combines updates of a selection of chapters from the first edition, with a large number of new contributions covering both conventional and state-of-the-art methodologies, as well as whole animal, cellular, and molecular approaches that have emerged or been extended to adipose tissue research during the past few years.

Unlike many other research areas in which the use of standard techniques and off-the-shelf kits is often sufficient and satisfactory, research involving adipose tissues and adipocytes requires considerable expertise. The methods compiled here are contributed by internationally recognized experts in their respective fields who have many years of experience not only in using these techniques, but also in troubleshooting and perfecting them. Consequently, each chapter contains a step-by-step description of the method and, more importantly, it provides invaluable tips and tricks to safeguard against potential mishaps and pitfalls. The chapters are arranged in a logical order, progressing from an overview of adipose tissue biology to detailed protocols covering all key aspects of adipose tissue research. *Adipose Tissue Protocols, Second Edition* will undoubtedly be an essential manual for newcomers to this burgeoning area, as well as a valuable addition to the laboratories and offices of more experienced researchers.

I would like to express my gratitude to all the authors who have made this book possible and thank Professor John Walker, the Series Editor, for the invitation to edit this second edition. My thanks are also due to Mr. Andrew Williams for his diligent assistance during the final stages of editing, proofreading, and collating. Last, but not least, a special thank you to Dr. Haiyan Guan for her help and support throughout the preparation of this book.

Kaiping Yang

Contents

Contributors

Danett K. Brake, BSc
Section of Leukocyte Biology, Children's Nutrition Research Center,
Department of Pediatrics and Department of Immunology, Baylor College
of Medicine, Houston, Texas

Ebba Brakenhielm, PhD
Rouen Medical University, Rouen, France

Bruce A. Bunnell, PhD
Department of Pharmacology, Center for Gene Therapy, Division of
Gene Therapy, Tulane National Primate Research Center, Covington, Louisiana

Denis Calise, MD
Zootechnical Facility, Institut Louis Bugnard, Toulouse, France

Barbara Cannon, PhD
The Wenner-Gren Institute, The Arrhenius Laboratories F3,
Stockholm University, Stockholm, Sweden

Yihai Cao, MD, PhD
Laboratory of Angiogenesis Research, Microbiology and
Tumor Biology Center, Karolinska Institutet, Stockholm, Sweden

Louis Casteilla, PhD
Institut Louis Bugnard, Toulouse, France

Jun Chen, BS
Image Reading Lab, Obesity Research Center, St. Luke's-Roosevelt
Hospital and Institute of Human Nutrition, Columbia University College
of Physicians and Surgeons, New York, New York

Xiaoli Chen, MD, PhD
Department of Food Science and Nutrition, University of Minnesota,
St. Paul, Minnesota

Saverio Cinti, MD
Institute of Normal Human Morphology, School of Medicine,
Marche Polytechnic University, Via Tronto, 10/A, 60020 Ancona, Italy

Béatrice Cousin, PhD
Institut Louis Bugnard, Toulouse, France

Bradley T. Estes, MS
Divison of Orthopaedic Surgery, Department of Surgery,
Duke University Medical Center, Durham, North Carolina

Andrea Frontini, BSc, MSc, PhD
Institute of Normal Human Morphology, School of Medicine,
Marche Polytechnic University, Via Tronto, 10/A, 60020 Ancona,
Italy

Gema Frühbeck, R Nutr, MD, PhD
Department of Endocrinology and Metabolic Research Laboratory,
Clínica Universitaria, University of Navarra, Pamplona, Spain

AnneMarie Gagnon, PhD
Departments of Medicine and of Biochemistry, Microbiology and Immunology,
University of Ottawa, Ottawa Health Research Institute, Ottawa, Ontario, Canada

Jeffrey M. Gimble, MD, PhD
Stem Cell Biology Laboratory, Pennington Biomedical Research Center,
Louisiana State University System, Baton Rouge, Louisiana

Antonio Giordano, MD, PhD
Institute of Normal Human Morphology, School of Medicine,
Marche Polytechnic University, Via Tronto, 10/A, 60020 Ancona,
Italy

Ramon Gomis, MD, PhD
Endocrinology and Nutrition Unit, Institut d'Investigacions Biomèdiques
August Pi i Sunyer (IDIBAPS) Hospital Clínic, Universitat de Barcelona,
Barcelona, Spain

Gijs H. Goossens, PhD
Department of Human Biology, Nutrition and Toxicology Research Institute
Maastricht (NUTRIM), Maastricht University, Maastricht, The Netherlands

Michel Grino, MD, PhD
Inserm U626, Université de la Mediterranée, Marseille, France

Haiyan Guan, MD
Children's Health Research Institute and Lawson Health Research Institute,
Departments of Obstetrics and Gynaecology and Physiology and Pharmacology,
The University of Western Ontario, London, Ontario, Canada

Farshid Guilak, PhD
Divison of Orthopedic Surgery, Department of Surgery,
Duke University Medical Center, Durham, North Carolina

Dorothy B. Hausman, PhD
Department of Foods and Nutrition, University of Georgia, Athens, Georgia

Gary J. Hausman, PhD
United States Department of Agriculture, Agricultural Research Service,
Athens, Georgia

Sonja Hess, Dr. rer. nat.
Proteomics and Mass Spectrometry Facility, National Institute of Diabetes
and Digestive and Kidney Diseases, National Institutes of Health,
Bethesda, Maryland

Colin R. Jefcoate, PhD
Department of Pharmacology, University of Wisconsin Medical School,
Madison, Wisconsin

Xian-Cheng Jiang, PhD
Department of Anatomy and Cell Biology, SUNY Downstate Medical Center,
Brooklyn, New York

Cristiana E. Juge-Aubry, MSc
University of Geneva Medical School, Geneva, Switzerland

Konstantin V. Kandror, PhD
Boston University School of Medicine, Boston, Massachusetts

Fredrik Karpe, MD, PhD
Oxford Centre for Diabetes, Endocrinology and Metabolism,
Nuffield Department of Clinical Medicine, University of Oxford,
Oxford, United Kingdom

Yong-Ho Lee, PhD
Division of Endocrinology, Diabetes and Metabolism, Department of Medicine,
University of Vermont College of Medicine, Burlington, Vermont

Xueqing Liu, PhD
Department of Pharmacology, University of Wisconsin Medical School,
Madison, Wisconsin

Constantine Londos, DDS, PhD
Laboratory of Cellular and Developmental Biology, National Institute
of Diabetes and Digestive and Kidney Diseases, National Institutes
of Health, Bethesda, Maryland

Daniela Malide, MD, PhD
Light Microscopy Core Facility, National Heart, Lung and Blood Institute
National Institutes of Health, Bethesda, Maryland

Ismo Mattila, MSc
VTT Technical Research Centre of Finland, Tietotie 2, FI-02044 VTT,
Espoo, Finland

Christoph A. Meier, MD
University of Geneva Medical School, Geneva, Switzerland; and Department
of Internal Medicine, Triemli Hospital, Zürich, Switzerland

Melina M. Musri, PhD
Endocrinology and Nutrition Unit, Institut d'Investigacions Biomèdiques
August Pi i Sunyer (IDIBAPS) Hospital Clínic, Universitat de Barcelona,
Barcelona, Spain

Saraswathy Nair, PhD
Department of Biological Sciences, University of Texas at
Brownsville and Texas Southmost College, Brownsville, Texas

Jan Nedergaard, PhD
The Wenner-Gren Institute, The Arrhenius Laboratories F3,
Stockholm University, Stockholm, Sweden

Matej Orešič, PhD
VTT Technical Research Centre of Finland, Tietotie 2, FI-02044 VTT,
Espoo, Finland

Hea Jin Park, PhD
Department of Animal and Dairy Science, University of Georgia,
Athens, Georgia

Marcelina Párrizas, PhD
Endocrinology and Nutrition Unit, Institut d'Investigacions Biomèdiques
August Pi i Sunyer (IDIBAPS) and Biochemistry and Molecular Genetics
Department, Hospital Clínic, Universitat de Barcelona, Barcelona, Spain

Luc Pénicaud, PhD
Institut Louis Bugnard, Toulouse, France

Paska A. Permana, PhD
Carl T. Hayden Veterans Affairs Medical Center, Phoenix, Arizona

Tuulikki Seppänen-Laakso, PhD
VTT Technical Research Centre of Finland, Tietotie 2, FI-02044 VTT,
Espoo, Finland

Wei Shen, MD
Obesity Research Center, St. Luke's-Roosevelt Hospital and Institute
of Human Nutrition, Columbia University College of Physicians
and Surgeons, New York, New York

Jun Shi, PhD
Boston University School of Medicine, Boston, Massachusetts

Shigeki Shimba, PhD
Department of Health Science, College of Pharmacy,
Nihon University, Tokyo, Japan

C. Wayne Smith, MD
Section of Leukocyte Biology, Children's Nutrition Research Center,
Department of Pediatrics, and Department of Immunology, Baylor College
of Medicine, Houston, Texas

Alexander Sorisky, MD
Departments of Medicine and of Biochemistry, Microbiology and Immunology,
University of Ottawa, Ottawa Health Research Institute, Ottawa, Ontario, Canada

Tapani Suortti, PhD
VTT Technical Research Centre of Finland, Tietotie 2, FI-02044 VTT,
Espoo, Finland

Sébastien Thalmann, MD
University of Geneva Medical School, Geneva, Switzerland

Srikant Viswanadha, DVM, PhD
Laboratory of Cellular and Developmental Biology, National Institute
of Diabetes and Digestive and Kidney Diseases, National Institutes
of Health, Bethesda, Maryland

Suqing Wang, PhD
Department of Pharmacology, University of Wisconsin Medical School,
Madison, Wisconsin

Kaiping Yang, PhD
Departments of Obstetrics and Gynaecology and Physiology and Pharmacology,
Children's Health Research Institute and Lawson Health Research Institute,
The University of Western Ontario, London, Ontario, Canada

Chapter 1
Overview of Adipose Tissue and Its Role in Obesity and Metabolic Disorders

Gema Frühbeck

Summary As the result of its apparent structural and histological simplicity, adipose tissue (AT) functions initially were limited to energy storage, insulation, and thermoregulation. Only decades later was the extraordinarily dynamic role of AT recognized, revealing its participation in a broad range of physiological processes, including reproduction, apoptosis, inflammation, angiogenesis, blood pressure, atherogenesis, coagulation, fibrinolysis, immunity and vascular homeostasis with either direct or indirect implications in the regulation of proliferation. The functional pleiotropism of AT relies on its ability to synthesize and, in some cases, secrete a large number of enzymes, hormones, growth factors, cytokines, complement factors, and matrix and membrane proteins, collectively termed adipokines. At the same time, white AT expresses receptors for most of these factors, warranting a wide cross-talk at both local and systemic levels in response to metabolic changes or other external stimuli. In this chapter, mounting evidence on the specific characteristics of AT from different depots is outlined in relation to fat distribution and comorbidity development. The current knowledge in this field is reviewed with a broad perspective ranging from classification, structure, and distribution to the key functional roles of AT with a particular focus on the role of adipokines and their involvement in the metabolic disorders accompanying obesity.

Key words Adipocytes; cytokines; adipokines, fat distribution; energy homeostasis; glucose and lipid metabolism; obesity; comorbidities.

1 Introduction

Storing and releasing lipids for oxidation by skeletal muscle and other tissues became so firmly established decades ago as the unique and simple role of white adipose tissue (WAT) that a persistent lack of interest hindered the study of its truely pleiotropic nature. With the identification of the extraordinarily dynamic behavior of adipocytes and the recognition of obesity as one of the major public

From: *Methods in Molecular Biology, Vol. 456:*
Adipose Tissue Protocols, Second Edition. Edited by: Kaiping Yang
© Humana Press, a part of Springer Science + Business Media, Totowa, NJ

health problems in modern society, unraveling the neuroendocrine systems that regulate energy homeostasis and adiposity has become a challenging research priority (1–6). Interestingly, obesity is not defined as an excess of body weight, but as an increased adipose tissue (AT) accretion to the extent that health may be adversely affected. WAT has become, therefore, the research focus of biomedical scientists for both pathophysiological and molecular reasons. Beyond the primary role of adipocytes in storing excess energy in the form of fat, it is now widely recognized that WAT lies at the heart of a complex network participating in the regulation of a variety of quite diverse biological functions (7–29).

2 Development

In homeotherms, WAT develops extensively with the proportion to body weight, varying greatly between species. Adipocytes differentiate from stellate or fusiform precursor cells of mesenchymal origin. Two main processes of adipose tissue formation can be distinguished (30). In humans, the primary fat formation takes place relatively early in the prenatal period around the 14th to 16th week, with gland-like aggregations of epitheloid precursor cells, called lipoblasts, or preadipocytes being formed in specific locations. Subsequently, a secondary fat formation takes place after the 23rd week of gestation as well as in the early postnatal period. In this case, the differentiation of fusiform precursor cells, which accumulate lipid to ultimately · coalesce into a single large fat droplet in each cell, leads to the dissemination of fat depots formed by unilocular white adipocytes in many areas of connective tissue. AT is partitioned by connective tissue septa into lobules.

Although the number of fat lobules remains constant, subsequent developmental changes are accompanied by a continous growth in the lobules' size. At the sites of early fat development, the adipocytes' multilocular morphology is predominant. The second trimester represents a critical period for the development of obesity in later life. Although relatively small, adipocytes are already present in the main fat depots at the beginning of the third trimester. A tight, temporo-spatial coordination of angiogenesis with fat cell cluster formation takes place during embryonic development. At birth, WAT reportedly accounts for approximately 16% of total body weight (with brown adipose tissue [BAT] constituting 2–5%). During the first year of life, an increase in total body fat of approx. 0.7 kg to 2.8 kg is observed.

The development of adipose tissue, called adipogenesis, varies according to sex and age. The adipose lineage originates from multipotent mesenchymal stem cells that develop into adipoblasts. Commitment of these adipoblasts gives rise to preadipose cells, also called preadipocytes, which are cells that have expressed early markers but have yet to express late markers or accumulate triacylglycerol stores (31–34). Multipotent stem cells and adipoblasts, which are found during embryonic development, are still present postnatally.

The existence of especially sensitive periods for changes in adipose tissue cellularity throughout life is well established. In particular, two peaks of accelerated adipose mass enlargement have been established, namely after birth and between 9 years and

13 years. Cell proliferation and differentiation are highest during the first year of life, being less pronounced in the years before puberty. Subsequently, the rate of cell proliferation slows down during adolescence and, in weight-stable individuals, remains fairly constant throughout adulthood. Under circumstances of maintained positive energy balance, adipose mass expansion takes place initially by an enlargement of the existing fat cells (hypertrophia). The perpetuation of this situation ends up in severe obesity, with an increase in the total fat cell number (hyperplasia). Therefore, excess energy storage starts initially as a hypertrophic process resulting from the accumulation of excess lipid in a normal number of unilocular adipose cells, which can reach four times their normal size. If the positive energy balance is prolonged over time, AT progresses by a hyperplasic or hypercellular expansion. Interestingly, the identification of the occurrence of apoptosis in WAT has changed the traditional belief that acquisition of fat cells is irreversible. Although childhood-onset obesity is characterized by a combination of fat cell hyperplasia and hypertrophy, in adult-onset obesity, hypertrophic growth predominates. In any case, the hyperplasic growth of adipocytes in adults does not happen until the existing fat cells reach a critical size. Noteworthy, new adipocyte formation has been reported in humans with WAT containing a significant proportion of cells with the ability of undergoing differentiation *(33,34)*.

In the course of adipogenesis, the morphological and functional changes taking place represent a shift in transcription factor expression and activity leading from a primitive, multipotent state to a final phenotype characterized by alterations in cell shape and lipid accumulation *(35)*. A number of redundant signaling pathways and transcription factors directly influence fat cell development by converging in the upregulation of peroxisome proliferator-activated receptor γ (PPARγ), which embodies a common and essential regulator of adipogenesis as well as of adipocyte hypertrophy. Among the broad array of transcription factors, CCAATT enhancer-binding proteins and the basic helix-loop-helix family (ADD1/SREBP-1c) also stand out in relation to their link with the existing nutritional conditions *(32,34)*. The transcriptional repression of adipogenesis includes both active and passive mechanisms, with the former directly interfering with the transcriptional machinery and the latter based on the binding of negative regulators to yield inactive forms of known activators. Interestingly, the simultaneous presence, at specific threshold concentrations, of some adipogenic factors may be a neccessary requirement to trigger terminal differentiation. In addition, the responsiveness of adipocytes to neurohumoral signals may vary according to peculiarities in the adipose lineage stage at the moment of exposure, with nutrients, hormones, cytokines, and growth factors influencing the dynamic changes related to adipose tissue mass as well as its pattern of distribution during this period and later in life.

3 Structure

Adipose tissue represents a special loose connective tissue containing lipid-laden adipocytes. Within the tissue, fat cells are individually held in place by delicate reticular fibers clustering in lobules bounded by fibrous septa surrounded by a rich capillary and innervation network. Adipocytes may comprise approx. 35% to 70% of adipose mass in adults, accounting only for approx. 25% of the total cell population. In fact, adipose tissue itself is composed of not only adipocytes but also by other cell types found in the stroma-vascular fraction, which include macrophages, fibroblasts, pericytes, blood cells, endothelial cells, and adipose precursor cells, among others (*see* **Fig. 1.1**). These diverse cell types account for the remaining 75% of the total cell population. This multicellularity represents a wide range of targets for an extensive autocrine-paracrine crosstalk (*see* **Figs. 1.2** and **1.3**).

The characteristic spherical form of adipocytes varies enormously in size (ranging from $20\,\mu m$ to $200\,\mu m$ in diameter, with variable volumes comprising from a few picoliters to approx. $3\,nL$). The coalescent lipid droplets contain a mixture of neutral fats, triglycerides, fatty acids, phospholipids, and cholesterol. Approximately 95% of the total lipid content stored is represented by triacylglycerols (constituted mainly by oleic and palmitic acids) and to a smaller degree by diacylglycerols, phopholipids, unesterified fatty acids, and cholesterol. Isotopic tracer studies have clearly shown that lipids are continuously being mobilized and renewed, even in individuals in energy

Fig. 1.1 Schematic representation of the multicellularity of adipose tissue with approximate quantitative contribution of adipocytes compared to the stromavascular fraction

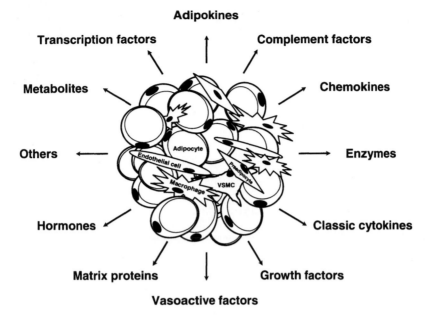

Fig. 1.2 Diagram of the multiple adipose-derived factors

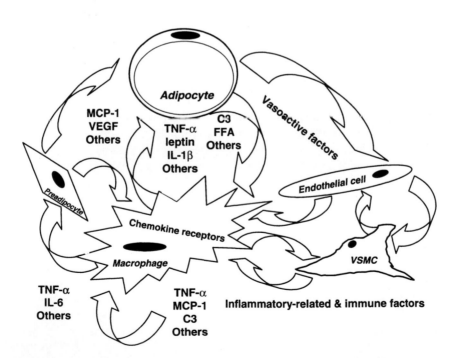

Fig. 1.3 Autocrine-paracrine crosstalk underlying the complex adiposity-inflammation-immune relation

balance. Fatty acid esterification and triglyceride hydrolysis take place continuously. The half-life of the depot lipids in rodents is about 8 days, meaning that almost 10% of the fatty acids stored in adipose tissue are replaced daily. A thin interface membrane separates the lipid droplet from the cytoplasmic matrix. Adipocytes are capable of changing their diameter 20-fold and their volumes by several thousand-fold to accommodate the increased lipid load. However, the enlargement of the fat cell size is not indefinite. The formation of new adipocytes from the precursor pool takes place once a maximum capacity is attained, which in humans averages 1000 pL. Given that about 90% of the cell volume is a lipid droplet, the nucleus becomes a flattened semilunar structure pushed against the edge of the cell with the thin cytoplasmic rim also being pressed to the periphery of the adipocyte. Mature white fat cells contain a single large lipid droplet and are described as unilocular. However, developing white adipocytes are transiently multilocular, containing multiple lipid droplets before these finally coalesce into a single large drop. In young fat cells the nucleus is round or oval, whereas in mature adipocytes it turns into a cup-shaped and peripherally displaced structure, accounting for only one fortieth of the cell volume. The cytoplasm is stretched to form a narrow sheath around the lipid drop. A thin external lamina called basal lamina surrounds the cell. A few mitochondria with loosely arranged membranous cristae can be observed around the nucleus. The cytoplasm has a small Golgi zone and is filled with free ribosomes but contains only a limited number of short profiles of the granular endoplasmic reticulum with occasional lysosomes.

4   Classification

In overweight and obese individuals, WAT can easily represent the largest endocrine organ of the whole organism. The anatomical distribution of single, concrete fat pads dispersed throughout the whole body and not connected to each other collides with a classic organ-specific localization. Despite representing one of the largest body compartments, a classification defining specific AT depots based on their anatomical location and physiological functionality has proven difficult (36). Therefore, the lack of an accepted taxonomy has, for many years, meant a limitation for researchers investigating AT topography and its functional correlates. In addition, the mounting evidence of the differences in the molecular biology of adipose tissue has rendered some classification systems obsolete.

Although many cell types contain small reserves of carbohydrate and lipid, AT is the body's most capacious energy reservoir. Because of the high energy content per unit weight of fat as well as its hydrophobicity, the storage of energy in the form of triglycerides is a highly efficient biochemical phenomenon (1 g of adipose tissue contains approx. 800 mg of triacylglycerol and only approx. 100 mg of water). It quantitatively represents the most variable component of the organism, varying from a few percent of body weight in elite athletes to more than half of the total body weight in morbidly obese patients. The normal range is about 10% to 20% body fat for males and around 20% to 30% for females, accounting for, approximately, a 2-month energy reserve. During pregnancy, most species accrue additional reserves of AT to help support the development of the fetus and to further facilitate the lactation period.

Total AT represents the sum of all fat depots, usually excluding the bone marrow as well as the small amounts of fat contained in the hands, feet, and head. The total fat mass is subcategorized into subcutaneous and internal adipose tissue *(36)*. The former comprises the fat layer found between the dermis and the aponeuroses and fasciae of the muscles. It can be further subdivided into superficial and deep subcutaneous AT. The superficial subcutaneous AT depot comprises the layer contained between the skin and a fascial plane in the lower trunk and gluteal-thigh area, whereas the deep subcutaneous compartment encompasses the layer found between the fascial plane in the lower trunk and gluteal-thigh areas and the muscle fascia. The internal AT compartment groups all the visceral and nonvisceral fat together with other rare or quantitatively very small depots. On the basis of recently acquired knowledge, AT can be classified according to topographical location of the depot and functional characteristics of the fat cells as summarized in **Table 1.1**.

Table 1.1 Main adipose tissue depots

Distribution and specific anatomic location
Subcutaneous: approx. 80% (superficial and deep adipose tissue layers)
Truncal
Cervical
Dorsal
Lumbar
Abdominal
Gluteofemoral
Mammary
Internal: approx. 20% of total body fat
Visceral
Intrathoracic (extraintrapericardial)
Intraabdominopelvic
Intraperitoneal
Omental (greater and lesser omentum)
Mesenteric (epiplon, small intestine, colon, rectum)
Umbilical
Extraperitoneal
Intraabdominal
Preperitoneal
Retroperitoneal (peripancreatic, periaortic, pararenal, perirenal -infiltrated with brown adipocytes)
Intrapelvic
Gonadal (parametrial, retrouterine, retropubic)
Urogenital (paravesical, para-retrorectal)
Nonvisceral
Intramuscular
Perimuscular
Intermuscular
Paraosseal
Other nonvisceral adipose tissue (orbital adipose tissue)
Ectopic (steatosis, intramyocardial, lipoma, lypodistrophy, etc.)

According to their diverse biochemical and functional characteristics, AT has been classically subclassified as WAT and BAT. Brown adipocytes are specialized in transferring energy from food into heat, thus playing an important role in body thermogenesis (37). The relation between brown and white fat during development is still being disentangled. In the newborn, BAT is well developed in the neck and interscapular region. BAT has a limited distribution in childhood and is present only to a small degree in adult humans, while exhibiting significant amounts in rodents and hibernating animals. Under normal physiological conditions, it is now assumed that adult humans exhibit low amounts of functional BAT.

In other larger mammals, the functional capacity of BAT decreases because of the relatively higher ratio between heat production from basal metabolism and the smaller surface area encountered in adult animals. In addition, clothing and indoor life have reduced the need for adaptive nonshivering thermogenesis. However, it has been recently shown that, especially in some depots, human WAT can be infiltrated with brown adipocytes expressing uncoupling protein-1 (UCP-1). Brown adipocytes can be detected among all white fat depots in variable amounts depending on species, localization, and environmental temperature. The transformation of characteristic brown adipocytes into white fat cells can take place rapidly in numerous species and depots during postnatal development.

The brown color of BAT is derived from a rich vascular network and abundant mitochondria and lysosomes. The individual multilocular adipocytes exhibit multiple small lipid droplets, which have not coalesced to a single drop as is the case in white fat cells. The spherical nuclei are centrally or eccentrically located within the cell. Compared with the unilocular white adipocytes, the cytoplasm of the multilocular brown fat cell is relatively abundant containing numerous mitochondria, which are involved in the oxidation of the stored lipid. Because of the fact that BAT exhibits a reduced potential to conduct oxidative phosphorylation, the energy produced in brown adipocytes is released in the form of heat in relation to the uncoupling activity of UCPs and is not captured in adenosine triphosphate. For this functional reason, BAT is extremely well vascularized so that the blood is warmed when it passes through the active tissue, helping to dissipate the energy produced.

5   Distribution and Depot-Specific Differences

WAT exhibits regional differences, as summarized previously (38–40). Unilocular fat is widely distributed in subcutaneous AT but exhibits quantitative regional characterisics that are influenced by age and sex. The sexual dimorphism is responsible for the particular body form of males and females, termed android and gynoid fat distribution. In females, subcutaneous fat is most abundant in the gluteofemoral regions as well as the breasts, whereas in males, the main subcutaneous depots include the nape of the neck, the area over the deltoid and triceps muscles, and the lumbosacral region. Additionally, extensive fat depots are found in the omentum, mesenteries, and the retroperitoneal area of both sexes. In well-nourished, sedentary

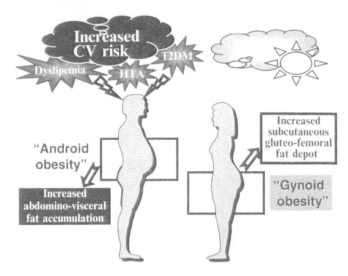

Fig. 1.4 Depot- and sex-specific adipose tissue accumulation and their differential impact on the development of the main comorbidities

individuals, the fat distribution persists and becomes more obvious with advancing age, with men tending to accumulate more fat in the visceral compartment. Depot-specific differences may be related not only to the metabolism of fat cells but also to their capacity to form new adipocytes. Moreover, regional differences also result from variations in molecular characteristics, such as hormone receptor distribution, adipokine secretory profile, and expression pattern, as well as from specific local environmental peculiarities in innervation and vascularization *(38–42)*.

Regional distribution of body fat is known to be an important indicator of metabolic and cardiovascular alterations in some individuals. The seminal observations that the topographic distribution of AT was considered relevant to understanding the relation of obesity with the development of glucose and lipid alterations were performed in the 1950s. Since then, numerous prospective studies have revealed that android or male-type obesity correlates more often with an elevated mortality and risk for the development of type 2 diabetes mellitus, dyslipidemia, hypertension, and atherosclerosis than gynoid or female-type obesity (*see* **Fig. 1.4**). The clustering of obesity-associated conditions is more evident in relation to increased central, intra-abdominal adiposity, which, at the same time, exerts a deleterious impact on both life expectancy and quality of life.

6   Functions and Pathophysiological Links

WAT is actively involved in cell function regulation through a complex network of endocrine, paracrine, and autocrine signals that influence the response of many tissues, including the hypothalamus, pancreas, liver, skeletal muscle, kidneys, endothelium,

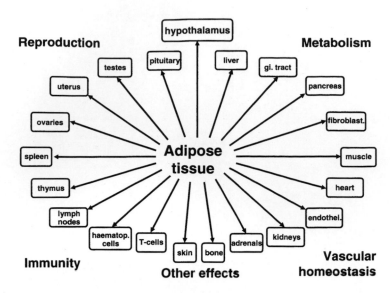

Fig. 1.5 Dynamic view of white adipose tissue based on the pleiotropic effects exerted on quite diverse organs and their respective physiological functions

and immune system, among others (*see* **Fig. 1.5**). AT serves the functions of being a store for reserve energy, an insulation against heat loss through the skin, and a protective padding for certain organs. A rapid turnover of stored fat can take place, such that AT can be used up almost completely during starvation. Furthermore, adipocytes are uniquely equipped to participate in the regulation of other functions such as reproduction, immune response, blood pressure control, coagulation, fibrinolysis, and angiogenesis, among others. Because of the fact that adipocytes express receptors for both pituitary hormones and hypothalamic releasing factors, AT should be contemplated as a fast-acting endocrine gland in close connection with the regulation of the central nervous system. In fact, the concept of "adipotropins" includes pituitary and hypothalamic hormones and releasing factors that have adipocytes as their direct targets via specific receptors *(43)*.

The identification of AT as a multifunctional organ has been brought about by the emerging body of evidence gathered during the last decades, which opposes the static view of AT as a merely passive organ devoted to the storage of excess energy in the form of fat. This pleiotropic nature is based on the ability of fat cells to secrete a large number of hormones, growth factors, enzymes, cytokines, complement factors, and matrix proteins, collectively termed as adipokines or adipocytokines (*see* **Table 1.2; Fig. 1.2**), at the same time as expressing receptors for most of these factors (*see* **Table 1.3**), which warrants an extensive crosstalk at a local and systemic level in response to specific external stimuli or metabolic changes (*see* **Fig. 1.3**). The vast majority of adipocyte-derived factors have been shown to be dysregulated in the alterations accompanying AT mass changes such as overfeeding (*see* **Fig. 1.6**), thus providing evidence for their implication in the etiopathology

Table 1.2 Main factors secreted by adipose tissue to the bloodstream

Molecule	Function/effect
Adiponectin/ACRP30/ AdipoQ/apM1/GBP28	Plays a protective role in the pathogenesis of T2DM and cardiovascular diseases
Adipsin	Possible link between the complement pathway and adipose tissue metabolism
Angiotensinogen	Precursor of angiotensin II; regulator of blood pressure and electrolyte homeostasis
Apelin	Elicits endothelium-dependent, nitric oxide-mediated vasorelaxation reducing blood pressure; potent and long-lasting positive inotropic activity
apoE	Surface constituent of plasma lipoproteins implicated in hepatic uptake of lipoprotein remnants; involved in cellular cholesterol efflux from macrophage foam cells
ASP	Influences the rate of triacylglycerol synthesis in adipose tissue
C-reactive protein	Biomarker of systemic inflammation
CSFs	Participates in hematopoietic, immunological and inflammatory processes
Estrogens	In addition to their role in the development and maintenance of female sex organs and control of the reproductive cycle and pregnancy, they are also involved in growth, blood flow, lipoproteins metabolism and coagulation
EGF	Endothelial growth factor with potent angiogenic and mitogenic effects
Factor B	Possible link between the complement pathway and adipose tissue metabolism
Fibronectin	Interacts with extracellular matrix and plays important roles in cell adhesion, migration, growth and differentiation
FFA	Oxidized in tissues to produce local energy serving as a substrate for the synthesis of triglycerides and structural molecules: involved in the development of insulin resistance
Glycerol	Structural component of the major classes of biological lipids and gluconeogenic precursor
IGF-I	Stimulates proliferation of a wide variety of cells and mediates many of the effects of growth hormone
IL-10	Suppresses macrophage function and inhibits the production of proinflammatory cytokines
IL-17D	Regulation of immune response
IL-1Ra	Acute-phase protein intervening in the counterregulation of inflammatory processes
IL-1β	Essential mediator of the inflammatory, pyrogenic, and anorectic response
IL-6	Implicated in host defense, glucose and lipid metabolism and regulation of body weight
IL-8	Involved in atherosclerosis and inflammatory processes
Leptin	Participates in the control of appetite and energy expenditure via signaling to the hypothalamus about body fat stores; extreme functional pleiotropism
MCP-1	Play a central role in inflammatory processes by regulating leukocyte migration into sites of tissue damage
MIF	Involved in proinflammatory processes and immunoregulation
MIP-1α	Regulates leukocyte migration in inflammatory processes

(continued)

Table 1.2 (continued)

Molecule	Function/effect
Monobutyrin	Stimulates angiogenesis and microvascular endothelial cell motility
NGF	Key signal in the development and survival of sympathetic neurons
NO	Important regulator of vascular tone; pleiotropic involvement in pathophysiological conditions
Osteonectin	Elicits changes in cell shape and influences the synthesis of extra-cellular matrix
PAI-1	Potent inhibitor of the fibrinolytic system
PGI_2 and $PGF_{2\alpha}$	Implicated in regulatory functions such as inflammation and blood clotting, ovulation, menstruation and acid secretion
Resistin	Putative role in insulin resistance May participate in inflammation
SAA3	Acute phase reactant produced in response to infection, injury, or inflammation
TGF-β	Regulates a wide variety of biological responses, including proliferation, differentiation, apoptosis, and development
Tissue factor	Major cellular initiator of the coagulation cascade
TNF-α	Interferes with insulin receptor signaling and is a possible cause of the development of insulin resistance in obesity
VAP-1/SSAO	Mediates leukocyte-endothelial cell interactions; produces reactive oxygen species through its amine oxidase activity
VEGF	Stimulation of angiogenesis
Visfatin	Binds to insulin receptor at a site distinct from insulin exerting a hypoglycemic effect
αl-acid glycoprotein	Acute phase protein involved in inflammatory response

and co-morbidities asssociated to obesity. The diversity of secreted molecules includes factors involved in lipid and glucose metabolism such as lipoprotein lipase (LPL), apolipoprotein E (apoE), cholesteryl ester transfer protein, PPAR, glucocorticoids, sex steroids, prostaglandins, adipsin, acylation stimulating protein, leptin, resistin, adiponectin/Acrp30/adipoQ, osteonectin, and cathepsins, among others. Secreted growth factors include insulin-like growth factor I (IGF-I), nerve growth factor (NGF), macrophage colony-stimulating factor, transforming growth factor-β (TGF-β), vascular endothelial growth factor (VEGF), heparin-binding epidermal growth factor, leukemia inhibitory factor, and bone morphogenetic proteins.

It is now evident that AT is the source of production of a multiplicity of secretory factors, which vary considerably in terms of structure, function, and cellular origin *(10–14)*. By definition, adipocytokines are cytokines produced by adipocytes. Although, strictly speaking, AT secretes a variety of factors, not all of them can be contemplated as adipocytokines. Therefore, the less strict term of adipokines has been coined to include a wider range of factors. Although some hormones and cytokines are mainly produced by adipocytes, some chemokines and proteins of the alternative complement system are mainly produced by cells of the stromovascular fraction. For instance, classical cytokines include TNF-α and interleukin-1, 6, 10, and 18 (IL-1, IL-6, IL-10, IL-18), whereas monocyte chemoattractant protein-1 (MCP-1), macrophage migration inhibitory factor (MIF), and IL-8 belong more to

Table 1.3 Relevant receptors expressed by adipose tissue involved in obesity and metabolic disorders

Receptor	Main effect of receptor activation on adipocyte metabolism
Hormone–cytokine receptors	
Adenosine	Inhibition of lipolysis
Adiponectin (AdipoR1 and AdipoR2)	Regulation of insulin sensitivity and fatty acid oxidation
Angiotensin II	Increase of lipogenesis
	Stimulation of prostacyclin production by mature fat cells
	Interaction with insulin in regulation of adipocyte metabolism
ANP	Oxidative glucose metabolism modulation (via NPR-A)
	Clearance of ANP (via NPR-C)
Chemokines	Regulation of leukocyte migration
CRH-R1 and R2	Implicated in energy homeostasis
EGF	Regulation of adipocyte differentiation
	Interference with the lipolytic effect of catecholamines
Endothelin	Stimulates leptin production
	Potentiates glucose transport
FGF	Regulation of adipocyte differentiation
Gastrin/CCK-B	Regulation of expression and secretion of leptin
GH	Induction of leptin and IGF-I expression
	Stimulation of lipolysis
GHS	Stimulates the differentiation of preadipocytes and antagonizes lipolysis
GIP	Fatty acid and triglyceride synthesis stimulation
	Amplification of insulin effects
GLP-1	Regulation of glycogen synthesis
	Enhancement of insulin-stimulated glucose metabolism
Glucagon	Stimulation of lipolysis
GPR39	Proposed receptor of obestatin supposedly involved in food intake control
IGF-I and -II	Inhibition of lipolysis
	Stimulation of glucose transport and oxidation
IL-1	Plays an important role in lipid metabolism by regulating insulin levels and lipase activity
IL-11	Potent inhibition of LPL activity and adipogenesis
IL-15	Effects on protein and lipid metabolism
IL-6	LPL activity inhibition
	Induction of lipolysis
Insulin	Inhibition of lipolysis and stimulation of lipogenesis
	Induction of glucose uptake and oxidation
	Stimulation of leptin expression
Leptin (OB-R)	Stimulation of lipolysis
	Autocrine regulation of leptin expression
MCR1:2 and 5	Regulation of adipocyte differentiation
	Regulation of lipolysis
NPY Y1 and Y5	Inhibition of lipolysis
	Induction of leptin expression
PDGF	Regulation of adipocyte differentiation
Prolactin	Decreases glucose transport and LPL activity
Prostaglandin	Strong antilypolitic effects (PGE_2)
	Modulation of preadipocyte differentiation ($PGF_{2\alpha}$ and PGI_2)
TGF-β	Potent inhibition of adipocyte differentiation

(continued)

Table 1.3 (continued)

Receptor	Main effect of receptor activation on adipocyte metabolism
TNF-α	Stimulation of lipolysis
	Regulation of leptin secretion
	Potent inhibition of adipocyte differentiation
	Involvement in development of insulin resistance
TSH	Regulation of adipocyte differentiation
	Regulation of lipolysis
VEGF	Stimulation of angiogenesis
Catecholamine-nervous system receptors	
Muscarinic	Inhibition of lipolysis
Nicotinic	Stimulation of lipolysis
β_1-AR	Induction of inositol phosphate production and PKC activation
β_2-AR	Inhibition of lipolysis
	Regulation of preadipocyte growth
β_{1-3}-AR	Stimulation of lipolysis
	Induction of thermogenesis
	Reduction of leptin mRNA levels
Nuclear receptors	
Androgen	Control of adipose tissue development (antiadipogenic signals)
	Modulation of leptin expression
Estrogen	Control of adipose tissue development (proadipogenic signals)
	Modulation of leptin expression
Glucocorticoids	Stimulation of adipocyte differentiation
PPARδ	Regulation of fat metabolism
	Plays a central role in fatty acid-controlled differentiation of preadipose cells
PPARγ	Induction of adipocyte differentiation and insulin sensitivity
Progesterone	Regulation of fat metabolism and distribution
RAR/RXR	Regulation of adipocyte differentiation
T_3	Stimulation of lipolysis
	Regulation of leptin secretion
	Induction of adipocyte differentiation
	Regulation of insulin effects
Vitamin D	Inhibition of adipocyte differentiation
Lipoprotein receptors	
HDL	Clearance and metabolism of HDL
LDL	Stimulation of cholesterol uptake
VLDL	Binding and internalization of VLDL particles. Involvement in lipid accumulation

Abbreviations: ACRP30, adipocyte complement-related protein of 30 kDa; ANP, atrial natriuretic peptide; apM1, adipose most abundant gene transcript 1; apoE, apolipoprotein E; ASP, acylation-stimulating protein; CCK-B, cholecystokinin-B; CRH-R1 and R2 corticotropin-releasing hormone receptors 1 and 2; CSF, colony-stimulating factor; EGF, epidermal growth factor; FFA, free fatty acids; FGF, fibroblast growth factor; GBP28, gelatin-binding protein 28; GH, growth hormone; GHS, growth hormone secretagogue; GIP, glucose-dependent insulinotropic peptide; GLP-1, glucagon-like peptide-1; GPR39, G protein-coupled receptor 39; HDL, high density lipoprotein.; HGF, hepatocyte growth factor; IGF, insulin-like growth factor; IL, interleukin; IL-1Ra, interleukin-1 receptor antagonist; LDL, low density lipoprotein; LPL, lipoprotein lipase; MCP-1, monocyte chemoattractant protein-1; MCR1:2 and 5, melanocortin receptors −1, −2, and −5; MIF, macrophage migration inhibitory factor; MIP-1α, macrophage inflammatory protein-1α; NGF,

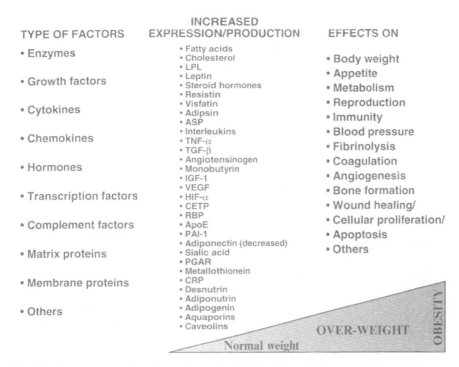

TYPE OF FACTORS	INCREASED EXPRESSION/PRODUCTION	EFFECTS ON
• Enzymes	• Fatty acids • Cholesterol • LPL	• Body weight
• Growth factors	• Leptin • Steroid hormones • Resistin	• Appetite • Metabolism
• Cytokines	• Visfatin • Adipsin • ASP	• Reproduction • Immunity
• Chemokines	• Interleukins • TNF-α • TGF-β	• Blood pressure • Fibrinolysis
• Hormones	• Angiotensinogen • Monobutyrin • IGF-1	• Coagulation • Angiogenesis
• Transcription factors	• VEGF • HIF-α • CETP	• Bone formation • Wound healing/
• Complement factors	• RBP • ApoE • PAI-1	• Cellular proliferation/
• Matrix proteins	• Adiponectin (decreased) • Sialic acid • PGAR	• Apoptosis • Others
• Membrane proteins	• Metallothionein • CRP • Desnutrin	
• Others	• Adiponutrin • Adipogenin • Aquaporins • Caveolins	

Fig. 1.6 Factors expressed and/or produced by WAT, which underlie the multifunctional nature of this endocrine organ, and their subsequent elevation with increased body weight

the subgroup of chemokines. The broad spectrum of adipokines contributes to the pleiotropism of WAT and underlies the extensive autocrine, paracrine, and endocrine cross-talk *(16,17,19–29)*.

The endocrine activity of WAT was postulated almost 20 yrs ago, alluding to the tissue's ability for steroid hormone interconversion. In the last years, especially since the discovery of leptin, the list of adipocyte-derived factors has been increasing at an outstanding pace. Another way of addressing the production of adipose-derived factors is by focusing on the functions they are implicated in. One of the

---

nerve growth factor; NO, nitric oxide; NPR-A, ANP receptor A; NPR-C, ANP receptor C; NPY-Y1 & -Y5, neuropeptide receptors Y-1 and −5; OB-R, leptin receptor; PAI-1, plasminogen activator inhibitor −1; PDGF, platelet-derived growth factor; PGE_2, prostaglandin E_2; $PGF_{2\alpha}$, prostaglandin $F_{2\alpha}$; PGI_2, prostacyclin; PPAR peroxisome proliferator-activated receptor; RAR, retinoic acid receptor; RXR, retinoid × receptor; SAA3, serum amyloid A3; T3, triiodothyronine; TGF-β, transforming growth factor-β; TSH, thyroid-stimulating hormone; TNF-α, tumor necrosis factor-α; VAP-1 (SSAO): Vascular adhesion protein-1 (semicarbazide-sensitive amine oxidase); VEGF, vascular endothelial growth factor; VLDL, very low density lipoprotein; β_1- and β_2-AR, β_1- and β_2-adrenergic receptors; β_{1-3} -AR, β_{1-3} adrenergic receptors.

best known aspects of WAT physiology relates to the synthesis of products involved in lipid metabolism *(44)* such as perilipin, adipocyte lipid binding protein (ALBP, FABP4 or aP2), cholesteryl ester transfer protein, and retinol binding protein (RBP). The lipid droplets contained in adipocytes are coated by structural proteins, such as perilipin, that stabilize the single fat drops and prevent triglyceride hydrolysis in the basal state. The phosphorylation of perilipin after adrenergic stimulation or other hormonal inputs induces a structural change of the lipid droplet that allows the hydrolysis of triglycerides. After hormonal stimulation, hormone sensitive lipase (HSL) and perilipin are phosphorylated and HSL translocates to the lipid droplet. ALBP, also termed aP2, then binds to the N-terminal region of HSL, preventing fatty acid inhibition of the enzyme's hydrolytic activity. Synthesis and secretion of RBP by adipocytes is induced by retinoic acid and shows that WAT plays an important role in retinoid storage and metabolism. In fact, RBP mRNA is one of the most abundant transcripts present in both rodent and human adipose tissue. Hepatic and renal tissues have been regarded as the main sites of RBP production, whereas the quantitative and physiological significance of the WAT contribution remains to be fully established.

AT also has been identified as a source of factors with immunological properties participating in immunity and stress responses, as is the case of ASP and metallothionein. More recently, the pivotal role of adipocyte-derived factors implicated in cardiovascular function control, such as angiotensinogen, adiponectin, PPARγ angiopoetin-related protein/fasting-induced adipose factor, and C-reactive protein (CRP), has been established. A further subsection of proteins produced by AT concerns other factors with an autocrine-paracrine function like PPAR-γ, IGF-I, monobutyrin, and the UCPs.

Energy balance regulation is an extremely complex process composed of multiple interacting homeostatic and behavioral pathways based on highly orchestrated interactions between nutrient selection, organoleptic influences, and neuroendocrine responses to diet, while, at the same time, being influenced by genetic and environmental factors. The final outcome resulting from the opposing impact of adipogenic versus anti-adipogenic factors yields the dynamic situation between fat accretion or reduction (*see* **Fig. 1.7**).

Inflammation constitutes one of the most prominent areas of development in obesity research. In this context, the proinflammatory-proliferative-atherosclerotic-vascular factors such as TNF-α, IL-6, IL-8, plasminogen activator inhibitor-1 (PAI-1), tissue factor, nitric oxide, angiotensinogen, metallothionein, and CRP represent an extraordinarily relevant group. The list of factors shown to be implicated, either directly or indirectly, in the regulation of vascular homeostasis through effects on blood pressure, inflammation, atherogenesis, coagulation, fibrinolysis, angiogenesis, proliferation, apoptosis, or immunity has increased at a magnificent rate *(7–24)*. In fact, parallels between adipocytes and immune cells have been drawn with preadipocytes being able to act like macrophages *(45)*. Acute phase proteins that have been recognized as adipokines include haptoglobin, serum amyloid A (SAA), and CRP. Other inflammation-related adipokines with vasoactive properties or growth factor potential include leptin, adiponectin, NGF, and VEGF *(8,11,13,23–26,46,47)*.

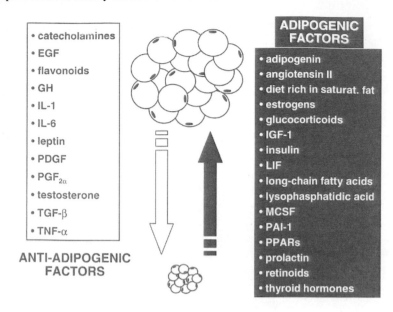

Fig. 1.7 Main factors impinging on the regulation of adipogenesis

The role of most adipokines has been extensively reviewed already *(7–24,48–51)*, with research now concentrated on unravelling the potential mechanistic basis for the link between inflammation and insulin resistance, focusing, in particular, on some of the adipokines that have emerged more recently in the proinflammatory and metabolic syndrome scenario. The teleological reason for the onset of inflammation in the setting of obesity has not been completely disentangled.

Currently, it is not clear whether elevated circulating concentrations of inflammatory factors serve as indicators of systemic inflammation or reflect a spillover of adipose-derived bioactive molecules in response to the hypoxia that takes place as a consequence of adipose mass expansion *(10,52)*. In this context, an increased expression of hypoxia inducible factor-1alpha (HIF-1α), a transcription factor that operates as a molecular sensor for low oxygen levels, seems to exert a key role. The fact that HIF-1α stimulates the production and subsequent release of inflammatory cytokines, chemokines, and angiogenic factors aimed at enhancing blood flow and vascularization *(53)*, points to the relevance of hypoxic conditions in triggering the inflammatory cascade.

Findings of the epidemiological association between inflammatory markers, such as circulating concentrations of fibrinogen and other acute-phase reactants, with obesity or type 2 diabetes mellitus (T2DM) were identified more than half a century ago *(54–56)*. However, at that time point no causal inferences in pathogenesis were established. In the last decade, more widespread epidemiological studies have provided additional information on the link between inflammation and insulin resistance at the same time as extending the early findings from the pathophysiological perspective *(48–62)*. Furthermore, the notion that WAT is a site of production of inflammation-related cytokines has quickly extended beyond TNF-α to incorporate IL-6, angiotensinogen, leptin, adiponectin, resistin, SAA, visfatin, MCP-1, retinol-binding protein 4, and toll-like receptor 4, among others *(63–82)*.

In the last years, the pathophysiological consequences of increased accumulation of fat in the liver has focused attention on the detrimental effects associated with being overweight and obese. Hepatic steatosis describes the phenomenon of excess triglyceride deposition in the liver. Absence of both significant alcohol consumption and other liver diseases are the criteria for establishing the diagnosis of nonalcoholic fatty liver disease, which can range from simple hepatic steatosis through fibrosis and nonalcoholic steatohepatitis to cryptogenic cirrhosis *(83)*. Patients with T2DM exhibit severe abnormalities in the disposal of liver triglycerides, leading to an increased hepatocellular lipid content, which is characteristic of insulin-resistant states *(84–86)*. Free fatty acids overflow and mitochondrial dysfunction and an imbalance of adipocytokines, represented by decreased concentrations of insulin-sensitizing factors, such as adiponectin, together with elevated levels of proinflammatory cytokines (mainly TNF-α, IL-6, IL-1β), contribute to the pathogenesis of nonalcoholic fatty liver disease. The subsequent activation of inflammatory pathways encompassing protein kinase C, nuclear factor-κB, and c-Jun-N-terminal kinase 1 can thereby participate in the progression of hepatic steatosis to nonalcoholic steatohepatitis and cirrhosis *(84–86)*. Therefore, the increased total adiposity together with the elevated intraparenchimatous deposition of fat exert a pivotal role in the development of a common obesity comorbidity.

7   Future Perspectives

The processes participating in controlling energy balance as well as the intermediary lipid and carbohydrate metabolism are intricately coupled by neurohumoral mediators. The coordination of the molecular and biochemical pathways involved underlies, at least in part, the large number of intracellular and secreted proteins produced by WAT with autocrine, paracrine, and endocrine effects. The identification that WAT secretes a plethora of pleiotropic adipokines at the same time as expressing receptors for a huge range of compounds has led to the development of a novel and much more broad-ranging insight into the function of AT at both the basic and clinical levels. Given the versatile and ever-expanding list of secretory proteins, receptors, and plasma membrane proteins found in AT, additional and unexpected consequences of physiological and clinical relevance are sure to emerge.

The advent of the microarray technology has dramatically changed the study of the pattern of gene expression by enabling the simultaneous analysis of thousands of genes in a single experiment. Interestingly, the high number and ample spectrum of genes found to be expressed in WAT together with the changes observed in samples of obese patients substantiates the view of an extraordinarily active and plastic tissue. The complex and complementary nature of the expression profile observed in obese AT reflects a pleiad of adaptive changes affecting crucial pathophysiological functions that deserve to be further explored through genomic, proteomic, and lipidomic approaches to gain more insight into the metabolic disorders accompanying obesity and their deleterious effects on health.

Acknowledgments This work was supported by FIS PI061458 from the Spanish Instituto de Salud Carlos III, Ministerio de Sanidad y Consumo and by the Department of Health of the Gobierno de Navarra (20/2005) of Spain.

References

1. Zhang Y, Proenca R, Maffei M, Barone M, Leopold L, Friedman JM (1994) Positional cloning of the mouse obese gene and its human homologue. Nature 372:425–432
2. Langin D, Lafontan, M (2000) Millennium fat-cell lipolysis reveals unsuspected novel tracks. Horm Metab Res 32:443–452
3. Trayhurn P, Beattie JH (2001) Physiological role of adipose tissue: white adipose tissue as an endocrine and secretory organ. Proc Nutr Soc 60:329–339
4. Frühbeck G, Gómez-Ambrosi J, Muruzábal FJ, Burrell MA (2001) The adipocyte: a model for integration of endocrine and metabolic signaling in energy metabolism regulation. Am J Physiol 280:E827–E847
5. Frühbeck G, Gómez-Ambrosi J (2001) Rationale for the existence of additional adipostatic hormones. FASEB J 15:1996–2006
6. Frühbeck G (2001) A heliocentric view of leptin. Proc Nutr Soc 60:301–318
7. Marette A (2002) Mediators of cytokine-induced insulin resistance in obesity and other inflammatory settings. Curr Opin Clin Nutr Metab Care 5:377–383
8. Blake GJ, Ridker PM (2002) Inflammatory bio-markers and cardiovascular risk prediction. J Intern Med 252:283–294
9. Frühbeck G, Gómez-Ambrosi J (2003) Control of body weight: a physiologic and transgenic perspective. Diabetologia 46:143–172
10. Trayhurn P, Wood IS (2004) Adipokines: inflammation and the pleiotropic role of white adipose tissue. Br J Nutr 92:347–355
11. Frühbeck G (2004) The adipose tissue as a source of vasoactive factors. Curr Med Chem (Cardiovasc Hematol Agents) 2:197–208
12. La Cava A, Alviggi C, Matarese G (2004) Unravelling the multiple roles of leptin in inflammation and autoimmunity. J Mol Med 82:4–11
13. Granger JP (2004) Inflammatory cytokines, vascular function, and hypertension. Am J Physiol Reg Integr Comp Physiol 286:R989–R990
14. Gimeno RE, Klaman LD (2005) Adipose tissue as an active endocrine organ: recent advances. Curr Opin Pharmacol 5:122–128
15. Hutley L, Prins JB (2005) Fat as an endocrine organ: relationship to the metabolic syndrome. Am J Med Sci 330:280–289
16. Trayhurn, P (2005) Endocrine and signalling role of adipose tissue: new perspectives on fat. Acta Physiol Scand 184:285–293
17. Yu Y-H, Ginsberg HN (2005) Adipocyte signaling and lipid homeostasis. Sequelae of insulin-resistant adipose tissue. Circ Res 96:1042–1052
18. Bell CG, Walley AJ, Froguel P (2005) The genetics of human obesity. Nat Rev Genet 6:221–234
19. Berg AH, Scherer PE (2005) Adipose tissue, inflamation and cardiovascular disease. Circ Res 96:939–949
20. Lau DCW, Dhillon B, Yan H, Szmitko PE, Verma, S (2005) Adipokines: molecular links between obesity and atherosclerosis. Am J Physiol Heart Circ Physiol 288:H2031–H2041

21. Nawrocki AR, Scherer PE (2005) Adipose tissue, adipokines, and inflammation. Drug Discovery Today 10:1219–1230
22. Fantuzzi G (2005) Adipose tissue, adipokines, and inflammation. J Allergy Clin Immunol 115:911–919
23. Van Gaal LF, Mertens IL, De Block CE (2006) Mechanisms linking obesity with cardiovascular disease. Nature 444:875–880
24. Kahn SE, Hull RL, Utzschneider KM (2006) Mechanisms linking obesity to insulin resistance and type 2 diabetes. Nature 444:840–846
25. Fain JN (2006) Release of interleukins and other inflammatory cytokines by human adipose tissue is enhanced in obesity and primarily due to the nonfat cells. Vitam Horm 74:443–477
26. Matsuzawa Y (2006) Therapy insight: adipocytokines in metabolic syndrome and related cardiovascular disease. Nat Clin Pract Cardiovasc Med 3:35–42
27. Trayhurn P, Bing C, Wood IS (2006) Adipose tissue and adipokines—energy regulation from the human perspective. J Nutr 136:1935S–1939S
28. Despres JP, Lemieux I (2006) Abdominal obesity and metabolic syndrome. Nature 444:881–887
29. Abizaid A, Gao Q, Horvath TL (2006) Thoughts for food: brain mechanisms and peripheral energy balance. Neuron 51:691–702
30. Pond CM (1999) Physiological specialisation of adipose tissue. Prog Lipid Res 38:225–248
31. Spiegelman BM, Hotamisligil GS, Graves RA, Tontonoz P (1993) Regulation of adipocyte gene expression and syndromes of obesity/diabetes. J Biol Chem 268:68236–6826
32. Rosen ED, Walkey CJ, Puigserver P, Spiegelman BM (2000) Transcriptional regulation of adipogenesis. Genes Dev 14:1293–1307
33. Gregoire FM (2001) Adipocyte differentiation: from fibroblast to endocrine cell. Exp Biol Med 226:997–1002
34. Rosen ED, MacDougald OA (2006) Adipocyte differentiation from the inside out. Nat Rev Mol Cell Biol 7:885–896
35. Unger RH (2003) The physiology of cellular liporegulation. Annu Rev Physiol 65:333–347
36. Shen W, Wang Z, Punyanita M, Lei J, Sinav A, Kral JG, Imielinska C, Ross R, Heymsfield SB (2003) Adipose quantification by imaging methods: a proposed classification. Obesity Res 11:5–16
37. Cannon B, Nedergaard, J (2004) Brown adipose tissue: function and physiological significance. Physiol Rev 84:277–359
38. Wajchenberg BL (2000) Subcutaneous and visceral adipose tissue: their relation to the metabolic syndrome. Endocrine Rev 21:697–738
39. Lafontan M, Berlan M (2003) Do regional differences in adipocyte biology provide new pathophysiological insights? Trends Pharmacol Sci 24:276–283
40. Fried SK, Ross RR (2004) Biology of visceral adipose tissue. In: Bray GA, Bouchard C (eds) Handbook of obesity. Etiology and pathophysiology, 2nd edn. Marcel Dekker, Inc., New York, pp 589–614
41. Gómez-Ambrosi J, Catalán V, Diez-Caballero A, Martínez-Cruz A, Gil MJ, García-Foncillas J, Cienfuegos JA, Salvador J, Mato JM, Frühbeck G (2004) Gene expression profile of omental adipose tissue in human obesity. FASEB J 18:215–217
42. Linder K, Arner P, Flores-Morales A, Tollet-Egnell P, Norstedt G (2004) Differentially expressed genes in visceral and subcutaneous adipose tissue of obese men and women. Prog Lipid Res 45:148–154
43. Schäffler A, Schölmerich J, Buechler C (2006) The role of 'adipotropins' and the clinical importance of a potential hypothalamic-pituitary-adipose axis. Nat Clin Pract Endocrinol Metab 2:374–383
44. Jump DB (2004) Fatty acid regulation of gene transcription. Crit Rev Clin Lab Sci 41:41–78
45. Cousin B, Munoz O, Andre M, Fontanilles AM, Dani C, Cousin JL, et al (1999) A role for preadipocytes as macrophage-like cells. FASEB J 13:305–312
46. Peeraully MR, Jenkins JR, Trayhurn P (2004) NGF gene expression and secretion in white adipose tissue: regulation in 3T3-L1 adipocytes by hormones and inflammatory cytokines. Am J Physiol Endocrinol Metab 287:E331–E339

47. Wang B, Jenkins JR, Trayhurn, P (2005) Expression and secretion of inflammation-related adipokines by human adipocytes differentiated in culture: integrated response to TNF-α. Am J Physiol Endocrinol Metab 288:E731–E740
48. Lehrke M, Reilly MP, Millington SC, Iqbal N, Rader DJ, Lazar MA (2004) An inflammatory cascade leading to hyper-resistinemia in humans. PloS Med 1:161–168
49. Frühbeck G (2006) Vasoactive factors and inflammatory mediators produced in adipose tissue. In: Fantuzzi G, Mazzone T (eds) Health and nutrition. Adipose tissue and adipokines in health and disease. Humana Press Inc., Totowa, NJ, pp 61–75
50. Chaldakov GN, Tonchev AB, Tuncel N, Atanassova P, Aloe L. Adipose tissue and mast cells. Adipokines as yin-yang modulators of inflammation. In: Fantuzzi G, Mazzone T, (eds) Health and nutrition. Adipose tissue and adipokines in health and disease. Humana Press Inc., Totowa, NJ, pp 147–154
51. Murdolo G, Smith U (2006) The dysregulated adipose tissue: a connecting link between insulin resistance, type 2 diabetes mellitus and atherosclerosis. Nutr Metab Cardiovasc Dis 16:S35–S38
52. Shoelson SE, Lee J, Goldfine AB (2006) Inflammation and insulin resistance. J Clin Invest 116:1793–1801
53. Wenger RH (2002) Cellular adaptation to hypoxia: O2-sensing protein hydroxylases, hypoxia-inducible transcription factors, and O2-regulated gene expression. FASEB J 16:1151–1162
54. Fearnly GR, Vincent CT, Chakrabarti R (1959) Reduction of blood fibrinolytic activity in diabetes mellitus by insulin. Lancet 2:1067
55. Ogston D, McAndrew GM (1964) Fibrinolysis in obesity. Lancet 14:1205–1207
56. Grace CS, Goldrick RB (1968) Fibrinolysis.and body build. Interrelationships between blood fibrinolysis, body composition and parameters of lipid and carbohydrate metabolism. J Atheroscler Res 8:705–719
57. Schmidt MI, Duncan BB, Sharrett AR, Lindberg G, Savage PJ, Offenbacher S, et al (1999) Markers of inflammation and prediction of diabetes mellitus in adults (Atherosclerosis Risk in Communities study): a cohort study. Lancet 353:1649–1652
58. Barzilay JI, Abraham L, Heckbert SR, Cushman M, Kuller LH, Resnick HE, et al (2001) The relation of markers of inflammation to the development of glucose disorders in the elderly: the Cardiovascular Health Study. Diabetes 50:2384–2389
59. Pradhan AD, Manson JE, Rifai N, Buring JE, Ridker PM (2001) C-reactive protein, interleukin 6, and risk of developing type 2 diabetes mellitus. JAMA 286:327–334
60. Festa A, D'Agostino R Jr., Tracy RP, Haffner SM (2002) Elevated levels of acute-phase proteins and plasminogen activator inhibitor-1 predict the development of type 2 diabetes: the insulin resistance atherosclerosis study. Diabetes 51:1131–1137
61. Freeman DJ, Norrie J, Caslake, MJ, Gaw A, Ford I, Lowe GD, et al (2002) C-reactive protein is an independent predictor of risk for the development of diabetes in the West of Scotland Coronary Prevention Study. Diabetes 51:1596–1600
62. Haffner S, Temprosa M, Crandall J, Fowler S, Goldberg R, Horton E, et al (2005) Intensive lifestyle intervention or metformin on inflammation and coagulation in participants with impaired fasting glucose tolerance. Diabetes 54:1566–1572
63. Linton MF, Fazio S (2003) Macrophages, inflammation, and atherosclerosis. Int J Obes Relat Metab Disord 27(Suppl 3):S35–S40
64. Castrillo A, Tontonoz P (2004) Nuclear receptors in macrophage biology: at the crossroads of lipid metabolism and inflammation. Annu Rev Cell Dev Biol 20:455–480
65. Clément K, Viguerie N, Poitou C, Carette C, Peñoux V, Curat CA, et al (2004) Weight loss regulates inflammation-related genes in white adipose tissue of obese subjects. FASEB J 18:1657–1669
66. Poitou C, Viguerie N, Cancello R, De Matteis R, Cinti S, Stich, V. et al (2005) Serum amyloid A: production by human white adipocyte and regulation by obesity and nutrition. Diabetologia 48:519–528
67. Charo IF, Taubman MB (2004) Chemokines in the pathogenesis of vascular disease. Circ Res 95:858–866
68. Malavazos AE, Cereda E, Morricone L, Coman C, Corsi MM, Ambrosi B (2005) Monocyte chemoattractant protein 1: a possible link between visceral adipose tissue-associated

inflammation and subclinical echocardiographic abnormalities in uncomplicated obesity. Eur J Endocrinol 153:871–877

69. Fukuhara A, Matsuda M, Nishizawa M, Segawa K, Tanaka M, Kishimoto K, et al (2005) Visfatin: a protein secreted by visceral fat that mimics the effects of insulin. Science 307:426–430

70. Berndt J, Kloting N, Kralisch S, Kovacs P, Fasshauer M, Schon MR, et al (2005) Plasma visfatin concentrations and fat depot-specific mRNA expression in humans. Diabetes 54:2911–2916

71. Yang Q, Graham TE., Mody N, Preitner F, Peroni OD., Zabolotny JM, et al (2005) Serum retinol binding protein 4 contributes to insulin resistance and type 2 diabetes. Nature 436:356–362

72. Cancello R, Clement, K (2006) Is obesity an inflammatory illness? Role of low-grade inflammation and macrophage infiltration in human white adipose tissue. BJOG 113:1141–1147

73. Arner P (2006) Visfatin—a true or false trail to type 2 diabetes mellitus. J Clin Endocrinol Metab 91:28–30

74. Graham TE, Yang Q, Blüher M, Hammarstedt A, Ciaraldi TP, Henry RR, et al (2006) Retinol-binding protein 4 and insulin resistance in lean, obese, and diabetic subjects. N Engl J Med 354:2552–2563

75. Cho YM, Youn BS, Lee H, Lee N, Min SS, Kwak SH, et al (2006) Plasma retinol-binding protein-4 concentrations are elevated in human subjects with impaired glucose tolerance and type 2 diabetes. Diabetes Care 29:2457–2461

76. Janke J, Engeli S, Boschmann M, Adams F, Bohnke J, Luft FC, et al (2006) Retinol-binding protein 4 in human obesity. Diabetes 55:2805–2810

77. Shi H, Kokoeva MV, Inouye K, Tzameli I, Yin H, Flier JS (2006) TLR4 links innate immunity and fatty acid-induced insulin resistance. J Clin Invest 116:3015–3025

78. Song MJ, Kim KH., Yoon JM, Kim JB (2006) Activation of Toll-like receptor 4 is associated with insulin resistance in adipocytes. Biochem Biophys Res Commun 346:739–745

79. Hotamisligil GS (2006) Inflammation and metabolic disorders. Nature 444:860–867

80. Tilg H, Moschen AR (2006) Adipocytokines: mediators linking adipose tissue, inflammation and immunity. Nat Rev Immunol 6:772–783

81. Lumeng CN, Bodzin JL, Saltiel AR (2007) Obesity induces a phenotypic switch in adipose tissue macrophage polarization. J Clin Invest 117:175–184

82. Lumeng CN, Deyoung SM, Bodzin JL, Saltiel AR (2007) Increased inflammatory properties of adipose tissue macrophages recruited during diet-induced obesity. Diabetes 56:16–23

83. Choudhury J, Sanyal AJ (2004) Clinical aspects of fatty liver disease. Semin Liver Dis 24:349–362

84. Roden M (2006) Mechanisms of disease: hepatic steatosis in type 2 diabetes—pathogenesis and clinical relevance. Nat Clin Pract Endocrinol Metab 2:335–348

85. Tilg H, Hotamisligil GS (2006) Nonalcoholic fatty liver disease: Cytokine-adipokine interplay and regulation of insulin resistance. Gastroenterology 131:934–945

86. Utzschneider KM, Kahn SE (2006) The role of insulin resistance in nonalcoholic fatty liver disease. J Clin Endocrinol Metab 91:4753–4761

Chapter 2
Choosing an Adipose Tissue Depot for Sampling

Factors in Selection and Depot Specificity

Louis Casteilla, Luc Pénicaud, Béatrice Cousin, and Denis Calise

Summary The importance and the role of adipose tissues are now largely expanded not only because the very high occurrence of obesity but also because the emerging view that adipose tissue could be a reservoir of therapeutic cells. A critical examination of the adipose tissue features according to their location shows that sampling is not as easy as previously thought and needs special attention to heterogeneity and differences. We discussed here these different points and give precise protocols to sample the different adipose tissues and manipulate them.

Key words White adipose tissue; brown adipose tissue; plasticity; heterogeneity; preadipocytes; adipose-derived stroma cells.

1 Introduction

Adipose tissues (ATs) were long considered as negligible and as simple filling tissues. The increase in knowledge concerning their role in energy balance and the increased occurrence of metabolic disorders, such as obesity and syndrome X, have focused the attention of the scientific community on these tissues. This evolution has been speeded by both the discovery of leptin and the development of transgenic and knockout techniques (*1,2*). The former and the following studies on adipokines emphasized the endocrine function and the involvement of adipose mass in most physiological functions (*3,4*). The latter made it possible to test hypotheses elaborated from in vitro findings in the organisms, but these approaches also revealed unexpected results and findings concerning the development of AT (*2*). More recently, the discovery that unexpected phenotypes can be obtained from stroma cells purified from adipose tissue has largely amplified the concept of adipose tissue plasticity (*5*). Although previous reports demonstrated the capability of preadipocytes to exhibit an osteogenic potential, the consideration of adipose tissue as a reservoir of stem cells was really undertaken after a publication by Zuk et al. (*6–8*). Such findings associated with the easy sampling of adipose tissue are rapidly

From: *Methods in Molecular Biology, Vol. 456:*
Adipose Tissue Protocols, Second Edition. Edited by: Kaiping Yang
© Humana Press, a part of Springer Science + Business Media, Totowa, NJ

attracting many new investigators. This chapter aims to give a general picture of the ATs presently available in mammals and to describe their sampling.

1.1 Concept and Principles: How to Classify ATs?

Three functionally different types of adipose tissues can be classically described in mammals: brown adipose tissue (BAT), white adipose tissue (WAT), and bone marrow adipose tissue (BMAT). The role of BMAT is poorly investigated and seems related to the control of hematopoiesis and osteoblastogenesis by acting as energy stores, but also via their paracrine activities (9,10). Beside these classic locations of adipose tissues, the emergence of adipocytes can be observed at unexpected locations, particularly in degenerative tissues. In this chapter, we will focus our discussions on white and brown fat tissues only.

Brown (BAT) and white adipose tissue (WAT) are characterized by different anatomical locations, morphological structures, functions, and regulation (11–14; see also Subheadings 3.2–3.4.). Both are called adipose because of the amount of fat stored in both types. BAT is so called because of its characteristic color, originating mostly from its abundant vascularization and cytochromes. Both ATs are able to store energy as triglycerides, although whereas white fat releases this energy according to the needs of the organism, brown fat converts it as heat. WAT is the main store of energy as lipids for the organism (14). BAT plays an important role in the regulation of body temperature in hibernating, as well as in small and newborn mammals (15). The developmental patterns of ATs are different and are species-dependent (see Subheadings 3.2–3.4). Adipocytes within a pad were long considered to belong to a single phenotype, i.e., either brown or white adipocytes.

Although studies on WAT were always concerned with energy metabolism, studies on BAT were first focused on nonshivering thermogenesis and thermoregulatory purposes (15). Later, the involvement of BAT in diet-induced thermogenesis led researchers to also investigate its role in various conditions associated with changes of energy balance (16–18).

1.2 Typical BAT

The main features specific to BAT are summarized in Table 2.1 and are compared with WAT. Its thermogenic function is assumed by the numerous mitochondria and by the presence of mitochondrial protein, uncoupling protein 1 (UCP1), in the brown adipocytes. UCP1 is specifically expressed in these cells and is located in the inner mitochondrial membrane. It is able to uncouple the mitochondria and enables heat production (17,19). Recently, the homologous proteins, UCP2 and UCP3, have been cloned (19) and can also be detected in this tissue. Some biochemical or molecular makers, including nuclear factors more or less specific to brown fat, are available, and these are also given in Table 2.1 (20,21).

Table 2.1 Main features differentiating WAT and BAT in rodents

	WAT	BAT
Location of main depots	Inguinal, retroperitoneal, gonadal (compare with **Table 2.2**)	Interscapular, perineal, axillary, paravertebral
Color	Ivory or yellow	Brown
Vascular sytem	++	+++
Innervation	Sympathetic (++)	Sympathetic (+++)
Adipose cells	Unilocular cells	Multilocular cells
Functions	Storage of energy as triglycerides	Storage of energy as triglycerides
	Fatty acids and glycerol release	Heat production
	Secretory tissue	Secretory tissue
Immune cells	+++	+/−
UCPs	UCP2 (++)	UCP1, UCP2 (+), UCP3
Deiodase type II	+	+++
GMP reductase	−	+++
Leptin	+++	At birth, not in adult
α-, β-Adrenoceptors	β_3 (++),α_2 (+)	β_3 (+++)
PGC1	+	+++

In most mammals, BAT develops during gestation and perinatal life *(15)*. It is prominent in the newborn or in young mammals in which nonshivering thermogenesis is necessary to counteract heat loss associated with birth and atmospheric life. It is mostly located around arterial vessels and vital organs. One exception is the piglet, which displays no brown fat, and is subject to thermoregulation dysfunction *(22)*. The development and quantity of BAT are associated with the degree of nonshivering thermognesis required by the organism to maintain its body temperature. This need corresponds to the balance between heat produced by the metabolic body mass and heat loss, which is correlated to body surface and the adequacy of insulation. With increasing age, as the rate of heat loss per unit body weight decreases, the tissue becomes indistinguishable from white fat. This point is crucial in humans and the existence of scattered brown adipocytes in adult is always debated and will be considered in the next paragraph *(18)*. Nevertheless, in hibernators and in some other small mammals (mice, rats, and so on), it regresses only partially and remains identifiable throughout life. In these species, adipose precursor cells are latent in the tissue, and can be recruited as necessary. This general presentation must be modified according to species and to the developmental stage of the newborn, as summarized in **Table 2.2** *(15)*. The most studied typical brown adipose deposits are the interscapular (IBAT) and perirenal BATs in rodents and large mammals, respectively. IBAT is located subcutaneously between the shoulders, and can easily be dissected (*see* **Subheading 3**). It is the only fat pad distinguishable at birth in laboratory rodents. Perirenal BAT is brown in large mammals during the perinatal period, its weight is greater, and it is impossible to sample or remove the whole pad without removing the kidney.

One of the strongest inducers of this type of AT is cold exposure. Acute exposure induces marked changes in metabolism and gene regulation but also stimulates proliferation and differentiation of the precursors into brown adipocytes, leading to

Table 2.2 Evolution of typical BAT according to species

	Immature (hamster)	Altricial (mouse, rat)	Precocial (rabbit, guinea pig, ruminants, primates)
Newborn	Underdeveloped	Nest-dependent	Well-developed
Amount at birth	1–2% body wt	1–2% body wt	2–5% body wt
Several days after birth	Poorly developed	11–12% of body wt	Decreasing
Several weeks after birth	Developing	Partial regression	Transformation into white fat
Adult	Developed	Present	Absent

the development of this tissue in days or weeks after exposure. Catecholamines or β-adrenoceptor agonists mimic the majority of these effects *(14,23,24)*.

1.3 WAT

This is the most abundant tissue of fat mass, and may account for more than half of body weight in severe obesity. It was considered as less vascularized and inner-vated than brown fat, but various reviews have questioned this opinion *(25–27)*. The importance of white fat in energy balance via its metabolism is well known. Besides this classic view, the wide range of products secreted by adipose cells emphasizes its secretory function and opens interesting fields for the understand-ing of the established links between the increase of fat mass and various associ-ated disorders, such as cardiovascular disease *(4,28–30)*. When white fat is compared with brown fat, it is noteworthy that no specific marker of white fat is presently available to positively identify it. In the adult, leptin could be a good marker for positive identification of white fat, but its strong expression at birth in brown fat makes this an open question *(31,32)*. Most of its development occurs after birth, and primarily results from hypertrophy of white adipose cells, which can reach 150 μm in diameter in some species. Nevertheless, a pool of preadipose cells is maintained throughout life in most species, including humans and can participate in this growth *(33)*.

1.4 Heterogeneity and Plasticity

The aforementioned classification must be qualified because of several findings: the presence of scattered brown adipocytes in white fat; the different properties of WATs according to location; and the putative conversion of one AT phenotype to the other.

1.4.1 Heterogeneity Within and Between Pads

Brown adipocytes have been observed in noncold-exposed rodents, as well as in several deposits considered as typical white fat in primates. The number of these cells can vary according to the location of fat pads and are most numerous in the periovarian fat of rodents, which can be compared with a patchwork of brown and white adipocytes (18,34,35).

It has long been known that the location of the development of adipose deposits during obesity differs according to gender and genetic determinants (36–38). Abdominal obesity is predominant in the male; subcutaneous (SC) fat mass is mostly involved in female obesity. Sex hormones play a major role in these differences (37). Increased intra-abdominal body fat mass is considered as an independent risk factor for health problems linked to obesity and is positively correlated with increased overall morbidity and mortality (39–41). These findings have been the basis for numerous investigations, including genetic approaches to differences of metabolic properties or precursor pools according to location of fat, and they reinforce the concept of heterogeneity but, in this case, between the fat pads. Taken together, these studies make it possible to distinguish SC from internal fat and upper from lower body fat.

However, this classification is not sufficiently clear and needs further definition: For instance, in humans, omental adipose fat is the most sensitive tissue in lipolysis, as well as in lipogenesis (42–44). This heterogeneity exists whatever the species (45,46). For example, abdominal pads also have greater interleukin 6 or plasminogen activator inhibitor secretion, in vitro differentiation capacity, thiazolidinedione sensitivity, and apoptosis than SC pads and a different redox metabolism (47–51). One exception seems to be leptin expression, which is higher in SC tissue (52,53). It is noteworthy that the regulation of this gene is depot-related (54). These depot-specific properties are partly genetically determined (38). Such heterogeneity also can be observed at molecular levels in rodents, as well as in humans (55,56). This could be attributed to the developmental origin of the different fat pads (57).

For a long time, adipose lineage cells have attracted most investigations and other cells constituting the whole adipose tissue were neglected. This view is changing because the emerging importance of the relationship between immune/inflammation and adipose tissue development (30,58). Thus, adipose tissue also contains immune cells, i.e. macrophages and lymphocytes, the amount of which will vary according to deposits (21).

1.4.2 Plasticity

Plasticity is the term which is used to indicate that some deposits are capable of converting from one type of AT to another. The transformation that has been described concerns the transformation of BAT into WAT-like AT, which takes place, as previously indicated, during postnatal development (15). The reversibility of this process differs according to species (18,34,59,60). The term of "convertible adipose tissue" was used to describe the deposits able to be reversibly transformed *(59)*. In fact, the same results can be obtained with all deposits, but with different intensity levels *(61,62)*. Marked development of brown adipocytes occurs among fat considered and studied as typical white fat, i.e., periovarian fat in rats, inguinal fat in mice, and numerous white fat pads in dogs. The proportion of the two phenotypes of adipose cells changes according to physiological (cold exposure, development, gestation-lactation cycles), pharmacological (β3-adrenergic agonist treatment), and pathophysiological conditions and genetic background *(34,60,63,64)*. The cellular mechanisms involved in this plasticity are not clearly deciphered. Although the overexpression of co-activator of peroxisome prolif- erators-activated receptor (PPAR) γ, PGC-1, drives the emergence of brown adipocyte phenotype from white adipocyte, convergent data, including genetic manipulations allowing irreversible labeling of brown adipocytes, show the independence of both line- ages *(65,66)*. Both hypotheses could be reconciled by considering that brown adi- pocytes could be transformed into a white adipocyte-like cells that we call dormant or masked brown adipocytes *(5)*. Whatever the answer, the data lead to conclusions about the heterogeneity within or between pads and the potential for transformation between the two phenotypes of adipose tissues, for which we have first proposed the term plas- ticity *(5)*. From these considerations, this notion is now extended. Indeed, it was recently established that preadipocytes can behave as endothelial- and macrophage-like cells and that adipose tissues host multipotent adipose-derived stroma cells (ADSCs) *(5,67,68)*. Again, these striking properties display site specific differences *(69,70)*.

The cell plasticity of adipose derived cells raises great hope raises great hope in regenerative medicine because adipose tissue can be easily harvested in adults and could represent an abundant source of regenerative cells. So far, when all data are carefully collected, it appears that adipose tissues can be considered as a subtle and complex mix- ture of cells, the differentiation potential of which can vary according to the deposit.

In any event, the investigator must be cautious and take into account this aspect of AT biology when: mice or rats are used as a model for humans; fat pads have to be pooled to obtain sufficient sample quantity; or it is only possible to remove an aliquot of AT to interpret the results as the index of the whole fat pad or the whole fat mass.

2   Materials

No specific materials are needed except sharpened and pointed surgery tools.

3   Methods

3.1   Choice of Species

The criteria of choice are numerous and are grouped here into three levels (**Table 2.3**). The first is the scientific aim, and the choice of species will be strictly dependent on it. In other cases, the decision is less clear, and each aspect may need discus-

Table 2.3 Parameters to be taken into account in the study of AT

	Parameters
Questions	Model for human physiology or pathology; BAT or WAT studies; physiological investigations; in vivo developmental studies; plasticity studies; transgenic model; photoperiod effect; in vitro studies.
Extrinsic	Species; breed; age; sex; diet; room temperature; photoperiod; gestation-lactation cycles
Intrinsic	Importance of metabolic pathways; BAT vs WAT phenotype; plasticity (*see* **Subheading 2.4.2.**); SC vs internal; accessibility of innervation or vascular system (depending on studies)

Table 2.4 Advisable species according to the investigations

Investigation fields	Advisable species
Model for humans	Primates, not rodents
BAT or WAT studies	All species
Physiological investigations	Rats
In vivo developmental studies	All species
Plasticity studies	Mouse, rat, rabbit
Transgenic and knockout models	Mouse
Photoperiodic effect	Sensitive species: hamster
Diet-obesity	Rats according to breed; dogs with high-fat diet; primates

Table 2.5 Main features differentiating ATs in rats, mice, and humans

	Rats and mice	Humans
Location of fat pad:		
Interscapular	+	−
Periovarian	+	−
Epididymal	+	
Persistence of brown fat in adults	+++	+/−
Convertible features	Mice > rats	
BAT → WAT	+	+++
WAT → BAT	+++	?
	Ing > PO > RP > Ep	
ADSC plasticity	Ing > PO > RP > Ep	
Main site of lipogenesis	AT	Liver
Glucose transport sensitive to insulin	+++	+
Catecholamine-stimulated lipolysis	$\beta_1, \beta_2, \beta_3$ (+)	β_1, β_2 (+) α_2 (−)

PO, periovarian; RP, retroperitoneal; Ep, epididymal.

sion. Nevertheless, an aid to decision can be suggested as illustrated in **Table 2.4**, which shows that rodents are valuable and convenient models in most cases, except for human studies. The chief reasons are given in **Table 2.5**. From these data, it is clear that classic laboratory rodents are not a good model for humans in metabolic or developmental studies. When metabolic features are considered, no important difference exists between adipocytes from non-human and human primates (*23*). Therefore, the only physiological models available as human models are primates. For developmental studies, large animals and rabbits seem to present the same features and can be used at least until weaning. After this time, the great difference in metabolism excludes the use of these animal species as human models.

3.2   Choice of Pad

When the species has been decided upon, the location of the fat pad to be studied must be chosen. Two aspects must be considered: once again, the scientific aim and the amount of tissue needed. Both aspects are summarized in **Tables 2.6** and **2.7** for rats or mice. The coarse ratio between IBAT and the three other sites described in **Table 2.7** is quite different in these two closely related rodent species, which suggests that IBAT is relatively more important in the mouse, the species used for transgenic studies, than in the rat. Whatever the AT, fine dissection is required, because of the developed vascular system and numerous lymph nodes.

3.3   Sampling IBAT in Mice or Rats

1. After euthanasia (*71*), the animals are placed on the abdomen, the head toward the investigator.

Table 2.6   Choice of fat pads in the rat or mouse according to the aim of investigation

Aim	IBAT	PO	RP	Inguinal	Ep
Sc vs abdominal				+	+
Plasticity	+	Rat		Mouse	
Denervation	+	+	+		
Vascular system	+			+	+
Isolated adipocytes	+			+	+
Primary culture	+			+	+

IBAT, interscapular brown adipose tissue; PO, periovarian adipose tissue; RP, retroperitoneal adipose tissue; Inguinal, Inguinal adipose tissue; Ep, epididymal adipose tissue.

Table 2.7   Weights of major sites of AT in young adult rats (9- to 10-wk old) and mice (7- to 8-wk-old)

	IBAT	Inguinal (g)	Gonadal (g)	Retroperitoneal (g)	IBAT/WAT
Mice	0.14	0.35	0.2–0.4	Negligible	Approx 20%
Rats	0.3	2.5	1.2	1	Approx 7%

IBAT, interscapular brown adipose tissue; WAT, white adipose tissue.

2. The shoulder region is abundantly rinsed with 70% EtOH to wet the coat and to avoid having hairs on the samples.
3. The skin just behind the head is grasped with tongs, lifted, and incised with scissors.
4. The skin is widely incised from this point to the middle of the black, and the field is opened.
5. The butterfly-shaped IBAT is revealed.

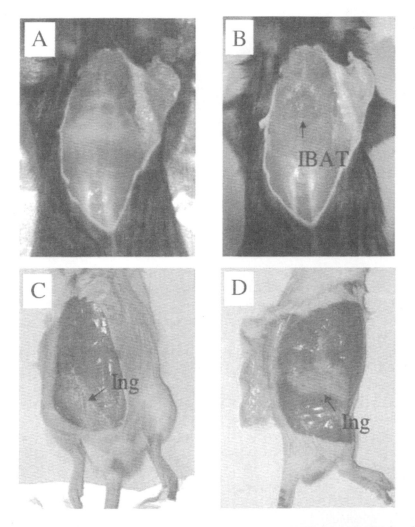

Fig. 2.1 IBAT and inguinal adipose tissue in rats. (**A**) Aspect of IBAT before removing the white part. (**B**) The white part of the pad has been carefully dissected. Brown fat appears as a butterfly between shoulders (**C,D**) Front and side views, respectively, of inguinal fat

6. Rub the fat pad with a paper tissue to discard the white part just above the IBAT and then carefully dissect the pad.
7. A binocular microscope can be used but, with some practice, this is not necessary; or remove the fat pad, and afterwards carefully dissect the butterfly of brown fat. In all cases, care must be taken to avoid the muscle closely associated with the brown fat.
8. The sample is ready and the parts of the pad can be separated as required. If RNA is to be extracted, freeze the tissue by immersion in liquid nitrogen and store at −80°C. It is better to freeze it at once, rather than to freeze it after putting it into a container, to prevent it sticking to the walls.

3.4   General Considerations for Sampling White Fat

WATs are organized in lobules, and the various pads can be found together within connective tissues, particularly in obese animals. Therefore, before cutting with scissors, it is sometimes better to separate the different parts by hand, taking care to remove only the whole fat of interest. AT can be frozen by immersion in liquid nitrogen, the same for brown fat. After freezing, the tissue can be reduced to powder to facilitate and homogenize the sample.

3.4.1   Sampling Inguinal AT in Mice or Rats

1. The procedure is the same as that previously described for IBAT but, in this case, the rodent is placed on its back with the tail toward the investigator.
2. The abdomen is rinsed with EtOH, and the skin is widely incised.
3. After removing the pad, dissect and discard the lymph nodes present among the fat. For females, take care not to confuse the fat pad and the mammary gland, which is involuted.
4. If sampling is done to study gene expression, depending on the size of the pad, it may be preferable to reduce the pad to powder, in order to use only the amount required for the study.

3.4.2   Sampling Gonadal AT in Mice or Rats (see Fig. 2.2)

1. Open the abdominal wall.
2. Extract the genitals (ovaries or testes, according to the sex) from the abdominal cavity.
3. Remove carefully, by dissecting the fat tissue or handling the gonadal tract with one hand, and separate fat from other tissues by gently pulling them with the other hand.

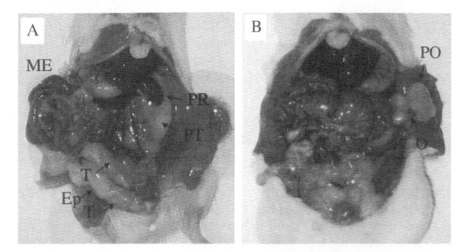

Fig. 2.2 Internal AT in male or female older rats. (**A**) The abdominal cavity of a 200-g weight male rat is opened. The left part corresponds to the natural appearance of abdominal cavity. In the right part, the testis (T) and the associated epididymal adipose pad (E) have been put in a prominent position. (**B**) Appearance of abdominal cavity of an older rat (body weight: 400 g) (**C**) Intestines are put on the right side and the left ovary (O) and periovarian pad (PO) placed on the opened abdominal wall. (**D**) The genital part was removed, and it is then possible to better distinguish the retroperitoneal (RP) and the perirenal fat (PR)

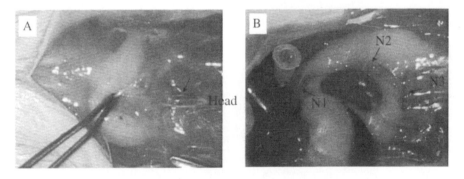

Fig. 2.3 Innervations of fat tissues. (**A**) Upper panel: The IBAT is lifted by fingers and the nervous fibers can be easily identified (arrow). (**B**) Lower panel: a, b, and c correspond to the three nervous fibers innervating the retroperitoneal pad

3.5 Denervation Studies

3.5.1 IBAT *(see Fig. 2.3)*

1. Proceed as in **Subheading 3.3.** to reach the BAT.
2. Carefully separate AT from muscle above the shoulders.

3. Carefully start to raise the IBAT; nerve fibers can now be seen arising from under each shoulder muscle.
4. Cut them at two points and remove the fragment to block regeneration.
5. Suture skin.

3.5.2 Retroperitoneal AT

1. The retroperitoneal fat pad is innervated by three nerve fibers, and the contralateral pad can be used as control. In this case, use the left or right one at random.
2. Proceed as in **Subheading 3.6.**, but the opening must be as small as possible. The goal is to maintain the animal alive after surgery.
3. Cut the three nerves at two points and remove the fragments.
4. Close the abdominal wall, then the skin.

3.6 Investigations via Vascular System

Almost all fat pads are individually vascularized, and it is possible to use such a feature to investigate some parameters (blood flow, arteriovenous differences, etc.) via the vascularization system or to inject particles (e.g., virus) with good preservation of the anatomy and the cellular interactions of the pad.

3.6.1 IBAT (see Fig. 2.4)

1. To easily obtain cannula with different diameters, use yellow tips for pipettes. Bring the center of the tip, which you hold at each end, near a flame. When the tip begins to melt and becomes translucent, rapidly stretch it to obtain catheters. The diameter depends on length of stretch. Cut one end obliquely.
2. Lay the anesthetised animal on one side, and carefully dissect the arterial vessel
3. Put in position the surgical silk, before doing a small incision in the artery.
4. After introducing the catheter into the artery, one can inject any solution (drugs, viral particles, etc.).

3.6.2 White Pad

A very good description of surgical and technical procedure was made by Scow *(72)* for the periovarian pad. Similar experiments can be performed with most fat pads.

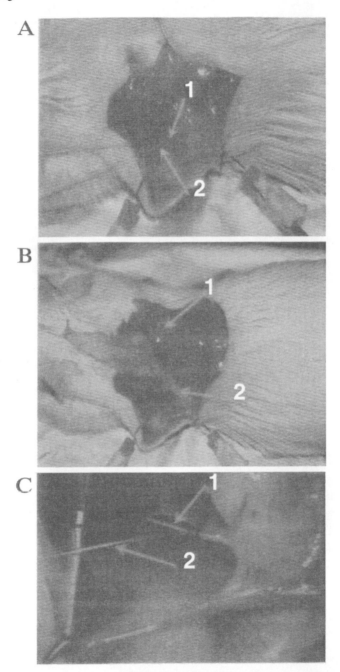

Fig. 2.4 Catheterization of arterial vessel of IBAT. (**A**) The rat is opened on the side and , after careful dissection, the arterial vessel of IBAT can be visualized (arrow 1). A catheter (arrow 2) is introduced into the vessel, and a surgical silk positioned to secure it in place. (**B**) After catheterization, the IBAT is washed with physiological buffer, and the IBAT is dissected and exposed. The washed part of the pad (arrow 2) appears clearer than the other part (arrow 1). (**C**) Enlargement of the catheterization. 1, arterial vessel; 2, catheter

References

1. Friedman JM, Halaas JL (1998) Leptin and the regulation of body weight in mammals. Nature 395:763–770
2. Valet P, Tavernier G, Castan-Laurell I, et al (2002) Understanding adipose tissue development from transgenic animal models. J Lipid Res 43:835–860
3. Kershaw EE, Flier JS (2004) Adipose tissue as an endocrine organ. J Clin Endocrinol Metab 89:2548–2556
4. Trayhurn P (2005) Endocrine and signalling role of adipose tissue: new perspectives on fat. Acta Physiol Scand 184:285–293
5. Casteilla L, Dani C (2006) Adipose tissue-derived cells: from physiology to regenerative medicine. Diabetes Metab 32:393–401
6. Dorheim MA, Sullivan M, Dandapani V, et al (1993) Osteoblastic gene expression during adipogenesis in hematopoietic supporting murine bone marrow stromal cells. J Cell Physiol 154:317–328
7. Gimble JM, Nuttall ME (2004) Bone and fat: old questions, new insights. Endocrine 23:183–188
8. Zuk PA, Zhu M, Ashjian P, et al (2002) Human adipose tissue is a source of multipotent stem cells. Mol Biol Cell 13:4279–4295
9. Gimble JM (1990) The function of adipocytes in the bone marrow stroma. New Biol 2:304–312
10. Gimble JM, Robinson CE, Wu X, et al (1996) The function of adipocytes in the bone marrow stroma: an update. Bone 19:421–428
11. Napolitano L (1965) The fine structure of adipose tissues. In: Renold AE, Cahill GF (eds): Adipose Tissue (Handbook of Physiology). American Physiological Society, Washington, DC, pp 109–124
12. Vague J, Fenasse R (1965) Comparative anatomy of adipose tissue. In: Renold AE, Cahill GF (eds): Adipose Tissue (Handbook of Physiology). American Physiological Society, Washington, DC, pp 25–36
13. Néchad M (1986) Structure and development of brown adipose tissue. In: Trayhurn P, Nicholls DG (eds) Brown adipose tissue. Arnold E., London, pp 1–30
14. Himms-Hagen J (1990) Brown adipose tissue thermogenesis: interdisciplinary studies. FASEB J 4:2890–2898
15. Nedergaard J, Connolly E, Cannon B (1986) Brown adipose tissue in the mammalian neonate. In: Trayhurn P, Nicholls DG (eds) Brown adipose tissue. Arnold E., London, pp 152–213
16. Rothwell N, Stock M (1986) Brown adipose tissue and diet-induced thermogenesis. In: Trayhurn P, Nicholls DG (eds) Brown adipose tissue. Arnold E., London, pp 269–298
17. Ricquier D, Bouillaud F (2000) Mitochondrial uncoupling proteins: from mitochondria to the regulation of energy balance. J Physiol 529:3–10
18. Cinti S (2006) The role of brown adipose tissue in human obesity. Nutr Metab Cardiovasc Dis 16:569–574
19. Nicholls DG, Locke RM (1984) Thermogenic mechanisms in brown fat. Physiol Rev 64:1–64
20. Casteilla L, Penicaud L, Cousin B, et al (2001) Choosing an adipose tissue depot for sampling. Factors in selection and depot specificity. Methods Mol Biol 155:1–19
21. Caspar-Bauguil S, Cousin B, Galinier A, et al (2005) Adipose tissues as an ancestral immune organ: site-specific change in obesity. FEBS Lett 579:3487–3492
22. Herpin P, Bertin R, Le Dividich J, et al (1987) Some regulatory aspects of thermogenesis in cold-exposed piglets. Comp Biochem Physiol A 87:1073–1081
23. Lafontan M, Berlan M (1993) Fat cell adrenergic receptors and the control of white and brown fat cell function. J Lipid Res 34:1057–1091
24. Giacobino JP (1995) Beta 3-adrenoceptor: an update. Eur J Endocrinol 132:377–385
25. Bartness T, Song CK (2007) Sympathetic and sensory innervation of white adipose tissue. J Lipid Res 48:1655–1672

26. Rosell S, Belfrage E (1979) Blood circulation in adipose tissue. Physiol Rev 59:1078–1104
27. Crandall DL, Hausman GJ, Kral JG (1997) A review of the microcirculation of adipose tissue: anatomic, metabolic, and angiogenic perspectives. Microcirculation 4:211–232
28. Ailhaud, G (2006) Adipose tissue as a secretory organ: from adipogenesis to the metabolic syndrome. C. R. Biol. 329:570–577; discussion 653–575
29. Gimeno RE, Klaman LD (2005) Adipose tissue as an active endocrine organ: recent advances. Curr Opin Pharmacol. 5:122–128
30. Tilg H, Moschen AR (2006) Adipocytokines: mediators linking adipose tissue, inflammation and immunity. Nat Rev Immunol 6:772–783
31. Dessolin S, Schalling M, Champigny O, et al (1997) Leptin gene is expressed in rat brown adipose tissue at birth. FASEB J 11:382–387
32. Cancello R, Zingaretti MC, Sarzani R, et al (1998) Leptin and UCP1 genes are reciprocally regulated in brown adipose tissue. Endocrinology 139:4747–4750
33. Ailhaud G, Grimaldi, P, Negrel, R (1992) Cellular and molecular aspects of adipose tissue development. Annu Rev Nutr 12:207–233
34. Cousin B, Cinti S, Morroni M, et al (1992) Occurrence of brown adipocytes in rat white adipose tissue: molecular and morphological characterization. J Cell Sci 103(Pt 4):931–942
35. Viguerie-Bascands N, Bousquet-Melou A, Galitzky J, et al (1996) Evidence for numerous brown adipocytes lacking functional beta 3-adrenoceptors in fat pads from nonhuman primates. J Clin Endocrinol Metab 81:368–375
36. Vague J, Rubin P, Jubelin J, et al (1974) The various forms of obesity. Triangle 13:41–50.
37. Bjorntorp P (1996) The regulation of adipose tissue distribution in humans. Int J Obes Relat Metab Disord 20:291–302
38. Bouchard C, Despres JP, Mauriege P (1993) Genetic and nongenetic determinants of regional fat distribution. Endocr Rev 14:72–93
39. Bjorntorp P (1997) Hormonal control of regional fat distribution. Hum Reprod 12(Suppl 1):21–25
40. Kissebah AH, Krakower GR (1994) Regional adiposity and morbidity. Physiol Rev 74:761–811
41. Despres JP, Lemieux I (2006) Abdominal obesity and metabolic syndrome. Nature 444:881–887
42. Tchernof A, Belanger C, Morisset AS, et al (2006) Regional differences in adipose tissue metabolism in women: minor effect of obesity and body fat distribution. Diabetes 55:1353–1360
43. Giorgino F, Laviola, L, Eriksson JW (2005) Regional differences of insulin action in adipose tissue: insights from in vivo and in vitro studies. Acta Physiol Scand 183:13–30
44. Laviola L, Perrini S, Cignarelli A, et al (2006) Insulin signaling in human visceral and subcutaneous adipose tissue in vivo. Diabetes 55:952–961
45. Berthiaume M, Laplante M, Festuccia W, et al (2007) Depot-specific modulation of rat intraabdominal adipose tissue lipid metabolism by pharmacological inhibition of 11{beta}-hydroxysteroid dehydrogenase type 1. Endocrinology 148:2391–2397
46. Tavernier G, Galitzky J, Valet P, et al (1995) Molecular mechanisms underlying regional variations of catecholamine-induced lipolysis in rat adipocytes. Am J Physiol 268, E1135–1142
47. Fried SK, Bunkin DA, Greenberg AS (1998) Omental and subcutaneous adipose tissues of obese subjects release interleukin-6: depot difference and regulation by glucocorticoid. J Clin Endocrinol Metab 83:847–850
48. Laplante M, Festuccia WT, Soucy G, et al (2006) Mechanisms of the depot specificity of peroxisome proliferator-activated receptor gamma action on adipose tissue metabolism. Diabetes 55:2771–2778
49. Niesler CU, Siddle, K, Prins JB (1998) Human preadipocytes display a depot-specific susceptibility to apoptosis. Diabetes 47:1365–1368
50. Festuccia WT, Laplante M, Berthiaume M, et al (2006) PPARgamma agonism increases rat adipose tissue lipolysis, expression of glyceride lipases, and the response of lipolysis to hormonal control. Diabetologia 49:2427–2436

51. Galinier A, Carriere A, Fernandez Y, et al (2006) Site specific changes of redox metabolism in adipose tissue of obese Zucker rats. FEBS Lett 580:6391–6398
52. Masuzaki H, Ogawa Y, Isse N, et al (1995) Human obese gene expression. Adipocyte-specific expression and regional differences in the adipose tissue. Diabetes 44:855–858
53. Montague CT, Prins JB, Sanders L, et al (1997) Depot- and sex-specific differences in human leptin mRNA expression: implications for the control of regional fat distribution. Diabetes 46:342–347
54. Russell CD, Petersen RN, Rao SP, et al (1998) Leptin expression in adipose tissue from obese humans: depot-specific regulation by insulin and dexamethasone. Am J Physiol 275, E507–515
55. Montague CT, Prins JB, Sanders L, et al (1998) Depot-related gene expression in human subcutaneous and omental adipocytes. Diabetes 47:1384–1391
56. Atzmon G, Yang XM, Muzumdar R, et al (2002) Differential gene expression between visceral and subcutaneous fat depots. Horm Metab Res 34:622–628
57. Gesta S, Bluher M, Yamamoto Y, et al (2006) Evidence for a role of developmental genes in the origin of obesity and body fat distribution. Proc Natl Acad Sci U S A 103:6676–6681
58. Hotamisligil GS (2006) Inflammation and metabolic disorders. Nature 444:860–867
59. Loncar D (1991) Convertible adipose tissue in mice. Cell Tissue Res 266:149–161
60. Champigny O, Ricquier D, Blondel O, et al (1991) Beta 3-adrenergic receptor stimulation restores message and expression of brown-fat mitochondrial uncoupling protein in adult dogs. Proc Natl Acad Sci U S A 88:10774–10777
61. Cousin B, Casteilla L, Dani C, et al (1993) Adipose tissues from various anatomical sites are characterized by different patterns of gene expression and regulation. Biochem J 292(Pt 3),873–876
62. Cinti S (2005) The adipose organ. Prostaglandins Leukot Essent Fatty Acids 73:9–15
63. Guerra C, Koza RA, Yamashita H, et al (1998) Emergence of brown adipocytes in white fat in mice is under genetic control. Effects on body weight and adiposity. J Clin Invest 102:412–420
64. Xue B, Coulter A, Rim JS, et al (2005) Transcriptional synergy and the regulation of Ucp1 during brown adipocyte induction in white fat depots. Mol Cell Biol 25:8311–8322
65. Moulin K, Arnaud E, Nibbelink M, et al (2001) Cloning of BUG demonstrates the existence of a brown preadipocyte distinct from a white one. Int J Obes Relat Metab Disord 25:1431–1441
66. Moulin K, Truel N, Andre M, et al (2001) Emergence during development of the white-adipocyte cell phenotype is independent of the brown-adipocyte cell phenotype. Biochem J 356:659–664
67. Gimble JM, Guilak F (2003) Differentiation potential of adipose derived adult stem (ADAS) cells. Curr Top Dev Biol 58:137–160
68. Planat-Benard V, Silvestre JS., Cousin B, et al (2004) Plasticity of human adipose lineage cells toward endothelial cells: physiological and therapeutic perspectives. Circulation 109:656–663
69. Villena JA, Cousin B, Penicaud L, et al (2001) Adipose tissues display differential phagocytic and microbicidal activities depending on their localization. Int J Obes Relat Metab Disord 25:1275–1280
70. Prunet-Marcassus B, Cousin B, Caton D, et al (2006) From heterogeneity to plasticity in adipose tissues: site-specific differences. Exp Cell Res 312:727–736
71. Peeters, LLH, Martensson L, Gilbert M, Pénicaud L (1984) The pregnant guinea pig, rabbit and rat as unstressed catheterized models. In: Nathanielz PW (ed) Animal Models in Fetal Medicine. Perinatology, Los Angeles, CA, pp 74–108
72. Scow RO (1965) Perfusion of isolated adipose tissue: FFA release and blood flow in rat parametrial fat body. In: Renold AE, Cahill GF (eds) Adipose tissue (handbook of physiology). American Physiological Society, Washington, DC, pp 435–454

Chapter 3
Application of Imaging and Other Noninvasive Techniques in Determining Adipose Tissue Mass

Wei Shen and Jun Chen

Summary In vivo adipose tissue quantification is an important tool to characterize phenotypes of obesity, especially in the human. The amount and distribution of adipose tissue is associated with many of the adverse consequences of obesity. Recent studies suggest that adipose tissue is not a single homogeneous compartment. Regional adipose tissue depots vary in biological functions and individual adipose tissue compartments have stronger associations with metabolic conditions than does total adipose tissue mass. Currently there is intense and increasing interest in regional adipose tissue compartments. Computed tomography and magnetic resonance imaging often are used to quantify adipose tissue volumes or cross-sectional adipose tissue areas. Other modalities, including dual-energy absorptiometry and magnetic resonance spectroscopy, provide whole-body or regional fat measures instead of adipose tissue mass quantification.

Key words Magnetic resonance imaging; computed tomography; adipose tissue; visceral adipose tissue; subcutaneous adipose tissue; intermuscular adipose tissue; body composition; abdominal obesity.

1 Introduction

Both the total amount and the distribution of adipose tissue (AT) is influenced by many factors, including sex, age, genotype, diet, physical activity, hormones, and drugs. In vivo AT quantification is important for investigating physiological and pathological conditions involving AT changes, such as obesity, insulin resistance, and lipodystropy.

Imaging methods, including computed tomography (CT) and magnetic resonance imaging (MRI), are considered the most accurate means available for in vivo quantification of AT, especially for human studies. The ability to measure different AT depots, such as subcutaneous AT (SAT) and visceral AT (VAT) with the use of these methods has greatly advanced our understanding of the relationships between

From: *Methods in Molecular Biology, Vol. 456:*
Adipose Tissue Protocols, Second Edition. Edited by: Kaiping Yang
© Humana Press, a part of Springer Science+Business Media, Totowa, NJ

adipose distribution and health risk. Although access and cost remain obstacles to routine use and technical skills are required, CT and MRI are now used extensively in body composition research.

The estimation of AT components with MRI is essentially the same as for CT. The choice between the two methods for studies is usually based on cost and availability of scanners as well as accuracy and reliability of image analysis (*see* **Note 1**). Without a concern for ionizing radiation, MRI is used by most investigators for multislice, whole-body, and longitudinal studies, especially in children or women in childbearing years. On the other hand, CT generates relatively consistent tissue attenuation values within image or among images and has been shown to provide slightly more reliable and repeatable data than MRI. To minimize radiation dose, CT has been mostly used as single-slice method in studying abdominal AT distribution.

Although commonly used interchangeably, AT and fat are different components according to the five-level body composition model (*1*). Although AT is a specialized loose connective tissue that is extensively laden with adipocytes, the molecular level or chemical component of fat is usually lipid in the form of triglycerides. AT contains ~80% fat and the remaining ~20% is water, protein, and minerals. Fat also exists in other tissues and the most widely used method for quantifying fat in vivo is dual-energy x-ray absorptiometry (DEXA). Magnetic resonance spectroscopy (MRS) methods also have been used for fat quantification; either regional (i.e., in skeletal muscle or liver) or at a whole body level (*see* **Note 2**). This chapter focuses on human AT quantification by CT and MRI methods. Animal CT and MRI methods and in vivo fat measurement methods (i.e., DEXA and MRS) are briefly summarized in the appendix.

2   Image Modalities and Software

2.1   *Instruments*

In human AT quantification studies, CT and MRI images are mostly acquired on clinical scanners. Major manufactures of CT scanners and MRI scanners include, but are not limited to, General Electric, Siemens, Philips, Picker, Toshiba, and Hitachi.

2.1.1   CT

The basic CT system consists of an x-ray tube that emits x-rays and a receiver that detects the attenuated x-rays having passed through tissues. During a scan, the x-ray tube and the receiver rotate in a perpendicular plane to the subject and the image is reconstructed with mathematical techniques. The x-ray attenuation is expressed as a measure of attenuation relative to air and water in Hounsfield units (HU), which gives contrast to the image and reflects the composition of different tissues. The CT number for AT pixels ranges from about −190 to −30 HU (*2,3*).

2.1.2 MRI

MRI is based on the interaction between hydrogen nuclei (protons), which is abundant in all tissue, and the magnetic fields of the MRI scanner. Protons in tissues behave like tiny magnets and align themselves with the magnetic field (usually no greater than 3.0 Tesla) in a known direction inside the magnet of an MRI unit. A pulsed radiofrequency field is then applied to the body tissues and "flips" hydrogen protons. When the radio frequency pulse is turned off, the protons gradually return to their original positions, in the process releasing energy in the form of a radio frequency signal. The signal is then detected and processed by the MRI system to generate the images. MRI signal contrast is based on differences in relaxation times between different tissues.

2.2 Image Analysis Software

Many software packages, either commercially available or developed in-house, have been reported for image analysis of CT and MRI. An example of commercially available software is sliceOmatic (Tomovision Inc., Montreal, Canada) and an example of free Image analysis software is Image J (http://rsb.info.nih.gov/ij/). In-house developed software usually runs in environments such as Interactive Data Language (IDL) or MATLAB. CT images can sometimes be analyzed with built in software package on the CT scanner console.

3 Image Acquisition

3.1 CT

CT is mostly used in single-slice studies and, with its relatively short acquisition time (i.e., several seconds), it is not considered a task for most subjects. Subjects are usually scanned in supine position. An example of parameters used are 120-kVp tube voltage, 220-mA tube current, 512 × 512 matrix (*4*).

3.2 MRI

MRI requires longer scanning time than CT, especially when the whole body is scanned. Before entering the MRI scanner room, subjects are always asked to remove all metal particles from their body (*see* **Note 3**). Subjects are scanned in either a supine or prone position and are instructed to lie still during the scan. Velcro bands may be used to maintain subjects' arms stretched above their heads without motion during upper body scans. When abdomen and chest are scanned, subjects are given instructions to hold their breath to reduce artifacts of respiratory motion. The most commonly used MRI sequence for AT quantification is T1-weighted spin echo sequence. An example of the parameters for T1-weighted sequences on a 1.5 Tesla scanner are repetition time/echo time (TR/TE) 210/17 ms, and matrix 256×256 (*see* **Note 4;** (*5*)). With these parameters, it takes approx. 20 to 30 min to acquire every 5 cm of whole body scan, including positioning time (*6*). The acquisition time can vary when subjects are scanned with other sequences such as fast spin echo, gradient echo, water suppressed, and Dixon method sequences.

Because whole-body scans have to be acquired in several series, between-series gaps are either prescribed equal to between-slice gaps or documented for the purpose of accurate volume calculation. It is also important to set anatomical landmark at "0" point and to double check that images of all series have been acquired completely.

CT and MRI data are usually saved in The Digital Imaging and Communications in Medicine (DICOM) format and are then transferred (FTP, CD, floppy etc.) to a separate workstation equipped with image analysis software (*see* **Note 5**).

3.3 Protocols

Based on the coverage of body regions, imaging protocol can be categorized into whole-body imaging protocol, multislice imaging protocol, and single-slice imaging protocol.

3.3.1 Whole-Body Protocol

CT is rarely used in whole body quantification because of concerns of high radiation. Using whole-body MRI data, researchers have the distinct advantages of detecting fat redistribtuion from one anatomical region to another or disproportional changes of AT across the whole body (*7*). Whole-body scans generally consist of multiple slices at predefined intervals (e.g., 3–5 cm) as contiguous scans require too much analyzing time. With subsampling, the coefficient of variation increases with interslice gap at a rate of 1.16%.cm (*6*). The extent to which variation in inter-image interval influences the derivation of subdepot of AT volume has been tested with the Visible Woman data set and the selection of the slice gap would depend on the accuracy requirements of the study (*8*).

3.3.2 Multislice Protocol

Some studies only focus on trunk AT and thus acquire multislice images at the trunk region either continuously or with a predefined interval. The anatomical coverage of the trunk region varies from study to study. Some studies included the whole trunk and consequently the quantified VAT consists of AT distributed in the three body cavities: intrathoracic (ITAT), intra-abdominal (IAAT), and intrapelvic (IPAT) *(3)*. Most investigators report visceral AT as IAAT, which is usually defined between superior border of the liver and L5-S1 intervertebral spaces *(9)*. Others define VAT as the sum of IAAT and IPAT ranged anatomically from the femoral heads or the bottom of pelvis to the liver dome *(6)*.

3.3.3   Single-Slice Protocol

Because of the cost of whole-body scans and concerns over exposure to radiation, the vast majority of studies have based their observations on a single-slice imaging, even though it conveys less information than multislice or whole-body scans. The single-slice CT and MRI studies are usually obtained at the level between the fourth and fifth lumbar vertebrae (L4-L5 level). However, this level has been recently proven not to be the location of the highest correlation between VAT volume and single-slice VAT area in large sample subjects *(5,10,11)*. VAT area at 5 to 10 cm above L4-L5 or at T12-L1 or L1-L2 level correlates the strongest with the VAT volume measure as well as health risks *(12,13)*. It is less of a concern for anatomical location in quantifying abdominal subcutaneous AT, as areas of SAT at all levels show high predictive values for estimating SAT volume *(10)*. On the other hand, if whole body instead of abdominal subcutaneous AT is of interest, area at 5 cm below L4-L5 is the strongest correlate (r = 0.96) with whole body subcutaneous AT *(14)*. It is important to recognize that single-slice studies only provide an area when reporting "adipose tissue," in contrast to the volumes reported in multiple-slice studies.

4   Image Analysis

Methods quantifying AT tissue area (cm^2) on each CT or MRI image can be divided into three categories based on the degree of automation, including: 1) manual methods, 2) semi-automated methods, and 3) fully automated segmentation. In manual segmentation, the technician traces the boundary of the interested tissue with a light pen, mouse, or track-ball controlled cursor *(15)*. Because it is time consuming and highly subjective to human input, purely hand tracing methods have been used much less in recent studies. Many efforts have been made towards fully automatic segmentation of AT and one successful example is automatic correction of inhomogeneity of magnetic field *(16)*. So far, claimed fully automatic methods are either of relatively low accuracy or have yet to be fully validated in diverse samples. Therefore, this chapter will only discuss the procedure for semiautomatic methods, which is currently the most widely used technique (*see* **Note 6**).

4.1 Semiautomatic Methods

The advantage of the semiautomatic technique is that it takes advantage of time-saving and precise computer automated procedures, whereas at the same time allowing for manual correction to achieve accuracy. Most of these steps are facilitated by using a transparent overlapping feature that allows the color tag to be superimposed on top of the original gray-level image.

4.1.1 Threshold-Based Tools

These tools are the automatic components of the semiautomatic procedure. All threshold-based tool-generated segment images are either completely or partially based on threshold differences between AT and other tissues. In general, threshold based automatic procedures identify AT "cleaner" in CT images than in MRI, especially in abdominal regions. Respiratory motion, intestinal movements, and blood flow cause artifacts in the abdomen to a greater extent in MRI than in CT (*see* **Note 7**). MRI images are also more difficult to segment than CT due to magnetic field heterogeneity *(7)*. Listed below are the most commonly used threshold based tools. A cross-sectional abdominal CT and a cross-sectional abdominal MRI are presented in **Fig. 3.1a** and **Fig. 3.2a**, respectively.

4.1.1.1 Histogram/Segmentation

These algorithms identify all pixels within a predefined range of intensities believed to be representative of AT. For CT images, the range of intensities of AT is usually −190 to −30 HU *(3)* but can vary, especially in subjects with pathological conditions such as lipodistrophy. For MRI, the range of AT intensity can possibly vary among scanners, protocols, subjects or even slices. Therefore, whether a study can use the same signal intensity range across subjects should be carefully evaluated and is not recommended in multi-center studies.

4.1.1.2 Region Growing/Seed Growing Function

Similar to histogram/segmentation, a threshold range needs to be defined for Region Growing. A seed is then selected within AT and regions that fall in the threshold range connected to the seed are then tagged (*see* **Fig. 3.1b**). Under this procedure, AT patches that are not connected to each other cannot be tagged all at once, and several seeds need to be selected. When inhomogeneity of magnetic field is present, region growing needs to be applied several times with different threshold ranges to tag anatomically connected AT (*see* **Fig. 3.2b**). Compared with

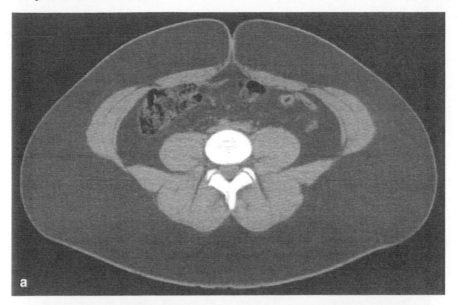

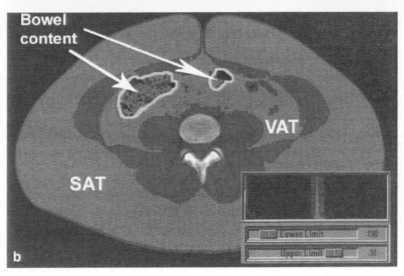

Fig. 3.1 Analysis of abdominal CT image. (**a**) Cross-sectional abdominal CT image. (**b**) Almost all AT is tagged by region growing applied only once with HU range set as −190 to −30. However, bowel contents (with boundaries highlighted) need to be removed with manual correction. SAT, subcutaneous adipose tissue; VAT, visceral adipose tissue

histogram/segmentation, region growing requires more skill and experience. On the other hand, by excluding non-AT that overlaps signal intensity with AT, region growing yields a more accurate segmentation and requires less manual editing than histogram/segmentation.

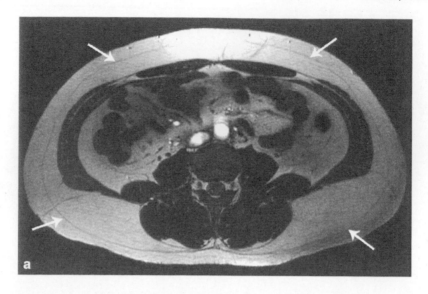

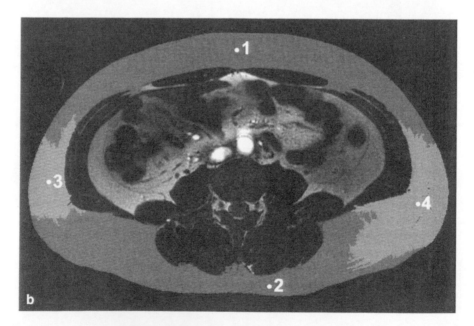

Fig. 3.2 Analysis of abdominal MRI. (**a**) Cross-sectional abdominal MRI with arrows indicating the thin fascial plane separating superficial and deep subcutaneous adipose tissue. (**b**) When inhomogeneity of magnetic field is present, region growing is applied four times with different threshold ranges to tag anatomically connected adipose tissue. Four seeds are indicated as 1, 2, 3 and 4, respectively.
(continued)

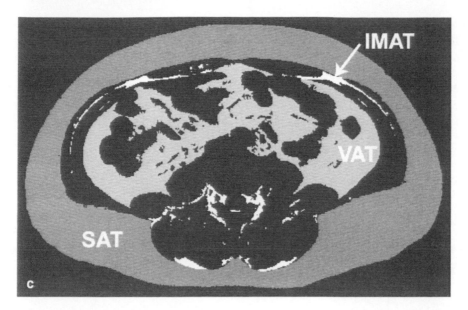

Fig. 3.2 (continued) (**c**) The final segmentation of different adipose tissue depots including sub-cutaneous adipose tissue (SAT), visceral adipose tissue (VAT), and intermuscular adipose tissue (IMAT)

4.1.1.3   Morphology

Morphology segmentation is done by automatically computing the local threshold contrast. This algorithm subdivides an image into many regions based on a local threshold. Each region should include no more than one tissue type. All regions corresponding to AT are then filled with tags. This procedure works well when contrast between AT and non-AT is high. When the contrast is low, some divided regions can contain both AT and non-AT.

The choice among histogram/segmentation, region growing/speed growing function, morphology and other automatic techniques depends on image quality, subject characteristics, and the subdepots of AT of interest. Sometimes it may require a combination of several automatic algorithms to correctly and efficiently identify AT.

4.1.2   Manual Correction

Manual correction is usually a necessary step following the automatic procedures regardless of which automatic procedures have been chosen. Erasing tools are used to remove color-tag from non-AT, such as bowel content and bone structures (*see* **Fig. 3.1b**). AT missed by automatic procedures is also manually tagged by tools such as paint brush. If histogram/segmentation is the major automatic tool used, special attention should be given to inflated or deflated regional boundaries of AT caused by inhomogeneity of magnetic field. Manual delineation is also used to separate AT components from each other (i. e., subcutaneous, visceral and intermuscular AT) and to assign different color tags for each component.

4.2   Subdivisions of AT

Total AT depots include three major components: subcutaneous, visceral, and intermuscular AT (*see* **Fig. 3.2c**; *see* **Note 8**). A detailed total body and regional AT classification system for imaging studies has been proposed by Shen et al. *(17)* (*see* **Note 9**).

4.2.1   Subcutaneous AT Segmentation

The subcutaneous AT of the lower trunk and the gluteal-thigh region can be divided into superficial and deep portions by a thin fascial plane (*see* **Fig. 3.1a**; *see* **Note 10**). As the fascial plane is not always visible on imaging and the majority of deep subcutaneous AT is located in the posterior half of the abdomen (18), anterior and posterior abdominal subcutaneous AT are sometimes used as approximates of superficial and deep subcutaneous abdominal AT *(19)*. A line can be drawn dissecting the abdomen into anterior and posterior depots using the anterior edge of the vertebrae *(20)*.

4.2.2   Visceral AT

On a cross-sectional image, the anatomical boundary for visceral AT is the inner abdominal muscle wall, which excludes inter-muscular and para-vertebral ATs *(17)*. Anatomically, visceral AT can be further divided into intraperitoneal and retroperitoneal AT *(15)* (*see* **Note 10**). Because the peritoneum can rarely be identified on cross-sectional images, an arbitrary method is adopted in which a straight line is drawn across the anterior border of vertebra and the psoas muscles, continuing on a tangent towards the posterior borders of the ascending and descending colon and extending to the abdominal wall *(21,22)*. For images in which the kidneys appear, an oblique line is drawn from the anterior border of the aorta and inferior vena cava to the anterior border of the kidney extending to the abdominal wall. On all images, extra-peritoneal AT is defined as the VAT located posterior to the lines drawn.

4.2.3 Intermuscular AT

Intermuscular AT is the AT distributed within the muscle region. Intermuscular AT is separated from subcutaneous and visceral AT by fascial planes (*see* **Fig. 3.2a, 3.2c** *(23,24)*).

4.3 Volume Extrapolation

The segmented AT area of a single image can be calculated by multiplying the number of included pixels by the area of an individual pixel. If multiple CT or MRI images are obtained, AT volumes are then calculated using geometrical models based on the tissue areas in the images and the distance between adjacent images. Adapted from the "parallel trapezium" model created by Kvist et al. *(3)*, a "two-column" geometrical model has been proposed by Shen et al. for estimating tissue volumes with cross-sectional images *(8)*:

$$V = (t + h)\sum_{i=1}^{N} A_i$$

where t is the thickness of each image, h is the distance (gap) between consecutive images, A_i is the tissue area (cm^2) of each image, and N is the total number of images.

The truncated cone model *(25)* has also been used but was found to slightly underestimate tissue volumes using the Visible Women *(26)* as in reference *(8)*.

CT and MRI volume measures for ATs can be converted to mass units by multiplying the volume with the AT density, which is assumed constant at 0.92 g/cm^3 *(27)*.

4.4 Reproducibility

Although true reproducibility would include testing duplicated acquisitions separated by days and repeated analysis separated by weeks, most investigators test reproducibility by duplicated analysis on the same CT or MRI images. The latter can be viewed as an evaluation of image readings. Reproducibility usually includes intra-observer and inter-observer reproducibility and both should be tested and reported for quality control purposes and for the readers' knowledge of accuracy. Interobserver reproducibility is tested by two or more individuals analyzing the same scans, whereas intraobserver reproducibility is tested by the same reader reading the same scans at different time points. Reproducibility is usually reported as %CV or intraclass correlation (*see* **Note 11**).

The coefficients of variation (CV) for repeated subcutaneous AT measurements by CT and MRI are similar and in the range of ~2% *(28,29)*. The CVs for visceral AT measurements by MRI are ~6–18% *(4,28,29)* and by CT ~2% *(30)*. The lower CV of CT is usually ascribed to a shorter image acquisition time and CT is thus less vulnerable to image artifacts produced by peristaltic intestinal movement *(29)* (*see* **Note 12**).

5   Notes

1. CT and MRI measures of abdominal subcutaneous AT are highly correlated and comparable. In general, CT requires shorter acquisition time than MRI for the same resolution, but recent advances in MRI technology have reduced the acquisition time dramatically and have made MRI much more accessible for AT quantification.
2. Refer to components evaluated by CT and MRI as "AT" instead of the chemical term "fat." Body fat or tissue fat is measured with the DXA or MRS method. This differentiation will leave no question as to what body constituent is actually being evaluated.
3. Great caution should be taken when scanning subjects with metal implants. Even without safety concerns, signal loss and distortion of image can impair the validity of results.
4. Maximum field of view (FOV; i.e., 50 cm for conventional CT and 48 cm for most MRI scanners) should be used to acquire images if obese subjects are studied. Some very obese subjects may be too large to fit into the scanners. Furthermore, as the actual field of view of the scanner is smaller than the bore diameter, out of field of view can still occur when the subject can fit in the scanner. Out of FOV tends to happen at the widest body regions such as hip and shoulder. When out of FOV happens, there is always an underestimation of SAT.
5. Make sure the images acquired on scanners are saved in a format compatible with the image analysis software.
6. Training in cross-sectional anatomy is essential even for analysts with a medical background.
7. Image analysis technicians should be trained to recognize artifacts including those caused by respiratory motion, blood flow, and intestinal movement, etc. Correctly identifying artifacts increases accuracy and precision of MRI estimates, especially in VAT segmentation.
8. "Total body" AT measured by imaging methods may be different from the actual volume of AT determined by dissection and histological analysis. Because bone marrow AT does not change with energy storage, it is usually not included in quantification of total-body AT in imaging methods. AT in the head, feet, and hands is difficult to distinguish from bone marrow AT, and these tissues are usually labeled as non-AT.

9. Because investigators differ in their interests and definitions of various AT compartments and there is considerable variation among study acquistion and analysis protocols, we suggest that any AT depot under study should be accurately named and characterized, with both the number and location of the (CT or MRI) images clearly defined and documented.

10. It is unclear whether segmenting SAT into deep and superficial compartments, or VAT into intraperitoneal and extraperitoneal compartments improves understanding of the relationships observed between AT and metabolic risk factors. Because the approximation sub-depot segmentation methods cause uncertainties of the accuracy, technological advances in accurately quantifying sub-depots of AT may improve our understanding of the biological role of these depots.

11. High reproducibility does not necessarily mean accuracy if image analysis is performed using automated methods only. Judgment and corrections by experienced readers are highly recommended for image analysis. It should be noted that, without exception, the CVs presented for CT represent repeated analysis using a single HU range is always low, especially when no manual correction (i.e., removal of bowel contents) is applied. HU range for AT may vary in subjects with disease conditions such as edema and lipodystrophy etc.

12. Reproducibility of AT is high in obese subjects or large adipose depots but is low in lean subjects, children, and small AT depots. It is always recommended to establish reproducibility for each study.

Appendix

Adipose Tissue Quantification With MRI and CT in Animals

The principles for human and animal AT acquisitions and analyses are the same. Animals such as primates, rabbits, dogs, guinea pigs, and rats can be scanned by human scanners *(31)*. Smaller coil may be used to facilitate signal detection and reduce scanning time of MRI. To acquire adequate resolution in a reasonable scanning time, small rodents (i.e., mice) are scanned by high magnetic field scanners (i.e., 7–9 Tesla MRI scanner) *(32)*. Animals are anesthetized before being scanned. The procedures of image analysis of animals are primarily the same as that of humans except for animal anatomy training.

MRS Methods

Proton MRS methods quantify fat content in a defined volume but are not usually used for AT mass quantification. MRS methods are mostly used to quantify liver and skeletal muscle fat content (i.e., intramyocellular lipid) in both human and animal studies *(33)*. MRS spectroscopy can also provide estimates of tissue free fatty acid composition *(34)*. Most clinical scanners either have MRS sequences installed or have the option to install a MRS package.

Quantitative magnetic resonance (QMR) method quantifies whole body fat, and both animal and human QMR instruments are different from MRI scanners. QMR method does not provide images and therefore cannot provide regional AT distribution information. The advantage of QMR method is that the instrument requires only simple maintenance and no anesthesia is required for animal studies *(35)*.

DEXA Measurement of Fat Content

DEXA estimates masses of whole body and regional fat rather than AT. The regional fat estimation includes arms, legs and trunk etc, but DEXA cannot differentiate visceral fat from subcutaneous fat or intermuscular fat. DEXA instruments have human and animal models and the mostly used brands are GE Lunar and Hologic. DEXA machines utilize X-ray of two different energy levels and body composition is calculated by the attenuation of the X-rays having passed through tissues *(36)*. The amount of radiation exposure of DEXA is very small and usually not a concern, except in pregnant women. The analysis of DEXA scan can only be done with the software packages provided by the same manufacturer.

References

1. Wang ZM, Pierson RNJ, Heymsfield SB (1992) The five level model: a new approach to organizing body composition research. Am J Clin Nutr 56:19–28
2. Chowdhury B, Kvist H, Andersson B, Bjorntorp P, Sjostrom L (1993) CT-determined changes in adipose tissue distribution during a small weight reduction in obese males. Int J Obes Relat Metab Disord 17:685–91
3. Kvist H, Sjostrom L, Tylen U (1986) Adipose tissue volume determinations in women by computed tomography: technical considerations. Int J Obes Relat Metab Disord 10:53–67
4. Mitsiopoulos N, Baumgartner RN, Heymsfield SB, Lyons W, Gallagher D, Ross R (1998) Cadaver validation of skeletal muscle measurement by magnetic resonance imaging and computerized tomography. J Appl Physiol 85:115–122
5. Shen W, Punyanitya M, Wang Z, Gallagher D, St-Onge MP, Albu J, Heymsfield SB, Heshka S (2004) Visceral adipose tissue: relations between single-slice areas and total volume. Am J Clin Nutr 80:271–278
6. Thomas EL, Saeed N, Hajnal JV, Brynes A, Goldstone AP, Frost G, Bell JD (1998) Magnetic resonance imaging of total body fat. J Appl Physiol 85:1778–1785
7. Ross R, Leger L, Morris D, Guise JD, Guardo R (1992) Quantification of adipose tissue by MRI: relationship with anthropometric variables. J Appl Physiol 72:787–795
8. Shen W, Wang Z, Tang H, Heshka S, Punyanitya M, Zhu S, Lei J, Heymsfield SB. (2003). Volume estimates derived in vivo by imaging methods: model comparisons with visible woman as the reference. Obes Res 11:217–225
9. Ross R (1997) Effects of diet- and exercise-induced weight loss on visceral adipose tissue in men and women. Sports Med 24:55–64

10. Demerath EW, Shen W, Lee M, Choh A, Czerwinski SA, Siervogel RM, Towne B (2007). Estimation of total visceral adipose tissue with a single MR image. Am J Clin Nutr 85:362–368

11. Abate N, Garg A, Coleman R, Grundy SM, and Peshock RM (1997) Prediction of total subcutaneous abdominal, intraperitoneal, and retroperitoneal adipose tissue masses in men by a single axial magnetic resonance imaging slice. Am J Clin Nutr 65:403–408

12. Shen W, Punyanitya M, Chen J, Gallagher D, Albu J, Pi-Sunyer X, Lewis CE, Grunfeld C, Heymsfield SB, Heshka SR (2007) Visceral adipose tissue: relationships between single slice areas at different locations and obesity-related health risks. Int J Obes, 31:763–769

13. Kuk JL, Church TS, Blair SN, Ross R (2006) Does measurement site for visceral and abdominal subcutaneous adipose tissue alter associations with the metabolic syndrome? Diabetes Care 29:679–684

14. Shen W, Punyanitya M, Wang Z, Gallagher D, St-Onge MP, Albu J, Heymsfield SB, Heshka S (2004) Total body skeletal muscle and adipose tissue volumes: estimation from a single abdominal cross-sectional image. J Appl Physiol 97:2333–2338

15. Abate N, Burns D, Peshock RM, Garg A, Grundy SM (1994) Estimation of adipose tissue mass by magnetic resonance imaging: validation against dissection in human cadavers. J Lipid Res 35:1490–1496

16. Yang GZ, Myerson S, Chabat F, Pennell DJ, Firmin DN (2002) Automatic MRI adipose tissue mapping using overlapping mosaics. MAGMA 4:39–44

17. Shen W, Wang ZM, Punyanita M, Lei J, Sinav A, Kral JG, Imielinska C, Ross R, Heymsfield SB (2003) Adipose tissue quantification by imaging methods: a proposed classification. Obes Res 11:5–16

18. Kelley DE, Thaete FL, Troost F, Huwe T, Goodpaster BH (2000) Subdivisions of subcutaneous abdominal adipose tissue and insulin resistance. Am J Physiol Endocrinol Metab 278, E941–E948

19. Misra A, Garg A, Abate N, Peshock RM, Stray-Gundersen J, Grundy SM (1997) Relationship of anterior and posterior subcutaneous abdominal fat to insulin sensitivity in nondiabetic men. Obes Res 5:93–99

20. Ross R, Aru J, Freeman J, Hudson R, and Janssen I (2002) Abdominal adiposity and insulin resistance in obese men. Am J Physiol Endocrinol Metab 282: E657–E663

21. Ross R, Rissanen J, Pedwell H, Clifford J, Shragge P (1996). Influence of diet and exercise on skeletal muscle and visceral adipose tissue in men. J Appl Physiol 81:2445–2455

22. He Q, Engelson ES, Albu JB, Heymsfield SB, Kotler DP (2003) Preferential loss of omental and mesenteric adipose tissue during growth hormone therapy of HIV-associated lipodystrophy. J Appl Physiol 94:2051–2057

23. Gallagher D, Kuznia P, Heshka S, Albu J, Heymsfield SB, Goodpaster B, Visser M, Harris TB (2005) Adipose tissue in muscle: a novel depot similar in size to visceral adipose tissue. Am J Clin Nutr 81:903–010

24. Janssen I, Fortier A, Hudson R, Ross R (2002) Effects of an energy-restrictive diet with or without exercise on abdominal fat, intermuscular fat, and metabolic risk factors in obese women. Diabetes Care 25:431–438

25. Ross R (1996) Magnetic resonance imaging provides new insights into the characterization of adipose and lean tissue distribution. Can J Physiol Pharmacol 74:778–785

26. US National Library of Medicine (1995) Visible Human Project. Visible Human CD-ROM, Version 1.1

27. Snyder WS, Cooke MJ, Manssett ES, Larhansen LT, Howells GP, Tipton IH (1975) Report of the Task Group on Reference Man, Pergamon, Oxford

28. Elbers JM, Haumann G, Asscheman H, Seidell JC, Gooren LJ (1997) Reproducibility of fat area measurements in young, non-obese subjects by computerized analysis of magnetic resonance images. Int J Obes Relat Metab Disord 21:1121–1129

29. Seidell JC, Bakker CJ, van der Kooy K (1990). Imaging techniques for measuring adipose-tissue distribution—a comparison between computed tomography and 1.5-T magnetic resonance. Am J Clin Nutr 51:953–957

30. Thaete FL, Colberg SR, Burke T, Kelley DE (1995) Reproducibility of computed tomography measurement of visceral adipose tissue area. Int J Obes Relat Metab Disord 19:464–467
31. Ross R, Leger L, Guardo R, De Guise J, Pike BG (1991) Adipose tissue volume measured by magnetic resonance imaging and computerized tomography in rats. J Appl Physiol 70:2164–2172
32. Changani KK, Nicholson A, White A, Latcham JK, Reid DG, Clapham JC (2003) A longitudinal magnetic resonance imaging (MRI) study of differences in abdominal fat distribution between normal mice, and lean overexpressers of mitochondrial uncoupling protein-3 (UCP-3). Diabetes Obes Metab 5:99–105
33. Szczepaniak LS, Babcock EE, Schick F, Dobbins RL, Garg A, Burns DK, McGarry JD, Stein DT (1999) Measurement of intracellular triglyceride stores by H spectroscopy: validation in vivo. Am J Physiol 276:E977–E989
34. Thomas EL, Taylor-Robinson SD, Barnard ML, Frost G, Sargentoni J, Davidson BR, Cunnane SC, Bell JD (1997) Changes in adipose tissue composition in malnourished patients before and after liver transplantation: a carbon-13 magnetic resonance spectroscopy and gas-liquid chromatography study. Hepatology 25:178–183
35. Tinsley FC, Taicher GZ, Heiman ML (2004) Evaluation of a quantitative magnetic resonance method for mouse whole body composition analysis. Obes Res. 12:150–160
36. Pietrobelli A, Formica C, Wang Z, Heymsfield SB (1996) Dual-energy X-ray absorptiometry body composition model: review of physical concepts. Am J Physiol 271: E941–E951

Chapter 4
Generation of Adipose Tissue-Specific Transgenic Mouse Models

Xian-Cheng Jiang

Summary Adipose tissue plays a critical role in energy homeostasis, not only in storing triglycerides, but also in responding to nutrient, neural, and hormonal signals, and secreting adipokines that control feeding, thermogenesis, immunity, and neuroendocrine function. It is conceivable that adipose tissue-specific gene expression would influence the aforementioned functions. A feasible approach to prepare adipose tissue-specific transgenic mouse models is necessary for such studies. Here, we report the preparation of adipose tissue-specific cholesteryl ester transfer protein transgenic mice. The general principle might apply to the establishment of other adipose tissue-specific transgenic mice models.

Key words Adipocytes; fat-specific fatty acid-binding protein (aP2) promoter/enhancer; adipose tissue-specific transgenic mouse; CETP; plasma lipoprotein metabolism.

1 Introduction

There has been a paradigm shift from the notion of adipose tissue merely as a storage site for energy to one where adipose tissue plays an active role in energy homeostasis and various processes *(1)*. The predominant type of adipose tissue, commonly called "fat" in mammals, is white adipose tissue (WAT). WAT comprises of mostly adipocytes surrounded by loose connective tissue that is highly vascularized and innervated, and contains macrophages, fibroblasts, adipocyte precursors, and various other cell types *(2)*. The increase in WAT mass in obesity is associated with profound histological and biochemical changes characteristic of inflammation *(3,4)*. Obesity also results in an increase in fibrinogen, plasminogen activator inhibitor-1, and various coagulation factors *(5)*. Visceral adiposity in the human has been associated with cardiovascular disease *(6)*. Increasing WAT formation in the subcutaneous region was associated with improvement in glucose and lipid metabolism and reduced incidence of atherosclerosis *(7)*.

From: *Methods in Molecular Biology, Vol. 456:*
Adipose Tissue Protocols, Second Edition. Edited by: Kaiping Yang
© Humana Press, a part of Springer Science+Business Media, Totowa, NJ

Many researchers *(8–11)*, including ourselves, have established adipose tissue-specific transgenic mice to study the function of adipocyte-derived genes. A common feature of these approaches is that fatty acid-binding protein (aP2), a fat-specific promoter/enhancer, is used. aP2 promoter/enhancer has also been used to prepare aP2-Cre recombinase transgenic mice, which can be utilized for adipose tissue-specific gene knockout mice preparation *(12)*. The strategy for adipose-specific gene knockout mice preparation is outlined in **Fig. 4.1**. In this chapter, we will discuss the preparation of aP2-cholesteryl ester transfer protein (CETP) transgenic (Tg) mice. We believe that our experience with aP2-CETP Tg mice preparation will help other researchers who are interested in establishing adipose tissue-specific transgenic mice.

CETP is a hydrophobic plasma glycoprotein that mediates the transfer and exchange of cholesteryl esters (CEs) and triglycerides between plasma lipoproteins and plays an important role in HDL metabolism *(13)*. Mice do not express CETP, and human CETP transgenic mice, with predominant liver expression, show reduced

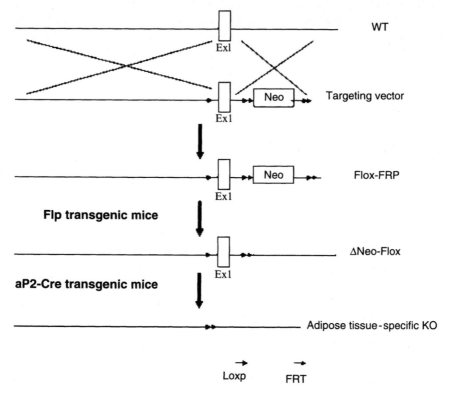

Fig. 4.1 The strategy for adipose tissue-specific gene knockout mouse preparation. We assume that Exon 1(Ex1) contains the start codon and deplete Ex1 could eliminate the gene expression. *(1) To prepare Flox-FRT mice:* first, a Loxp site in 5' flanking region, then, a neomycin resistance gene (Neo) cassette, flanking with one pair of Loxp and one pair of FRT, in intron 1, are inserted. *(2) To prepare ΔNeo-Flox mice.* To delete Neo cassette, heterozygous Flox-FRT mice are crossed with Flp transgenic mice *(29,30)* from Jackson Laboratories (Maine). *(3) Adipose-specific knockout mouse preparation.* Homozygous of ΔNeo-Flox mice cross with aP2-Cre transgenic mice to produce adipose-specific knockout mice

HDL-C levels *(14)*. However, the liver is not the major source of CETP in all mammalian species. In hamsters, for example, adipose tissue shows the highest levels of CETP expression and CETP mRNA is almost undetectable in the liver *(15)*. Adipose tissue appears to be a highly conserved site of CETP expression across species. However, its function in adipose tissue is still unknown. It has been reported that plasma CETP concentrations are positively correlated with adipose tissue CETP mRNA levels in hamsters *(16)*. In monkeys, a correlation between adipose CETP mRNA and CETP levels was also observed *(17)*. Moreover, human adipose tissue maintained in organ culture synthesizes and secretes CETP *(18)*. Other studies have shown that plasma CETP activity in humans correlates with the degree of adiposity *(19,20)* and that weight reduction is associated with a decrease in plasma CETP activity *(19)*. A hamster study also indicated that adipose tissue releases CETP activity during incubation in vitro and is subject to hormonal and nutritional regulation *(21)*. Furthermore, the release of CETP activity from cultured hamster adipose tissue increased following a period of fasting *(22)*. Despite these correlative findings, there is no direct evidence that adipose-derived CETP enters the circulation or that it influences plasma lipoprotein levels. In order to investigate these issues, we established an adipose tissue-specific CETP transgenic mouse line (*see* **Notes 1** and **2**).

2   Materials

2.1   Creation of aP2-CETP-pBK Vector

1. 5.4-kb mouse aP2 enhancer/promoter region.

2.2   aP2-CETP Transgenic Mouse Preparation

2.2.1   Animal Setting

1. Ten C57BL/6J or FVB mice (4–6 wk, the optimal age for injecting pregnant mare's serum) for harvesting fertilized eggs (*see* **Note 3**).
2. Ten C57BL/6J or FVB male mice (8–20 wk) for mating.
3. Ten to fifteen Swiss Webster female mice (7–15 wk) for preparing pseudopregnant females.
4. Ten to fifteen Vasectomized male mice (sterile stud male mice, 8–24 wk) for mating with Swiss Webster females.

2.2.2   Harvesting and Washing Eggs

1. M2 Medium (Sigma): use at room temperature.
2. M16 Medium (Sigma): use at 37°C.

2.2.3   Microinjecting DNA Into Pronuclei

1. Microinjection system (Eppendorf).

2.3   aP2-CETP Transgenic Mouse Screening

1. Tail Tip DNA Digestion Solution: 100 mM NaCl; 10 mM Tris-HCl, pH 8.0; 2.5 mM EDTA; 0.5% sodium dodecyl sulfate; 0.1 mg/mL Proteinase K.

2.4   CETP Activity Assay

1. Assay buffer: 10 mM Tris-HCl, 0.9% Sodium Chloride, 2 mM EDTA, pH 8.0.
2. A fluorescence plate reader (7620 Microplate Fluorimeter; Cambridge Technology, Watertown, MA).

3   Methods

3.1   Creation of aP2-CETP-pBK Vector

1. Digest a 5.4-kb mouse aP2 enhancer/promoter region using *Kpn* I and *Not* I from a aP2- pBluescript SK vector (*see* **Fig. 4.2**; *(23)*).
2. Ligate the fragment to upstream region of CETP-pBK vector that contains a 6.8-Kb human CETP minigene (*see* **Note 4**; **Fig. 4.2**; *(24)*).

3.2   aP2-CETP Transgenic Mouse Preparation

3.2.1   Animal Setting

1. On day 1 at 1:00 pm perform intraperitoneal injection of pregnant mare's serum at 5 units/C57BL/6J or FVB female mouse (to mimic follicle-stimulating hormone and induce superovulation).
2. On day 3 at 1:00 pm, perform intraperitoneal injection of human chorionic gonadotropin at 5 units/C57BL/6J or FVB female mouse (to mimic LH and induce superovulation). Set up mating, C57BL/6J female with C57BL/6J male, or FVB female with FVB male for harvesting fertilized eggs, and

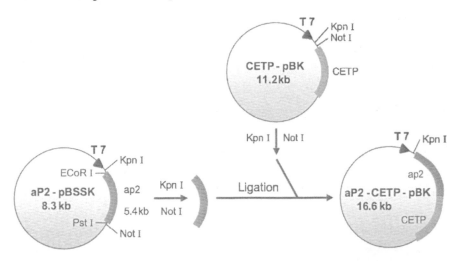

Fig. 4.2 Creation of aP2-CETP-pBK vector. A 5.4-kb mouse aP2 enhancer/promoter region was digested by *Kpn* I and *Not* I from aP2- pBluescript SK vector and then ligated to upstream of CETP-pBK vector which contains an 6.8 kb human CETP minigene

Swiss Webster female with vasectomized male for preparing pseudopregnant female.

3. On day 4 at 9 am, check the plugged Swiss Webster females that will be used as pseudopregnant egg recipient females and collect mated FVB or C57BL/6J females for egg harvesting.

3.2.2 Harvesting and Washing Eggs

1. Open the abdominal cavity, find, and cut both sides of coils of oviducts
2. In M2 medium, break the oviducts to release the eggs, add and incubate with hyaluronidase (300 µg/mL) for 30 min to remove cumulus cells.
3. Wash the eggs by transferring the eggs to a dish containing fresh M2 medium three times. Then, transfer the clean eggs to M16 medium to culture for two hours at 37°C. The eggs are used for injection.

3.2.3 Microinjecting DNA into Pronuclei

A aP2-CETP-pBK vector was digested by *Kpn* I and became a linear DNA fragment. The final concentration was 5 µg/mL in 10 mM Tris and 0.1 mM EDTA.

1. Set up the micromanipulators with holding pipette and injection needle. Set up injection chamber containing M2 medium covered with mineral oil.

2. Transfer 10 to 20 eggs into the chamber each time and inject the optimal of DNA into the pronuclei of the eggs (a swollen pronuclei should be se
3. Move the injected eggs back to M16 medium at 37°C until transferring oviduct of pseudopregnant females.

3.2.4 Oviduct Transfer

This step can be done right after microinjection or the day after injection.

1. Weigh the pseudopregnant egg recipient mouse and then anesthetize by ir it intraperitoneally with 0.016 mL/g of 2.5% avertin.
2. On the lateral side of the mouse, make a small incision of skin and underner cle, and then pull out the ovary, oviduct, and part of the uterus. Find the inf lum (the opening to the oviduct, usually hidden within coils of oviduct).
3. Insert the transfer pipet containing eggs into the infundibulum and blow pipet until the eggs enter the oviduct (20–30 eggs for each side of ovidu
4. Put back the ovary, oviduct, and uterus. Close the incision with a suture.

The pups are delivered 20–21 d after oviduct transfer. Genotyping can be per when the young mice reach 4–5 wk old.

3.3 aP2-CETP Transgenic Mouse Screening

1. Cut 0.5 cm of mouse tail and put into 0.5 mL of DNA Digestion Solution rotate overnight).
2. Centrifuge the tube and carefully transfer the supernatant to a new tube.
3. Add 0.5 mL of phenol to extract the DNA once.
4. Add 0.5 mL of $CHCl_3$ to extract the DNA once.
5. Add 1 mL of EtOH (you can see the DNA at this time), centrifuge the high speed for 10 min.
6. Wash the pellet once with 70% EtOH.
7. Dissolve the pellet with 60 µL of TE and incubate in 56°C for at least 2
8. The DNA solution can be used to do PCR and Southern blot.

3.4 CETP Activity Assay

The CETP activity was measured by a fluorescence method (Roar Biomedic New York, NY), which is comparable with the radiolabeled method (18).

1. Add 4 µL of donor, 4 µL of acceptor, 3 µL of plasma, and 89 µL of assa and incubate for 1 h.
2. Read emission at 460 nm and excitation 530 nm.

4 Notes

1. CETP mRNA was predominantly expressed in adipose tissue in our two lines of aP2-CETPTg mice. There were also low levels of expression in heart and muscle (*see* **Fig. 4.3**). This might be attributable to the contamination of adipose tissue in both organs. For instance, the epicardium contains substantial adipose tissue that could contribute to the total RNA prepared from heart. It has been reported that the aP2 promoter/enhancer also functions in macrophages *(25,26)*. However, there was no detectable CETP mRNA in the macrophages from aP2-CETPTg mice (*see* **Fig. 4.3**). Although the expression levels are negligible compared with adipose tissue, it is still not known why lung also can express low levels of aP2-CETP transgene (*see* **Fig. 4.3**).

2. To evaluate the impact of CETP expression on adipose tissue lipid levels, we extracted total lipids from both wild-type and aP2-CETPTg mouse adipose tissues and measured triglyceride, cholesterol, and phospholipids. We found that both triglyceride and cholesterol levels were significantly decreased (50% and 40%, $p < 0.001$ and $p < 0.01$, respectively) compared with the wild type, whereas no significant change was found in phospholipid levels. It has been reported that when triglyceride and cholesterol stores are depleted, adipocytes reduce their size *(27,28)*. To see whether this is the case in the aP2-CETPTg adipocyte, we dissected the adipose tissue and stained it with hematoxylin and eosin. The images were analyzed to determine the cross-sectional surface area of each adipocyte. aP2-CETPTg mice had a greater number of small adipocytes than wild-type mice (*see* **Figs. 4.4A** and **4.4B**) . This difference in frequency distribution (*see* **Fig. 4.4C**) was reflected in a 44% decrease in mean surface area of adipocytes from aP2-CETPTg mice (*see* **Fig. 4.4D**).

3. Eggs from FVB mice have larger nuclei than that from C57BL/6J. It is easier to inject the DNA fragment into FVB eggs than into C57BL/6J eggs.

4. It is better to use the gene, or minigene, to create the transgenic mice. We did not obtain aP2-CETP Tg mice when we used CETP cDNA to prepare the vector.

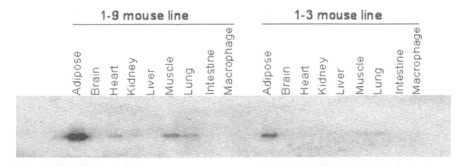

Fig. 4.3 Northern blot analysis of total RNA from various mouse tissues. Twenty μg/lane of total RNA from various mouse tissues was probed with a 444-base pair fragment (nucleotides 726–1170) that was random primer-labeled

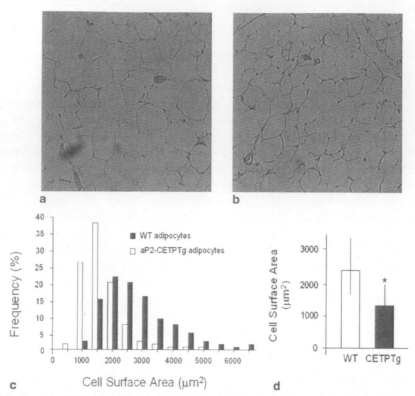

Fig. 4.4 Decrease of adipocyte size in aP2-CETPTg mice. Hematoxylin and eosin-stained sections of adipose tissue from wild-type (WT) (**A**) and aP2-CETPTg (**B**) mice, respectively. Images captured at 10x magnification. (**C**) Frequency distribution of adipocyte cell surface area from three WT and four aP2-CETPTg mice, respectively. More than 250 cells were measured for each mouse. The distribution includes 915 cells from WT and 1007 cells from aP2-CETPTg mice, respectively. (**D**) Mean surface area of adipocytes. Values are mean ± -SD, n = 3–4/group. p < 0.001

References

1. Flier JS (2004) Obesity wars: molecular progress confronts an expanding epidemic Cell 116:337–350
2. Ahima RS (2006) Adipose tissue as an endocrine organ. Obesity Suppl 5:242S–249S
3. Weisberg SP, McCann D, Desai M, Rosenbaum M, Leibel RL, and Ferrante AW Jr (2003) Obesity is associated with macrophage accumulation in adipose tissue. J Clin Invest 112:1796–1808
4. Xu H, Barnes GT, Yang Q, et al (2003) Chronic inflammation in fat plays a crucial role in the development of obesity-related insulin resistance. J Clin Invest 112:1821–1830
5. Skurk T, Hauner H (2004) Obesity and impaired fibrinolysis: role of adipose production of plasminogen activator inhibitor-1. Int J Obes Relat Metab Disord 28:1357–1364
6. Belanger C, Luu-The V, Dupont P, Tchernof A (2002) Adipose tissue intracrinology: potential importance of local androgen/estrogen metabolism in the regulation of adiposity. Horm Metab Res 34:737–745

7. Morton NM, Paterson JM, Masuzaki H, et al (2004) Novel adipose tissue-mediated resistance to diet-induced visceral obesity in 11 beta-hydroxysteroid dehydrogenase type 1 deficient mice. Diabetes 53:931–938

8. Shepherd PR, Gnudi L, Tozzo E, Yang H, Leach F, Kahn BB (1993) Adipose cell hyperplasia and enhanced glucose disposal in transgenic mice overexpressing GLUT4 selectively in adipose tissue. J Biol Chem 268:22243–22246

9. Kopecky J, Clarke G, Enerback S, Spiegelman B, Kozak LP (1995) Expression of the mitochondrial uncoupling protein gene from the aP2 gene promoter prevents genetic obesity. J Clin Invest 96:2914–2923

10. Horton JD, Shimomura I, Ikemoto S, Bashmakov Y, Hammer RE (2003) Overexpression of sterol regulatory element-binding protein-1a in mouse adipose tissue produces adipocyte hypertrophy, increased fatty acid secretion, and fatty liver. J Biol Chem 278: 36652–36660

11. Jurczak MJ, Danos AM, Rehrmann VR, Allison MB, Greenberg CC, Brady MJ (2007) Transgenic overexpression of protein targeting to glycogen markedly increases adipocytic glycogen storage in mice. Am J Physiol Endocrinol Metab 292:E952–E963

12. Barlow C, Schroeder M, Lekstrom-Himes J, Kylefjord H, Deng CX, Wynshaw-Boris A, Spiegelman BM, Xanthopoulos KG (1997) Targeted expression of Cre recombinase to adipose tissue of transgenic mice directs adipose-specific excision of loxP-flanked gene segments. Nucleic Acids Res 25:2543–2545

13. Tall AR (1993) Plasma cholesteryl ester transfer protein. J Lipid Res 120:1255–1274

14. Agellon LB, Walsh A, Hayek T, Moulin P, Jiang XC, Shelanski SA, Breslow JL, Tall AR (1991) Reduced high density lipoprotein cholesterol in human cholesteryl ester transfer protein transgenic mice. J Biol Chem 266:10796–1080

15. Jiang XC, Moulin P, Quinet E, Goldberg IJ, Yacoub LK, Agellon LB, Compton D, Schnitzer-Polokoff R, Tall AR (1991) Mammalian adipose tissue and muscle are major sources of lipid transfer protein mRNA. J Biol Chem 266:4631–4639

16. Quinet EM, Huerta P, Nancoo D, Tall AR, Marcel YL, McPherson R (1993) Adipose tissue cholesteryl ester transfer protein mRNA in response to probucol treatment: cholesterol and species dependence. J Lipid Res 34:845–852

17. Quinet E, Tall A, Ramakrishnan R, Rudel L (1991) Plasma lipid transfer protein as a determinant of the atherogenicity of monkey plasma lipoproteins. J Clin Invest 87:1559–1566

18. Radeau T, Lau P, Robb M, McDonnell M, Ailhaud G, McPherson R (1995) Cholesteryl ester transfer protein (CETP) mRNA abundance in human adipose tissue: relationship to cell size and membrane cholesterol content. J Lipid Res 36:2552–2561

19. Dullaart RP, Sluiter WJ, Dikkeschei LD, Hoogenberg K, Van Tol A (1994) Effect of adiposity on plasma lipid transfer protein activities: a possible link between insulin resistance and high density lipoprotein metabolism. Eur J Clin Invest 24:188–194

20. Arai T, Yamashita S, Hirano K, Sakai N, Kotani K, Fujioka S, Nozaki S, Keno Y, Yamane M, Shinohara E. (1994) Increased plasma cholesteryl ester transfer protein in obese subjects. A possible mechanism for the reduction of serum HDL cholesterol levels in obesity. Arterioscler Thromb 14:1129–1136

21. Remillard P, Shen G, Milne R, Maheux P (2001) Induction of cholesteryl ester transfer protein in adipose tissue and plasma of the fructose-fed hamster. Life Sci 69:677–687

22. Shen GX, Angel A (1995) Regulation of cholesteryl ester transfer activity in adipose tissue: comparison between hamster and rat species. Am J Physiol 269:99–107

23. Zhou H, Li Z, Hojjati MR, Jang D, Beyer TP, Cao G, Tall AR, Jiang XC (2006) Adipose tissue-specific CETP expression in mice: impact on plasma lipoprotein metabolism. J Lipid Res 47:2011–2019

24. Jiang XC, Agellon LB, Walsh A, Breslow JL, Tall A (1992) Dietary cholesterol increases transcription of the human cholesteryl ester transfer protein gene in transgenic mice. Dependence on natural flanking sequences. J Clin Invest 90:1290–1295

25. Yvan-Charvet L, Even P, Bloch-Faure M, Guerre-Millo M, Moustaid-Moussa N, Ferre P, Quignard-Boulange A (2005) Deletion of the angiotensin type 2 receptor (AT2R) reduces

adipose cell size and protects from diet-induced obesity and insulin resistance. Diabetes 54:991–999

26. Fu Y, Luo N, Lopes-Virella MF, Garvey WT (2002) The adipocyte lipid binding protein (ALBP/aP2) gene facilitates foam cell formation in human THP-1 macrophages. Atherosclerosis 165:259–269

27. Chen HC, Farese RV Jr. (2002) Determination of adipocyte size by computer image analysis. J Lipid Res 43:986–989

28. Le Lay S, Ferre P, Dugail I (2004) Adipocyte cholesterol balance in obesity. Biochem Soc Trans 32:103–106

29. Farley FW, Soriano P, Steffen LS, Dymecki SM (2000) Widespread recombinase expression using FLPeR (flipper) mice. Genesis 28:106–110

30. Jones JR, Shelton KD, Magnuson MA (2005) Strategies for the use of site-specific recombinases in genome engineering. Methods Mol Med 103:245–257

Chapter 5
Angiogenesis in Adipose Tissue

Ebba Brakenhielm and Yihai Cao

Summary Angiogenesis is required for the growth and expansion of both healthy and pathological tissues. The plasticity of the adipose tissue is reflected by its remarkable ability to expand or to reduce in size throughout the adult lifespan. We, and others, have recently shown that expansion of fat mass is dependent on angiogenesis, and suppression of angiogenesis might provide a novel therapeutic approach for prevention and treatment of obesity. Here, we outline two technical procedures for assessment of angiogenesis in adipose tissues.

Key words Angiogenesis; growth factors; adipogenesis; obesity; vascular density.

1 Introduction

Both white and brown adipose tissues (WAT and BAT) display relatively high blood perfusion rates compared with many other organs, and they are extensively vascularized *(1–4)*. The complex crosstalk between *angiogenesis* and *adipogenesis* has only recently begun to unravel. Whereas angiogenesis involves the migration, proliferation, and differentiation of endothelial cells, adipogenesis entails the differentiation of preadipocytes into adipocytes and their subsequent growth in size through lipogenesis *(5,6)*.

During the formation of the primitive fat organ in embryos, the vascular bed has been found to develop before adipocyte differentiation *(2,7,8)*. Similarly, during postnatal adipose tissue expansion, angiogenesis has been observed to precede or to coincide with adipogenesis, when the preformed adipocytes increase in size (hypertrophy), followed by an increase in adipocyte cell numbers (hyperplasia) *(7,9–12)*. Recently, it has become clear that the growth and expansion of adipose tissue is angiogenesis-dependent as antiangiogenic therapy has been demonstrated in animal models to block adipogenesis and even to regress preexisting adipose

From: *Methods in Molecular Biology, Vol. 456:*
Adipose Tissue Protocols, Second Edition. Edited by: Kaiping Yang
© Humana Press, a part of Springer Science+Business Media, Totowa, NJ

tissues *(13–18)*. Further support for the essential role of the vasculature is obtained by the demonstration that adipose tissue growth can be promoted by proangiogenic protein therapy, such as fibroblast growth factor (FGF)-2 *(19–21)*.

The endogenous proangiogenic properties of adipose tissue have been recognized for some time. Approximately 20 yrs ago, it was discovered that preadipocytes and 3T3 fibroblast-differentiated adipocytes in culture secrete endothelial cell-specific mitogenic factors, and that omental and subcutaneous adipose tissues produce factors that promote angiogenesis in vivo *(22–26)*. Subsequently, it has been demonstrated that both WAT and BAT produce and secrete many different types of proangiogenic factors, such as vascular endothelial growth factor (VEGF)-A and hepatocyte growth factor (HGF), the two key angiogenic factors produced by adipocytes *(27–29)*. Other examples of adipose-tissue derived factors with pro-angiogenic effects include monobutyrin (1-Butyrylglycerol), VEGF-B, VEGF-C, angiopoietin (Ang)-1, Ang-2, FGF-2, placental growth factor (PlGF), leptin, estrogen, interleukin (IL)-6, transforming growth factor (TGF)-α, TGF-β, tissue factor (TF), matrix metalloproteinase (MMP)-2 and MMP-9, SPARC/osteonectin, and cathepsins *(15,30–48)*. The adipose tissue also produces endogenous antiangiogenic factors, such as adiponectin, thrombospondin (TSP)-1, and TSP-2 *(35,49)*. Thus, the regulation of angiogenesis in adipose tissue may depend on the local balance between proangiogenic and antiangiogenic factors. In this chapter, we describe methods related to the characterization of vascular changes occurring during adipose tissue remodeling, with a specific focus on the detection and quantification of angiogenesis.

In the first protocol, the tissue vascularity is determined by an immunohistochemical approach with specific staining for endothelial cells, which allows for the determination of vascular density (vessels per mm^2 or vessel / adipocyte ratio) and vascular profiles (percentage of small vs. large diameter vessels or average blood vessel size). The changes that occur in vascular density during adipose tissue expansion or reduction, particularly when considering the parameter "vessel numbers per area," would depend not only on the emergence of new blood vessels but also on changes occurring simultaneously in other cell populations in the tissue. Notably, in expanding adipose tissue, active angiogenesis is accompanied by the event of increasing adipocyte cell sizes, which might effectively reduce the area available for vessels, resulting in a reduction of relative vessel density (vessels/area) observed in the tissue despite an increase in total tissue vasculature *(3,37,50)*. Despite this possibility, we have detected elevated "vessel numbers per area" in subcutaneous WAT from both genetically obese *ob/ob* mice (characterized by both hypertrophy and hyperplasia) as well as in high caloric diet fed *wt* mice (characterized mainly by hypertrophy), as compared with lean *wt* mice *(14)*. This finding is consistent with the notion of stimulation of angiogenesis in the tissue *(see* **Fig. 5.1a**).

During expansion of adult adipose tissue, the number of adipocytes per area would generally decrease because of the increase in adipocyte cell size. Thus, the ratio of "vessel/adipocytes" may remain unchanged or even increase during adipogenesis. However, it is possible that antiangiogenic therapy would increase the number of adipocytes per area (caused by a reduction of adipocyte sizes), leading to a decrease in the "vessel/adipocyte ratio." We, and others, have demonstrated that the vessel/adipocyte ratio is approx. 0.5–1.2 in mouse subcutaneous WAT depending on the homeostatic or expanding status of the tissue *(14,37)*. Indeed, we found that inhibition of angiogenesis in several mouse models of obesity leads to a significant decrease (about 10-fold) in adipocyte cell sizes, and a significant increase (about 50%) of adipocyte cell numbers per area. Interestingly, a slight decrease (15%) of vessel numbers per area has been observed, resulting in a very significant decrease (50%) in the vessel/adipocyte ratio (*see* **Fig. 5.1a–d;** *(14)*).

A third possible measurement to consider is the sizes of newly formed vessels, which could potentially be increased because of new sprouting and vessel dilation. Thus, the average blood vessel size (vessel area, μm^2) per sample might be initially increased during adipose tissue expansion. Recent studies show, in a high caloric diet-induced obesity model in mice, that blood vessel sizes are increased (approx. 50–60%) after moderate expansion of the adipose tissue *(37)*. This phenomenon is accompanied by a decrease (50–60%) in adipocyte numbers per area, due to a notable concomitant increase (~2-fold) in adipocyte sizes, leading to a slight decrease (20–40%) in vessel numbers per area. Taken together, concomitant angiogenesis and adipogenesis result in a significant increase (50–70%) of vessel/adipocyte ratio *(37)*. However, when adipocyte hypertrophy becomes pronounced, large blood vessels might divide to form several small vessels (by bridging or intussuceptive growth) *(5,51)*. Indeed, smaller luminal diameters have been observed in the

Fig. 5.1 Assessment of adipose tissue vascularity. The vascular density was determined in subcutaneous WAT samples from lean or obese animals by immunohistochemical detection of CD31 in 12 μm frozen tissue sections. Both genetically obese *ob/ob* mice and *wt* mice fed a high caloric diet for 16 wk displayed elevated vascular densities compared with lean control animals (**a**). Antiangiogenic therapy with TNP-470 of *ob/ob* and *wt* mice fed a high caloric diet for 12–16 wk resulted in a significant reduction of adipose tissue vascular density. Adipocyte density (**b**) and adipocyte sizes (**c**) were analyzed in lean *wt* mice fed a normal diet or in obese *wt* mice fed a high caloric diet with or without antiangiogenic therapy with TNP-470. Analysis of the normalized blood vessel density, i.e., vessel/adipocyte ratio, revealed that weight gain is associated with an increase in vessel numbers relative to adipocyte cell numbers (**d**). In contrast, antiangiogenic therapy and weight loss are associated with a significantly reduced vessel / adipocyte ratio. *p < 0.05, ***p < 0.001 Figure modified from Brakenhielm et al. *(14)*

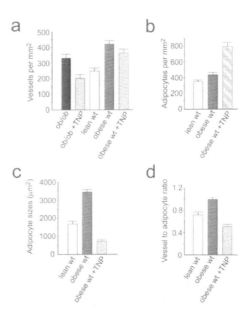

growing adipose tissue of obese animals, suggesting that extensive adipocyte hypertrophy is associated with decreases in vessel lumen diameters *(7,50)*. Thus, during an evaluation of angiogenic responses or vascular characteristics in adipose tissue, the following factors should be considered: 1) the homeostatic status of the adipose depot; 2) the balance between adipocyte hypertrophy and adipocyte hyperplasia, and 3) the balance between vessel numbers and vessel sizes.

In the second protocol, we describe an immunohistochemical approach to detect actively dividing endothelial cells in blood vessels of the adipose tissue. Similar to detection of proliferating cells in other organs and tissues, several commonly used proliferative markers can be used for the detection of dividing endothelial cells by immunohistochemistry. These include evaluation of the incorporation of tritiated thymidine or 5-bromo-2′-deoxyuridine (BrdU), or analysis of the expression of cell cycle-associated proteins such as cyclin D1, Ki67, or proliferating cell nuclear antigen *(52,53)*. BrdU, an analog of thymidine, is incorporated into DNA during the S-phase of the cell cycle, and might be subsequently detected immunologically with specific antibodies. To study the proliferative potential of different tissues, animals are injected with BrdU, sacrificed after a certain time, and the tissues are removed and fixed before microscopic analysis.

To determine the percentage of vessels with proliferating endothelial cells during adipogenesis, it is necessary to consider that endothelial proliferation might be most pronounced in the early stages of adipose tissue expansion. Therefore, BrdU-positive vessels in already well-established adipose tissue might be considerably less frequent than in the early onset of obesity characterized by rapidly expanding adipose tissue. In contrast, because adult adipose tissue expansion is mainly associated with adipocyte hypertrophy, and less often with adipocyte hyperplasia, and further because of the large sizes of adipocytes, it is rare to find BrdU-positive adipocytes in thin tissue sections even during pronounced adipogenesis.

A technical concern for this method is that the time of exposure and dosage of BrdU in vivo will determine the sensitivity of the assay. Long-term exposure and higher dosages of BrdU might yield results that are more positive. For this purpose, it is generally recommended that proliferation labeling at several different time points be performed. Another technical concern regarding this method is that to detect incorporated BrdU immunologically, the DNA must be denatured to allow penetration of the antibody. Denaturation of DNA is often accomplished by treatment of tissues with HCl, which might negatively affect cell morphology. For this reason, tissue sections from prefixed samples embedded in paraffin are often used instead of tissue sections from frozen and lightly postfixed samples. A drawback of this approach is that many antibodies might not recognize as efficiently their antigens in the sections following extensive cross-linking of cellular proteins. To retain cellular morphology even in more lightly fixed samples, an alternative to the acid-based DNA denaturation procedure has been developed consisting of the addition of special nucleases *(54)*.

To analyze the results, the number of BrdU-positive and -negative adipose tissue blood vessels in each section is determined manually by counting at least five representative sections per animal. The average percentage of vessels positively stained (both BrdU and endothelial phenotype marker-positive) is then calculated, and the results expressed as the percentage of double-labeled microvessels per total number of vessels (BrdU labeling index). BrdU labeling indexes of 10–30% of the endothelial cells in epididymal WAT sections have been observed in genetically obese *ob/ob* mice during weight gain *(2,13)*. In contrast, during weight loss, induced by antiangiogenic therapy, the number of proliferating endothelial cells is significantly decreased *(13)*.

2 Materials

2.1 Reagents and Equipment

1. Cryotome or microtome for production of thin (4- to 12-µm thick) sections.
2. Poly-L-lysine or other coated glass slides with strong adherence to the tissue sections.
3. Humidified incubation chamber for slides, for example, a sandwich box lined with damp paper.
4. Slide washing jars and racks.
5. Coplin jar or similar vessel capable of holding 50 mL of staining solution and slides.
6. Tissues or cotton wool swabs for drying around specimens after washes.
7. Endothelial immunohistochemical detection: protocol 1, primary antibody: monoclonal rat antimouse CD31 (BD Pharmingen, clone MEC13.3, cat. no. 550274), secondary antibody: FITC-conjugated rabbit antirat IgG (Vector laboratories, cat. no. FI4001); protocol 2, primary antibody: polyclonal rabbit anti-human von Willebrand factor (vWF, factor VIII) IgG (Abcam, cat. no. ab6994 or Dako, cat. no. A0082); secondary antibody: biotinylated donkey antirabbit IgG (Jackson Immunoresearch, cat. no. 711–065–152); biotin detection: streptavidin conjugate (SA-Cy3; Jackson Immunoresearch, cat. no. 016–160–084).
8. BrdU labeling reagents: BrdU (e.g., Sigma, cat. no. 858811); primary mouse anti-BrdU IgG$_{2a}$ antibody clone BU-1 (e.g., Chemicon, cat. no. MAB3510); FITC-conjugated donkey antimouse IgG secondary antibody (Jackson Immunoresearch, cat. no. 715–095–151).
9. Nuclear counter-staining: membrane–permeable Hoechst's dye 33342 (Sigma, cat. no. B2261) or Hoechst's dye 33258 (Sigma, cat. no. 861405).

2.2 Buffers and Solutions

1. Phosphate buffered saline (1x PBS): 11.5 g of Na_2HPO_4 2.96 g of $NaH_2PO_4*2H_2O$, 5.84 g of NaCl in 1 L of distilled H_2O, pH 7.4. Store at 4°C until use or prepare a *10x* concentrated stock solution and keep at room temperature until use.

2. Carnoy's fixative II: 1 part glacial acetic acid, 3 parts chloroform, 6 parts 95% or absolute ethanol. Store in a cool and dark place. Make a fresh solution before each use.
3. Blocking buffer: *1x* PBS supplemented with 5–10% whole serum from the same species as the secondary antibody used. Prepare fresh before each use.
4. Antibody dilution buffer: *1x* PBS supplemented with 2–5% whole serum from the same species as the secondary antibody used. Prepare fresh before each use.
5. Mounting medium: Vectashield water-based mounting medium with anti-fade agents (Vector laboratories, cat. no. H1000), or Fluoromount G (Southern Biotechnology, cat. no. 0100–01) or 'home-made' medium consisting of a mix of 10 ml *1x* PBS and 90 ml glycerol with addition of 100 mg anti-fade *p*-phenylenediamine hydrochloride. Note that *p*-phenylenediamine hydrochloride might interfere with certain cyanine dyes (especially Cy2). Store at 4°C in the dark for up to 2 wk.

3   Methods

3.1   Analysis of Adipose Tissue Vascularity in Mouse

1. Snap freeze WAT pieces of less than 1 cm size on dry ice (*see* **Note 1**) in plastic or metal cryomoulds and store at −80°C until sectioning in an air-tight plastic bag to prevent dehydration (*see* **Note 2**).
2. Cut cryosections of 10- to 12-μm thickness on a cryotome set at −30°C and collect on coated glass slides (*see* **Note 3**).
3. Air-dry the tissue sections for 1–1.5 h at room temperature.
4. Fix the tissue by submerging the slides in 100% acetone for 10 min in a Coplin jar.
5. Wash the slides in distilled H$_2$O followed by three washes of 5 min each in 1x PBS (pH 7.4).
6. Block nonspecific signals by incubating the sections with blocking buffer. After the last wash, drain the glass slide free of excess wash solution by tapping against a paper and wipe gently around the specimen without touching the tissue borders. Add a drop of the blocking solution (100–150 μL per sample depending on the sizes of the tissue sections) to completely cover the tissue (*see* **Note 4**). Use this technique for all subsequent steps. Place the slides in a moisture chamber and incubate at room temperature for 30 min (*see* **Note 5**).
7. Remove the blocking buffer and replace with a solution containing primary antibody, a rat antimouse CD31 IgG diluted 1:100 (0.15 μg/mL) with an antibody dilution buffer made of PBS supplemented with 3% rabbit non-immune whole serum. Incubate the sections in a moisture chamber at room temperature for 1 h (*see* **Note 6**).
8. Wash the slides in PBS three times for 5 min each.

9. Add the secondary antibody solution, consisting of a FITC-conjugated rabbit antirat IgG diluted 1:50 in PBS supplemented with 3% rabbit non-immune whole serum, to the sections and incubate at room temperature in a moisture chamber for 1 h in the dark (*see* **Note 7**).

10. Wash the slides in PBS three times as above.

11. To counter-stain the tissue and to visualize cell nuclei, incubate the sections with a membrane–permeable Hoechst's dye 33342 diluted 1:2000 in PBS from a stock solution of 10 mg/ml prepared in distilled H_2O (may be stored at 4°C in the dark for approx. 1 mo) to make a working concentration of 5 μg/mL. Incubate the samples at room temperature for 20 min in the dark (*see* **Note 8**).

12. Wash the slides in PBS twice for 5 min each, followed by one wash in distilled H_2O.

13. Mount the sections in a water-based mounting medium (1–2 drops of mounting medium is placed directly on the section before covering with a thin cover glass). Store the slides at 4°C in the dark until examination within a couple of days. For long-term storage, the slides might be sealed using nail polish at the edges of the glass to prevent drying of the mounting medium. Some water-based mounting media also allows freezing at −20°C, which will allow storage of the slides for several weeks without significant fading of the positive signal.

14. Observe the tissue under a fluorescence microscope and take digital pictures at 10x and 20x magnifications.

15. Analyze the images with the aid of Adobe Photoshop or other image analysis software programs, and calculate the number of microvessels per image field, manually or digitally.

16. The vessel density is given as vessels per field or vessels per mm^2 (5–10 sections per animal are analyzed and results are then averaged). Another aspect of the vasculature during remodeling or growth can be observed by estimating vessel diameters (manually or with analysis software programs such as IPlab) in the tissue sections. For example, the vessel profile of the tissue can be determined by comparing the number of small (< 4 μm diameter), middle-sized (4–10 μm in diameter), and large (> 10 μm in diameter) vessels and plotting this number as percentage of small-, middle-, or large-sized vessels per total vessels in the field *(7,51)*. Alternatively, the total vascularization area per field may be calculated by a computer-assisted analysis, and this number divided by the total number of vessels per field to give an estimate of the average vessel size (vessel area, $μm^2$, calculated according to the formula Πr^2) in the section *(37)*. Additionally, the percentage of vascularity of the tissue may be estimated by calculating the individual vessel areas and dividing the sum of the vascular areas in the field with the total image field area.

17. As a way to control for adipocyte hypertrophy and hyperplasia, an absolute measurement may be obtained by calculating the normalized blood vessel density, i.e., the vessel / adipocyte ratio. This value is obtained simply by counting the number of adipocytes per area by computer-assisted image analysis of the sections (either after a hematoxylin/eosin counter-staining of the sections (*see* **Note 9**), or by examining the CD31-stained sections where the adipocytes

appear as hollow structures) and then dividing the total number of vessels with the total number of adipocytes per area.

3.2   Evaluation of Angiogenesis in Adipose Tissue

1. Inject animals with BrdU in vivo. Dissolve BrdU in sterile-filtered (pore size of 0.22 µm) PBS or saline solution at a concentration of 3 mg/ml immediately before use (*see* **Note 10**). Administer BrdU to mice by intraperitoneal injection at a dose of 30 mg/kg. Sacrifice mice two hours later, and excise subcutaneous and omental WAT.
2. Fix WAT pieces of less than 1 cm² in Carnoy's fixative II (*see* **Note 11**) for 4 h, transfer to 100% ethanol, and embed in paraffin according to standard protocols (*see* **Note 12**).
3. Cut tissue sections of 4–5 µm with a microtome and collect on coated glass slides.
4. Dry the tissue sections onto the glass slides by 'baking' the slides in an oven at 55°C for 2 hs.
5. Process the slides for dewaxing and rehydration as follows:

   a. Remove the paraffin by dipping the slides in 100% xylene (in a Coplin jar) for 10–20 min (perform all xylene washes in a fume hood!)
   b. Move slides to a fresh Coplin jar containing xylene for an additional 10 min wash or until no trace of paraffin is left on the slide
   c. Rinse the slides twice for 2 min in 100% EtOH
   d. Rinse the slides twice for 2 min in 95% EtOH
   e. Place slides in 80% EtOH for 2 min, followed by 5 minutes in distilled H₂O
   f. Rinse slides several times with fresh distilled H₂O followed by a 5-min wash in PBS

6. Permeabilize sections by incubation with 10 µg/ml proteinase K in 0.1 M Tris-HCl buffer (pH 7.4) at 37°C for 10–20 minutes
7. Wash the slides in PBS three times (*see* **Notes 13** and **14**).
8. Block the nonspecific background in the tissue using 5% non-immune whole donkey serum in PBS (*see* **Note 5**), as described in **Subheading 3.1.6.**
9. Remove the blocking buffer and replace with a solution containing the primary antibody, a mouse anti-BrdU IgG$_{2a}$ clone BU-1, diluted according to the manufacturer's instructions (generally 1:200–1:5000) with an antibody blocking buffer (*see* **Note 15**). Incubate the sections for 30–60 min at room temperature according to the manufacturer's instructions (*see* **Note 16**).
10. Wash the slides in PBS three times.
11. Stain endothelial cells by incubating the sections with a polyclonal rabbit anti-human von Willebrand factor (vWF) primary antibody diluted 1:250 with a PBS solution containing 5% nonimmune whole donkey serum at 4°C overnight in a moisture chamber.
12. Wash the slides with PBS three times.

13. Incubate the sections with a secondary antibody consisting of a biotinylated donkey antirabbit IgG diluted 1:200 with a PBS solution containing 5% nonimmune whole donkey serum at room temperature for 30 minutes in a moisture chamber.
14. Wash the slides in PBS three times.
15. Detect BrdU-positive cells using a FITC-conjugated donkey antimouse IgG secondary antibody diluted 1:200, and detect endothelial cells using a fluorescent streptavidin conjugate (SA-Cy3) diluted at 1:1000 to 1:2500 (centrifuge stock vial before each use to remove undissolved crystals) in PBS containing 5% nonimmune whole donkey serum. Incubate the slides in a moisture chamber at room temperature in the dark for 30 min.
16. Wash the slides in PBS three times.
17. To counter-stain the tissue and visualize cell nuclei, incubate the sections in Hoechst's dye 33258 diluted to 1 µg/mL in PBS. Incubate the samples at room temperature for 20 min in the dark in a moisture chamber.
18. Wash the slides in PBS twice for 5 min each, followed by one wash in distilled H_2O.
19. Mount the sections in a water-based mounting medium, as described in **Methods 3.1.3**.
20. Examine the tissue under a fluorescence microscope and take digital pictures at 10x, 20x, or 40x magnifications.
21. Analyze the images using Adobe Photoshop or other image analysis software programs, and calculate the number of microvessels per image field, manually or digitally.
22. Determine the numbers of stained and unstained adipose tissue blood vessels for each section slide manually by counting about 10 representative fields. Calculate the BrdU labeling indices (the percentage of cells positively stained per area or per total number of vessels) accordingly.

4 Notes

1. The freezing procedure for larger pieces of tissue is more efficient if freezing in 2-methylbutane (isopentane), pre-chilled to −140°C to −80°C in a pyrex/styrofoam beaker kept on liquid N_2, is performed rather than placing the cryomould directly on dry ice (−90°C to −70°C). Further, the tissue pieces might be embedded in Tissue-Tek O.C.T compound to facilitate homogenous freezing and cryosectioning of the samples. Place the samples, held in plastic or metal moulds, on top of the chilled isopentane until freezing. Make sure that the Tissue-Tek O.C.T-filled moulds freeze from the bottom rather than from the top, as it would cause less risk of the block 'cracking' due to freeze-induced shrinkage of the OCT compound. Once samples are frozen, transfer to dry ice and store at −80°C until sectioning.
2. Instead of immediately freezing WAT, it is recommended to fix the samples in paraformaldehyde-based fixative (3% PFA) to obtain better morphology. PFA is

prepared as follows: 3 g paraformaldehyde dissolved in 100 ml of 1X PBS pre-heated to 60°C in a water-bath. Mix on a magnetic stirrer at room temperature until the PFA is almost dissolved. Add one or two drops of NaOH (0.1–0.5 M) to completely dissolve the PFA. Store at 4°C. The solution may be stored maximally for one week to ensure an optimal fixation process. The tissue is incubated in the fixation solution for 12 h at 4°C, followed by a sucrose substitution gradient (15% sucrose in PBS at 4°C for 3–5 h followed by 30% sucrose in PBS for 12–24 h at 4°C) to prepare the tissue for freezing. After immersion in sucrose the excess liquid on the tissue is removed by dabbing the tissue against a paper towel before embedding it with Tissue-Tek OCT compound and freezing it as above. Note that some antibodies, such as the rat antimouse CD31 antibody suggested in this protocol (and many antibodies against CD1, CD3, CD4, CD7, CD8, and CD19 antigens), do not react with PFA-fixed antigen epitopes, which might lead to negative results or weaker signals. In this case, substituting the PFA fixation with a Zinc-based fixation is recommended for retaining epitope reactivity similar to snap frozen samples while improving tissue morphology. Fixation in formalin-free Tris-buffered Zinc fixative is made as follows:

a. Prepare 0.1 M Tris-HCl buffer, pH 7.4: 12.1 g Tris Base (TRIZMA), 81.5 ml 1.0 M HCl, 900 mL deionized H_2O.
b. Prepare Zinc-Fixative: 1000 ml 0.1 M Tris-HCl buffer, pH 7.4:0.5 g calcium acetate, 5.0 g of zinc acetate, 5.0 g of zinc chloride. Mix to dissolve. The final pH will be approximately 6.5–7.0. Do not readjust pH, as this would cause zinc to come out of the solution! Store the zinc fixative at room temperature until use, and prepare a fresh solution before each experiment.
c. Fix tissues by placing freshly dissected tissue pieces (no larger than 5–10 mm in size) in the zinc fixative and incubate for 6–24 h at room temperature.

3. After sectioning, slides may be stored frozen for a short period of time at −20°C (days) or −80°C (weeks) if sealed in a slide holder kept in an airtight freezer bag. Alternatively, they may be individually wrapped in aluminum foil prior to storage in freezers.
4. To minimize the amount of reagent required to cover the sections, a PAP hydrophobic barrier pen (available from Sigma) might be used to trace, in grease, the outer borders of the tissue section.
5. The nonimmune whole serum in the blocking buffer could be replaced with 3% BSA in PBS, but this is less effective in blocking of nonspecific binding of the secondary antibody. However, it reduces the introduction of free biotins to the tissue, which is advantageous to avoid if streptavidin or avidin-conjugated fluorochromes or horseradish peroxidase (HRP) enzyme conjugates are used for the detection of biotinylated primary or secondary antibodies, especially if an amplification system such as ABC or NEN TSA is used.
6. An overnight incubation at 4°C with a primary antibody might also be used. In this case, the use of a moister chamber is crucial so that the tissue sections are not dehydrated. Instead of CD31 many other antigens have also been used to detect endothelial cells such as CD34 (membrane antigen) or von Willebrand factor

(intracellular antigen). Further, arteriolar and arterial vessels might be detected using a marker of smooth muscle cells and pericytes, such as α-smooth muscle actin (α-SMA) and desmin. However, in our experience the CD31 antigen gives the most homogenous staining results of both small (capillary) and large (arteries and veins) branches of the vasculature. In addition, indirect stains based on markers found in the endothelial basement membrane (e.g., fibronectin, collagen type IV, laminin), or based on recognition of endothelial-associated glycoproteins by lectin molecules, e.g. Bandeiraea (Griffonia) simplicifolia isolectin B_4 (GSL I-B_4/ BSI lectin, for rodents), and Ulex europaeus agglutinin (UEA-I, for humans) might also be used, but these approaches are less specific and might vary depending on the type of organ and/or on the status of the vasculature. For studies in the rat, we have previously used a biotinylated mouse antirat CD31 antibody (1:50 dilution, BD Pharmingen, clone TLD-3A12, cat. no. 555026) on acetone-fixed frozen sections followed by an ABC amplification system (Vector laboratories, cat. no. PK6100) and visualization of the signals with 3,3~-Diaminobenzidine (DAB). For studies in humans we have used a polyclonal rabbit antihuman von Willebrand factor (1:200, Dako, cat. no. A0082) on PFA-fixed paraffin embedded sections after proteinase K antigen retrieval. Positive signals are revealed using a biotinylated secondary antibody followed by amplification using ABC system as above or using fluorescent SA conjugates.

7. Instead of detecting the CD31 antigen with a fluorescent secondary antibody one can also use an HRP-conjugated secondary antibody. This requires blocking of the endogenous peroxidase activity (10–15 min incubation in 0.3% H_2O_2 in H_2O) before adding the primary antibody. The HRP signal might preferentially be amplified using a HRP-based amplification kit, such as NEN TSA biotin system (NEN Lifescience, cat. no. NEL700001KT). Alternatively, a biotinylated rat antimouse CD31 primary antibody might be used (BD Pharmingen, clone MEC13.3, cat. no. 553371) at 1:100 dilution as above, followed directly by the NEN TSA amplification system and development of the signals using DAB as a substrate. Prior to adding the primary antibody, the endogenous avidin and biotin binding sites in the tissue should be blocked, e.g. using avidin/biotin blocking kit (Vector laboratories, cat. no. SP-2001; *see* **Note 14**). Further, the blocking and antibody dilution buffers used in the protocol should also be modified when working with detection of biotin in tissue samples using an amplification system. Preferentially, the non-immune whole serum, which may contain biotin and avidin, should be replaced with 3% BSA in PBS. Alternatively, a blocking reagent supplied with the NEN TSA kit should be used.

8. Instead of Hoechst's dye, other nuclear stains might also be considered, such as 4 -6-diamidino-2-phenylindole dihydrochloride (DAPI, at 2 μg/mL), or propidium iodine (PI, at 5 μg/mL) or ethidium bromide (at 0.5–1.0 μg/mL) after permeabilization of the tissue (by incubation for 10–15 min in 0.1% Triton-X100 or 0.3% saponin in PBS). Alternatively, for nonfluorescence immunohistochemistry, dyes such as hematoxyline might be used for counter-staining after fixation of the developed sections with paraformaldehyde for 10–20 mins (*see* **Note 9**).

9. Hematoxylin and eosin counterstaining is a standard immunohistochemical staining protocol most often used for paraffin-embedded formalin-fixed tissue sections, but it may also be performed on cryosections that have been post-fixed. Hematoxylin, a water-based blue colorant, binds negatively charged nucleic acids. Eosin, an alcohol-based pink-red colorant, stains all cellular proteins. Thus, slides to be stained in eosin must first be dehydrated in EtOH. Both these stains might also be used alone in a more dilute form (dilute the 1x solution to 1:4 in H_2O or EtOH, respectively) to serve as counter-stains for immunoperoxidase staining. In hematoxylin and eosin-stained sections, adipocytes display a characteristic "signet ring" appearance with a compact, flattened nucleus found at the cell periphery, and a thin rim of cytoplasm encircling an empty space, since all the intracellular lipid droplets have been removed during the tissue processing. Further, each adipocyte is surrounded by a basement membrane rich in collagen, which appears as eosinophilic wavy fibers with a diameter of 2 Hematoxylin and eosin $10 \mu m$. Reagents: a) Harris hematoxylin (1x stock; Sigma, cat. no. HHS12); b) eosin B (1x stock; Sigma, cat. no. 861006); c) acid alcohol: 76.6% EtOH, 1/300 (v/v) conc. HCl (230 mL 100% EtOH, 70 mL of H_2O, 1 mL of HCl); and d) ammonia solution: 0.3% (v/v) NH_4OH (1 L H_2O, 3 mL of 28% w/v NH_4OH stock). Procedure: Dip slides 20 times up and down in each solution unless indicated otherwise: a) hematoxylin, 2 min (x1); b) running tap water (x1); c) acid alcohol (x1); d) H_2O (x1); e) ammonia solution (x1); f) running tap water, 5 minutes (x1); g) 80% EtOH (x1); h) eosin, 15 s; i) 95% EtOH (x2); j) 100% EtOH (x2); k) 100% Xylene, 3 min (x3); l) the slides are mounted in a xylene-based mounting medium, such as Permount, and examined by light microscopy.

10. To increase the uptake of BrdU into the cells, and thus increase the sensitivity of the assay, animals might be simultaneously exposed to 5-fluoro-2 -deoxyuridine (FdU), an inhibitor of thymidilate synthetase (55), leading to a decrease in the competition by endogenous thymidine to BrdU incorporation. For this purpose, a 10:1 mixture of BrdU and FdU might be used (3 mg/mL BrdU and 0.3 mg/mL FdU dissolved in sterile PBS). Another way to increase the uptake of BrdU in vivo is to use a greater dose (100–400 mg/kg BrdU has been used for mice and rats) administrated by intraperitoneal injection once or twice a day, and/or waiting longer time before sacrificing the animals (4–72 h after BrdU injection). Alternatively, osmotic minipumps (Alzet) might be used to continuously deliver BrdU (400 mg/kg body weight per day) by subcutaneous infusion for 1–3 d. Alternatively, BrdU might be dissolved in the drinking water (0.8 mg/ml) and the animals allowed to ingest it for 2–6 d before sacrifice.

11. If one wishes to combine BrdU labeling with staining for antigens sensitive to fixation, such as CD31 or macrophage Mac-3, Carnoy's fixative should be replaced by zinc fixation. As an alternative, "Methacarn" or "Methyl Carnoy's," a Carnoy's fixative based on MeOH rather than on EtOH might also be used for improved antigenicity of the tissue: six parts absolute MeOH, 3 parts chloroform, and 1 part glacial acetic acid. Submerge tissue at room temperature in the solution for a few hours, then transfer to 100% MeOH, followed by paraffin

embedding (*see* **Note 12**) starting on step d. Alternatively, the freshly dissected tissue may be embedded in Tissue-Tek O.C.T and snap frozen (*see* **Note 1**) to prepare 4-µm cryosections. The tissue sections are fixed by immersing in absolute MeOH for 10 min at 2–8°C and allowed to air-dry prior to rehydration by washing twice in PBS for 5 min each, after which proceed immediately to **Subheading 3.2.6**.

12. Paraffin embedding protocols are usually run in automated machines (such as Citadel) with the following standard program for PFA-fixed samples: a) 70% EtOH 1 h (x1); b) 90% EtOH 1 hour (x1); c) 95% EtOH 1 h (x1); d) 100% EtOH 1 h (x1); e) 100% EtOH 2 h (x2); f) 100% xylene 1 h (x1); g) 100% xylene 1.5 h (x2); h) paraffin (56°C) 2 h (x1); i) paraffin (56°C) 3 h (x1). For samples fixed with an alcohol-based fixative like Carnoy's, skip steps *a–c* and start with step *d* immediately. For preservation of tissue integrity and antigen reactivity, paraffin wax temperature should preferentially not exceed 60°C. Another concern is that for large WAT or BAT samples the solvents and the paraffin might not infiltrate the tissue sufficiently to allow removal of all fat, thus preventing easy sectioning. In this case the embedding protocol might be prolonged by increasing the time of incubation in each step (30 min to 1 h for each step) and prolonging the last paraffin embedding step for up to 4 hours. However, if the tissue is incubated too long in the alcohol and xylene baths it might become over-dehydrated, which runs the risks of loosing antigenicity or becoming too brittle for sectioning. Thus, these conditions must be determined empirically by each user depending on the tissue samples used.

13. Depending on which anti-BrdU antibody clone is used, a DNA denaturation step might be needed. However, the mouse monoclonal anti-BrdU antibody clone BU-1 does not require pretreatment of cells, due to the presence of contaminating DNases in the antibody solution *(56)*. In contrast, other anti-BrdU antibody clones, such as BMC9318 (Chemicon, cat. no. MAB3424), BU-33 (Sigma, cat. no. B8434), and ZBU30 (Invitrogen, cat. no. 18–0103), require varying degrees of DNA denaturation (e.g., incubation of the slides in $1–4 M$ HCl for 30–60 min at 37°C, or incubating the slides in 1–4 mg/mL of pepsin in $0.01 M$ HCl for 30 min at room temperature). The acid is neutralized by immersing the slides twice for 5 min each in $0.1 M$ borate $(Na_2B_4O_7)$ buffer, pH 8.5. Alternatively, treatment with DNase might be used to "open up" the DNA to reveal the incorporated BrdU *(54)*: a) after deparaffinization and rehydration, permeabilize the tissue by incubation with 10 µg/mL proteinase K in $0.1 M$ Tris-HCl buffer (pH 7.4) at 37°C for 10–20 min; b) wash the slides 3× 5 min in prewarmed 37°C Tris-buffered saline (TBS: 40 mM Tris–HCl, 10 mM NaCl, 6 mM MgCl$_2$ 10 mM CaCl$_2$, pH 7.9); c) add DNase solution at 100–500 U/mL DNase I dissolved in TBS; d) incubate the slides for 10 min at 37°C; e) wash slides in PBS; f) proceed to the addition of blocking solution (*see* **Subheading 3.2.8.**)

14. If the negative control (omitted primary antibody) shows high background staining, it is necessary to perform an additional blocking step to remove

endogenous binding sites for avidin and biotin. A commercial avidin/biotin blocking kit (Vector laboratories, cat. no. SP-2001) might be used or, alternatively, the following protocol may be used: a) incubate slides for 10 min in 0.001% avidin in PBS, then rinse briefly in PBS; b) incubate the slides for 10 min in 0.001% biotin in PBS, then wash in PBS; c) proceed to the addition of blocking solution (*see* **Subheading 3.2.8.**)

15. If the Cell Proliferation Kit from Amersham is used, the antibody is used undiluted in their proprietary nuclease solution.

16. The time of incubation with the anti-BrdU antibody solution can be adjusted depending on the sensitivity required, but exceeding 2 h of incubation might lead to nonspecific staining. In any case, negative control slides must be included in the analysis to verify the specificity of the signal. For negative controls: sample tissues from an animal *not* exposed to BrdU and process it together with the other slides for BrdU labeling.

References

1. Di Girolamo M, Skinner NS Jr., Hanley HG, Sachs RG (1971) Relationship of adipose tissue blood flow to fat cell size and number. Am J Physiol 220:932–937
2. Crandall DL, Hausman GJ, Kral JG (1997) A review of the microcirculation of adipose tissue: anatomic, metabolic, and angiogenic perspectives. Microcirculation 4:211–232
3. Gersh I, Still MA (1945) Blood vessels in fat tissue: relation to problems of gas exchange. J Exp Med 81:219–232
4. Cinti S (2005) The adipose organ. Prostaglandins Leukot Essent Fatty Acids 73:9–15
5. Risau W (1997) Mechanisms of angiogenesis. Nature 386:671–674
6. Carmeliet P (2000) Mechanisms of angiogenesis and arteriogenesis. Nat Med 6:389–395
7. Hausman GJ, Richardson RL (1983) Cellular and vascular development in immature rat adipose tissue. J Lipid Res 24:522–532
8. Poissonnet CM, LaVelle M, Burdi AR (1988) Growth and development of adipose tissue. J Pediatr 113:1–9
9. Kimura Y, Ozeki M, Inamoto T, Tabata Y (2002) Time course of de novo adipogenesis in matrigel by gelatin microspheres incorporating basic fibroblast growth factor. Tissue Eng 8:603–613
10. Sypniewska G, Bjorntorp P (1987) Increased DNA synthesis in adipocytes and capillary endothelium in rat adipose tissue during overfeeding. Eur J Clin Invest 17:202–207
11. Krotkiewski M, Bjorntorp P, Sjostrom L, Smith U (1983) Impact of obesity on metabolism in men and women. Importance of regional adipose tissue distribution. J Clin Invest 72:1150–1162
12. Di Girolamo M, Mendlinger S, Fertig JW (1971) A simple method to determine fat cell size and number in four mammalian species. Am J Physiol 221:850–858
13. Rupnick MA, Panigrahy D, Zhang CY, Dallabrida SM, Lowell BB, Langer R, Folkman MJ (2002) Adipose tissue mass can be regulated through the vasculature. Proc Natl Acad Sci U S A 99:10730–10735
14. Brakenhielm E, Cao R, Gao B, Angelin B, Cannon B, Parini P, Cao Y (2004) Angiogenesis inhibitor, TNP-470, prevents diet-induced and genetic obesity in mice. Circ Res 94:1579–1588
15. Dallabrida SM, Zurakowski D, Shih SC, Smith LE, Folkman J, Moulton KS, Rupnick MA (2003) Adipose tissue growth and regression are regulated by angiopoietin-1. Biochem Biophys Res Commun 311:563–571

16. Liu L, Meydani M (2003) Angiogenesis inhibitors may regulate adiposity. Nutr Rev 61: 384–387
17. Yamaguchi M, Matsumoto F, Bujo H, Shibasaki M, Takahashi K, Yoshimoto S, Ichinose M, Saito Y (2005) Revascularization determines volume retention and gene expression by fat grafts in mice. Exp Biol Med (Maywood) 230:742–748
18. Fukumura D, Ushiyama A, Duda DG, Xu L, Tam J, Krishna V, Chatterjee K, Garkavtsev I, Jain RK (2003) Paracrine regulation of angiogenesis and adipocyte differentiation during in vivo adipogenesis. Circ Res 93: e88–e97
19. Tabata Y, Miyao M, Inamoto T, Ishii T, Hirano Y, Yamaoki Y, Ikada Y (2000) De novo formation of adipose tissue by controlled release of basic fibroblast growth factor. Tissue Eng 6:279–289
20. Vashi AV, Abberton KM, Thomas GP, Morrison WA, O'Connor A, J., Cooper-White JJ, Thompson EW (2006) Adipose tissue engineering based on the controlled release of fibroblast growth factor-2 in a collagen matrix. Tissue Eng 12:3035–3043
21. Kawaguchi N, Toriyama K, Nicodemou-Lena E, Inou K, Torii S, Kitagawa Y (1998) De novo adipogenesis in mice at the site of injection of basement membrane and basic fibroblast growth factor. Proc Natl Acad Sci U S A 95:1062–1066
22. Castellot JJ Jr., Karnovsky MJ, Spiegelman BM (1982) Differentiation-dependent stimulation of neovascularization and endothelial cell chemotaxis by 3T3 adipocytes. Proc Natl Acad Sci U S A 79:5597–5601
23. Castellot JJ Jr., Karnovsky MJ, Spiegelman BM (1980) Potent stimulation of vascular endothelial cell growth by differentiated 3T3 adipocytes. Proc Natl Acad Sci U S A 77:6007–6011
24. Folkman J (1982) Angiogenesis: initiation and control. Ann N Y Acad Sci 401:212–227
25. Goldsmith HS, Griffith AL, Kupferman A, Catsimpoolas N (1984) Lipid angiogenic factor from omentum. JAMA 252:2034–2036
26. Silverman KJ, Lund DP, Zetter BR, Lainey LL, Shahood JA, Freiman DG, Folkman J, Barger AC (1988) Angiogenic activity of adipose tissue. Biochem Biophys Res Commun 153:347–352
27. Zhang QX, Magovern CJ, Mack CA, Budenbender KT, Ko W, Rosengart TK (1997) Vascular endothelial growth factor is the major angiogenic factor in omentum: mechanism of the omentum-mediated angiogenesis. J Surg Res 67:147–154
28. Rehman J, Considine RV, Bovenkerk JE, Li J, Slavens CA, Jones RM, March KL (2003) Obesity is associated with increased levels of circulating hepatocyte growth factor. J Am Coll Cardiol 41:1408–1413
29. Saiki A, Watanabe F, Murano T, Miyashita Y, Shirai K (2006) Hepatocyte growth factor secreted by cultured adipocytes promotes tube formation of vascular endothelial cells in vitro. Int J Obes (Lond) 30:1676–1684
30. Dobson DE, Kambe A, Block E, Dion T, Lu H, Castellot JJ Jr., Spiegelman BM (1990) 1-Butyryl-glycerol: a novel angiogenesis factor secreted by differentiating adipocytes. Cell 61:223–230
31. Asano A, Kimura K, Saito M (1999) Cold-induced mRNA expression of angiogenic factors in rat brown adipose tissue. J Vet Med Sci 61:403–409
32. Asano A, Irie Y, Saito M (2001) Isoform-specific regulation of vascular endothelial growth factor (VEGF) family mRNA expression in cultured mouse brown adipocytes. Mol Cell Endocrinol 174:71–76
33. Silha JV, Krsek M, Sucharda P, Murphy LJ (2005) Angiogenic factors are elevated in overweight and obese individuals. Int J Obes (Lond) 29:1308–1314
34. Stacker SA, Runting AS, Caesar C, Vitali A, Lackmann M, Chang J, Ward L, Wilks AF (2000) The 3T3-L1 fibroblast to adipocyte conversion is accompanied by increased expression of angiopoietin-1, a ligand for tie2. Growth Factors 18:177–191
35. Voros G, Maquoi E, Demeulemeester D, Clerx N, Collen D, Lijnen HR (2005) Modulation of angiogenesis during adipose tissue development in murine models of obesity. Endocrinology 146:4545–4554

36. Teichert-Kuliszewska K, Hamilton BS, Deitel M, Roncari DA (1992) Augmented production of heparin-binding mitogenic proteins by preadipocytes from massively obese persons. J Clin Invest 90:1226–1231
37. Lijnen HR, Christiaens V, Scroyen I, Voros G, Tjwa M, Carmeliet P, Collen D (2006) Impaired adipose tissue development in mice with inactivation of placental growth factor function. Diabetes 55:2698–2704
38. Cao R, Brakenhielm E, Wahlestedt C, Thyberg J, Cao Y (2001) Leptin induces vascular permeability and synergistically stimulates angiogenesis with FGF-2 and VEGF. Proc Natl Acad Sci U S A 98:6390–6395
39. Bouloumie A, Drexler HC, Lafontan M, Busse R (1998) Leptin, the product of Ob gene, promotes angiogenesis. Circ Res 83:1059–1066
40. Cleland WH, Mendelson CR, Simpson ER (1983) Aromatase activity of membrane fractions of human adipose tissue stromal cells and adipocytes. Endocrinology 113:2155–2160
41. Ichinose Y, Asoh H, Yano T, Yokoyama H, Inoue T, Tayama K, Ueda T, Takai E (1995) Use of a pericardial fat pad flap for preventing bronchopleural fistula: an experimental study focusing on the angiogenesis and cytokine production of the fat pad. Surg Today 25: 811–815
42. Guan H, Arany E, van Beek JP, Chamson-Reig A, Thyssen S, Hill DJ, Yang K (2005) Adipose tissue gene expression profiling reveals distinct molecular pathways that define visceral adiposity in offspring of maternal protein-restricted rats. Am J Physiol Endocrinol Metab 288: E663–E673
43. Rehman J, Traktuev D, Li J, Merfeld-Clauss S, Temm-Grove CJ, Bovenkerk JE, Pell CL, Johnstone BH, Considine RV, March KL (2004) Secretion of angiogenic and antiapoptotic factors by human adipose stromal cells. Circulation 109:1292–1298
44. Samad F, Pandey M, Loskutoff DJ (1998) Tissue factor gene expression in the adipose tissues of obese mice. Proc Natl Acad Sci U S A 95:7591–7596
45. Bouloumie A, Sengenes C, Portolan G, Galitzky J, Lafontan M (2001) Adipocyte produces matrix metalloproteinases 2 and 9: involvement in adipose differentiation. Diabetes 50:2080–2086
46. Tartare-Deckert S, Chavey C, Monthouel MN, Gautier N, Van Obberghen E (2001) The matricellular protein SPARC/osteonectin as a newly identified factor up-regulated in obesity. J Biol Chem 276:22231–22237
47. Li J, Yu X, Pan W, Unger RH (2002) Gene expression profile of rat adipose tissue at the onset of high-fat-diet obesity. Am J Physiol Endocrinol Metab 282: E1334–E1341
48. Taleb S, Lacasa D, Bastard JP, Poitou C, Cancello R, Pelloux V, Viguerie N, Benis A, Zucker JD, Bouillot JL, Coussieu C, Basdevant A, Langin D, Clement K (2005) Cathepsin S, a novel biomarker of adiposity: relevance to atherogenesis. FASEB J 19:1540–1542
49. Brakenhielm E, Veitonmaki N, Cao R, Kihara S, Matsuzawa Y, Zhivotovsky B, Funahashi T, Cao Y (2004) Adiponectin-induced antiangiogenesis and antitumor activity involve caspase-mediated endothelial cell apoptosis. Proc Natl Acad Sci U S A 101:2476–2481
50. Crandall DL, Goldstein BM, Huggins F, Cervoni P (1984) Adipocyte blood flow: influence of age, anatomic location, and dietary manipulation. Am J Physiol 247:R46–R51
51. Cao R, Brakenhielm E, Pawliuk R, Wariaro D, Post MJ, Wahlberg E, Leboulch P, Cao Y (2003) Angiogenic synergism, vascular stability and improvement of hind-limb ischemia by a combination of PDGF-BB and FGF-2. Nat Med 9:604–613
52. Ohta Y, Ichimura K (2000) Proliferation markers, proliferating cell nuclear antigen, Ki67: 5-bromo-2~-deoxyuridine, and cyclin D1 in mouse olfactory epithelium. Ann Otol Rhinol Laryngol 109:1046–1048
53. Yu CC, Woods AL, Levison DA (1992) The assessment of cellular proliferation by immunohistochemistry: a review of currently available methods and their applications. Histochem J 24:121–131
54. Ye W, Mairet-Coello G, Dicicco-Bloom E (2006) DNAse I pre-treatment markedly enhances detection of nuclear cyclin-dependent kinase inhibitor p57Kip2 and BrdU double immunostaining in embryonic rat brain. Histochem Cell Biol 127:195–203

55. Ellwart J, Dormer P (1985) Effect of 5-fluoro-2'-deoxyuridine (FdUrd) on 5-bromo-2'-deoxy-uridine (BrdUrd) incorporation into DNA measured with a monoclonal BrdUrd antibody and by the BrdUrd/Hoechst quenching effect. Cytometry 6:513–520
56. Gonchoroff NJ, Greipp PR, Kyle RA, Katzmann JA (1985) A monoclonal antibody reactive with 5-bromo-2-deoxyuridine that does not require DNA denaturation. Cytometry 6:506–512

Chapter 6
Adipose Organ Nerves Revealed by Immunohistochemistry

Antonio Giordano, Andrea Frontini, and Saverio Cinti

Summary Brown and white adipose tissue have recently gained prominence as key players in obesity and related health problems, such as type-2 diabetes and cardiovascular disease. Brown adipose tissue-dependent nonshivering thermogenesis significantly affects the body's energy balance. Originally considered as a passive store of lipids, white adipose tissue has recently been found to secrete a number of hormones and cytokines and to be thus involved in the control of body metabolism and energy balance at multiple sites. These findings have renewed the interest in adipose organ biology, including its innervation by the autonomic nervous system and sensory nerves. Here, we describe our protocols for detecting different types of adipose tissue nerves by light microscopy using peroxidase immunostaining and by laser scanning confocal microscopy using immunofluorescence. With these techniques, the presence, distribution, and colocalization of autonomic and sensory nerves can be effectively investigated in subcutaneous and visceral adipose depots of normal and obese animals.

Key words White adipose tissue; brown adipose tissue; immunohistochemistry; immunofluorescence; confocal microscopy; innervation; noradrenaline; neuropeptides; rat; mouse.

1 Introduction

In mammals, the adipose organ is made up of several fat depots containing variable amounts of brown adipose tissue (BAT) and white adipose tissue (WAT) *(1,2)*. BAT and WAT share the ability to store lipids as triglycerides, but they use them for different purposes. Whereas BATs oxidize fat to produce heat *(3)*, WATs store lipids that are then used by other tissues in the intervals between meals *(4)*. BATs and WATs are densely innervated by sympathetic nerves *(5)*. Noradrenaline released by postganglionic sympathetic nerves stimulates BATs to "burn" lipids and produce heat, whereas stimulation of WATs induces lipolysis and release of free fatty acids

From: *Methods in Molecular Biology, Vol. 456:*
Adipose Tissue Protocols, Second Edition. Edited by: Kaiping Yang
© Humana Press, a part of Springer Science+Business Media, Totowa, NJ

into the bloodstream. Sensory nerves are sparser, less abundant in the parenchyma, and currently do not appear to be associated with any specific function in relation to adipocytes. Evidences suggest that these nerves may convey information about body fat levels to the brain (6) and that sensory neuropeptides may affect the differentiation of brown adipocytes (7).

Transmission electron microscopy (TEM) allows the examination of nerve bundles entering fat pads as well as individual axons running through the parenchyma among adipocytes. With this technique, some morphological aspects of adipose depot innervation, such as axon diameter, the presence of myelinated axons, type of synaptic vesicles, synaptic-like junctions on adipocytes and vessel wall cells can be described and quantified (1,2,8,9). Unfortunately, TEM is uninformative with regard to the neurotransmitter/neuropeptide content of nerve fibers, as it does not allow one to discriminate between autonomic (sympathetic or parasympathetic) and sensory nerves. This, however, is of paramount importance to characterize the functional role of individual nerve fibers in adipocyte biology. Reliable ultrastructural investigation of adipose tissue nerves requires excellent specimen conservation that is afforded only by strong, prolonged fixation and embedding in epoxy resins. As a result, tissue immunoreactivity is strongly impaired in TEM preparations, hampering detection by immunogold staining of antigens expressed at very low levels, such as those in nerves.

Light microscopy, although not achieving the morphological resolution of electron microscopy, does nevertheless allow visualization of nerves belonging to the autonomic or sensory component of the peripheral nervous system using immunohistochemistry. With this procedure, nerve fibers can be labeled with antibodies directed against molecules specifically involved in neurotransmission (e.g., tyrosine hydroxylase for sympathetic noradrenergic nerves, vesicular acetylcholine transporter for parasympathetic cholinergic nerves, substance P or calcitonin gene-related peptide for sensory nerves). Functionally characterized nerve fibers stained by immunohistochemistry appear as varicose-like cords with an irregular course among stromal and parenchymal adipose components (10–14). Their distribution and amount can then be adequately studied in different physiological and pathological conditions.

Confocal microscopy has added new potential to the study of adipose organ innervation. With the use of immunofluorescence staining, the technique permits double and triple staining of the same nerve (colocalization studies) and allows, for instance, the assessment of whether noradrenergic or cholinergic nerves also contain neuropeptides or whether some peptidergic nerves in the adipose depots lack the classic neurotransmitters (13). In addition, fluorescent structures are observed on very thin optical planes and the planes collected and reconstructed as 3D images, thus considerably increasing optical resolution. This makes it possible to study the spatial relationships between differently stained structures, for example neurotransmitter-releasing sites along axons and uncoupling protein 1-expressing brown adipocytes (our unpublished observations).

Here, we describe our routine protocols for detecting adipose tissue nerves using immunohistochemistry for light and confocal microscopy. Immunoperoxidase staining of counterstained sections followed by light microscopic observation (15) is the method of choice to visualize specific nerve types and their vascular and/or paren-chymal distribution in the mammalian adipose organ. This technique is also suitable for semiquantitative studies because it allows staining obtained under standard and experimental conditions to be compared. We recommend immunofluorescence with confocal microscopic analysis when the general distribution of a given nerve type is already known to the experimenter. Use of immunofluorescence in BAT and WAT paraffin sections yields poor results, at least when applied to nerves (which are small, follow an irregular course and are closely associated with blood vessels and/or adipocytes), because of the excessively small difference between signal (immunos-taining) and background. Because rodent paraffin-embedded adipose tissue exhibits a high degree of autofluorescence, immunofluorescence with confocal microscopy on cryosections, where possible (*see* **Subheading 3.2.5**), is more effective.

The general methods and techniques used by our group for studying BAT and WAT morphology have been published earlier *(16)*, and the reader is referred to that work, especially for the discussion of the importance of positive and negative con-trols in adipose tissue immunohistochemistry.

Investigations by morphological techniques do not involve analysis of numbers with mathematical instruments, but of images using qualitative techniques. Thus, image "interpretation" is a fairly subjective but crucial element. Correct morpho-logical interpretation depends on adequate methodological and technical sample preparation as much as observer experience and intellectual honesty.

2   Materials

2.1   Animal Perfusion and Tissue Preparation

1. Anesthetics: ketamine and xylazine.
2. Fixative: 4% paraformaldehyde in 0.1 M phosphate buffer (PB), pH 7.4.
3. 0.1 M PB, pH 7.4; 0.2 M Na$_2$HPO$_4$; and 0.2 M NaH$_2$PO$_4$ mixed with bidistilled water (1:1, v/v).
4. Phosphate-buffered saline (PBS): 0.008 M Na$_2$HPO$_4$, 0.0015 M KH$_2$PO$_4$, 0.027 M KCl, and 0.137 M NaCl.
5. Reagents for cryoprotection and embedding: sucrose, ethanol, xylol, paraffin (melting point 58°C), OCT medium.

2.2   Peroxidase Immunohistochemistry

1. Avidin–biotin-peroxidase complex (ABC) and *ABC Elite* (both from Vector Laboratories).
2. 0.3% H$_2$O$_2$ in methanol.

3. 3.3′-diaminobenzidine hydrochloride chromogen (DAB; Sigma).
4. Normal serum and secondary antibodies (Vector Laboratories).
5. Counterstain: hematoxylin.

2.3 Immunofluorescence

1. PB-NaCl buffer: 0.1 M PB, pH 7.4; and 0.9 % NaCl.
2. Normal donkey serum and fluorescent secondary antibodies (Jackson ImmunoResearch; *see* **Note 1**).
3. Fluorescent nuclear counterstain: TOTO-3 iodide (Molecular Probes).
4. Vectashield mounting medium (Vector Laboratories).

3   Methods

3.1 Immunohistochemistry on Paraffin-Embedded Sections

3.1.1 Perfusion and Fixation

1. With the animal under deep anesthesia (we use an intraperitoneal injection of 100 mg/kg ketamine in combination with 10 mg/kg xylazine), the thoracic cavity is opened and a blunt-tip needle of a size appropriate for the animal is inserted at the level of the apex of the left ventricle.
2. The right atrium must be excised immediately after beginning the perfusion, to allow the blood, and subsequently the excess fixative, to flow out (*See* **Note 2**). We use 4% paraformaldehyde in 0.1 M PB, pH 7.4, because it allows us to obtain excellent histological sections also for morphometric investigations as well as preservation of sufficient antigenicity for immunohistochemical experiments.
3. After perfusion, specimens are collected and immersed in the same fixative (where they should float freely in the jar) for 12–15 h at 4°C.
4. After fixation, specimens are thoroughly washed in PB to remove any residual fixative and stored in the same buffer at 4°C until embedding.

3.1.2 Dehydration and Embedding

After fixation, specimens must be embedded in a medium (paraffin) that will allow sectioning. For immunohistochemistry, the melting temperature of the paraffin used must not exceed 58°C, to avoid heat-dependent protein denaturation. Because paraffin is not miscible with water and because specimens cannot be transferred directly into melted paraffin after fixation, tissue fragments must first be dehydrated with increasing concentrations of ethanol, which gradually replaces water, and then cleared in a solvent (xylol) miscible with paraffin before impregnation.

1. Dehydrate with gradual ethanol series as follows: 30 min in 75%, 2 × 75 min in 95%, and 3 × 60 min in 100%, at room temperature (RT), stirring.
2. Clear specimens in xylol, 2 × 60 min.
3. Impregnate with paraffin overnight at 58°C.
4. Change the paraffin twice.
5. Orientate samples in embedding molds.

3.1.3 Sectioning

Samples are cut into 3- to 4-μm thick sections with a normal sliding or rotary microtome, placed in a warm bath (distilled water at 37–38°C), picked up on glass, and dried overnight at 40°C.

3.1.4 Peroxidase Immunostaining

We use an indirect immunological sequence. The specific (primary) antibody, raised against the antigen to be localized, binds directly to the tissue antigen in the section. Incubation with another antibody against the primary antibody ("bridge" serum), conjugated to one biotin molecule, is the second step in the sequence. Detection is with the ABC method, which is based on the ability of avidin (glycoprotein of egg albumin) to bind to four molecules of biotin (vitamin H). Incubation with ABC solution allows non-immunological crosslinking between the biotin conjugated to the bridge serum, itself bound to the primary antibody, and the avidin present in the complex, starting a chain reaction that progressively amplifies the signal. The peroxidase in the ABC is localized using a solution containing diaminobenzidine (DAB) as the substrate and hydrogen peroxide; the reaction product is brown-colored and easy to detect by light microscopy.

1. Deparaffinize sections and hydrate to water.
2. Processing through 12 incubation steps. Dewaxed sections are covered with aliquots of serum solution in a humid chamber; for the washing steps and the enzymatic reaction the slides are immersed in Hellendhal type jars.
3. Incubate in 0.3% H_2O_2 in methanol for 30 min, to block endogenous peroxidase.
4. Wash in $0.015 M$ PBS, pH 7.3, 2 × 15 min, gently stirring.
5. Incubate with normal serum (raised in the same animal as the secondary antibody) at 1:75 dilution in PBS, to reduce non-specific background staining.
6. Incubate with the primary (poly- or monoclonal) antibody raised against the antigen to be localized (the antibody should be tested to find the best working dilution) diluted in PBS, overnight at 4°C in a humid chamber.
7. Wash in PBS, 2 × 15 min.
8. Incubate with biotinylated secondary IgG antibody against the primary antibody raised in the same animal as the normal serum in step 2 (dilution 1:200), in PBS for 30 min at RT.
9. Wash in PBS, 2 × 15 min.

10. Incubate in ABC or ABC *Elite* for 1 h at RT (*see* **Note 3**).
11. Wash in PBS, 2×15 min.
12. Histoenzymatic visualization of peroxidase using DAB and H_2O_2 for 4–5 min in a dark chamber (0.075% DAB in 0.05 M Tris-HCl buffer, pH 7.6, and 0.03% H_2O_2).
13. Wash in tap water.
14. Weakly stain nuclei with hematoxylin and mount sections in Entellan or other synthetic resin. An example of nerve immunostaining on paraffin embedded adipose tissue is shown in **Fig. 6.1A**.

3.1.5 Morphometric Analysis

Quantitation of immunohistochemical data allows comparison of the degree of innervation of adipose depots from control animals and those subjected to pharmacological treatment *(17)*, diet modifications, or exposure to different temperatures *(12)*. For these studies, immunohistochemistry must be performed in identical conditions for all adipose tissue samples, taking special care that all sections are from specimens processed with the same type and time of fixation, share the same thickness and have been incubated with the primary and secondary antibodies at the same concentration for the same amount of time.

The total nerve supply in an adipose depot or the number of nerve fibers of a given type may then be obtained by calculating the density of nerve fibers expressing a pan-nerve marker (such as PGP 9.5 or S100) or an antigen specifically expressed by the nerve fiber being investigated. For example, the density of innervation of noradrenergic nerves is the area occupied by the specific brownish precipitate (indicating both the presence and the amount of tyrosine hydroxylase-immunoreactive structures), divided by the total area of the section. For each adipose section, 10 or 20 high-power fields are randomly selected and the area occupied by the specific staining and the total area of the section are measured using a morphological imaging system. We use a light microscope fitted with a color video camera, which displays the selected field directly on the computer screen. The computer is equipped with the morphological imaging system LUCIA, Version 4.5 (Nikon Instruments, Rome, Italy), which allows us to measure profiled areas in μm². Density of innervation must be measured in a number of sections, which allows for statistical analysis.

3.2 Immunohistochemistry on Cryosections

3.2.1 Perfusion and Fixation

This procedure does not substantially differ from the one described in **Subheading 3.1.1** Sometimes, when tissues are to be processed for immunofluorescence and in order to

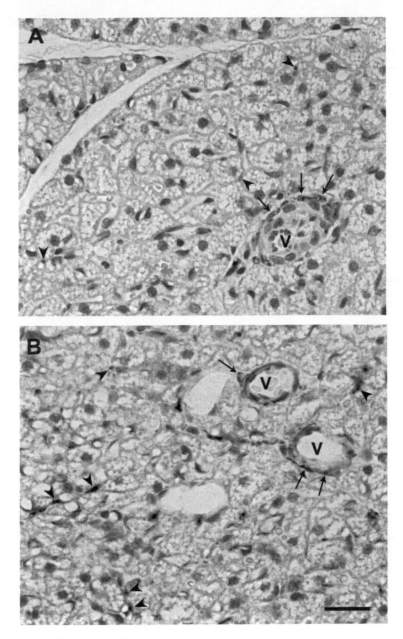

Fig. 6.1 Sympathetic innervation of brown adipose tissue. A 15-d-old male rat was perfused; the interscapular depot was isolated and post-fixed. Half was embedded in paraffin while the other half was frozen. Paraffin sections and cryosections underwent the same peroxidase immunostaining to detect noradrenergic nerves (primary antibody: sheep polyclonal anti-tyrosine hydroxylase from Chemicon, Temecula, CA, at a dilution of 1:200 v/v). Noradrenergic nerves are evident around blood vessels (V, arrows) and among adipocytes (arrowheads) both in the 3 μm thick paraffin section (**A**) and in the 6-μm thick cryosection (**B**). Frozen brown fat exhibits greater immunoreactivity than paraffin-embedded BAT (compare the density of innervation of the adipose parenchyma in **B** and **A**). In contrast, morphological resolution is much better preserved in the paraffin section (**A**), where most parenchymal adipocyte and capillaries are easily distinguished. Bar, 10 μm

improve perfusion (*see* **Note 4**), it may be useful to perfuse the animal first with physiological solution containing an anticoagulant (heparin) and then with the fixative.

3.2.2 Cryoprotection, Freezing, and Embedding

After fixation, specimens are cryoprotected in a solution of 30% sucrose in 0.1 *M* PB, pH 7.4. The procedure, which produces a degree of tissue dehydration, is essential to avoid specimen damage (fractures, holes, etc.) during the subsequent freezing step. BAT samples, especially those obtained from cold-exposed animals, which have a lower lipid content, often sink after 24–36 h of cryoprotection; this means that they are ready to be frozen. WAT samples never sink due to their low tissue density and we recommend to avoid cryoprotection exceeding 48 h. Specimens are then placed on one end of a marked card, embedded in OCT medium, and frozen by immersion for 1 min in a glass beaker containing 2-methylbutane (isopentane) pre-cooled in liquid nitrogen for about 2 min. Specimens are finally stored at −80°C and retain good antigenicity for about one year.

3.2.3 Sectioning

Sections from frozen specimens are obtained with a Leica CM1900 cryostat (Leica Microsystems, Vienna, Austria), which allows to control both the temperature of the chamber (which affects blade temperature) and that of the holder (which determines specimen temperature). Excellent BAT sections are obtained at chamber and holder temperatures of about −25°C. Sectioning is a compromise between specimen temperature and thickness of the section. BAT cryosections are usually thicker (5–7 μm) than paraffin sections (3–4 μm) and as a result allow less morphological resolution than paraffin sections (*see* **Note 5**). In contrast, obtaining good WAT sections is difficult. We have so far been unable to obtain WAT cryosections from the major depots, such as rodent epidydimal, retroperitoneal and subcutaneous depots of control animals, even at chamber temperatures of −35°C and holder (specimen) temperatures of −50°C. Only WAT adhering to other tissues or organs, like WAT surrounding BAT, mesenteric WAT and perirenal WAT may be successfully cut in control animals, but then tissue morphology is often so altered such that it precludes reliable morphological analysis and immunohistochemical procedures (*see* **Note 6**). After cutting, sections are placed on gelatine-coated or Superfrost slides (*see* **Note 7**) and air-dried overnight. Then they can be stored at −20°C, where they retain good antigenicity for several months.

3.2.4 Peroxidase Immunostaining

This procedure does not differ from the one described under 3.1.4. As a rule, better antigen preservation is afforded by cryosections than by paraffin sections. As a consequence, peroxidase immunostaining of cryosections often allows visualization of more nerves of a given type and the primary antibody is often used at higher dilution than is the case in paraffin sections (*see* **Note 8**). An example of nerve immunostaining on BAT cryosections is shown in **Fig. 6.1B**.

3.2.5   Indirect Immunofluorescence (Double or Triple Labeling)

Immunofluorescence is performed on cryosections.

1. Wash thoroughly in PB-NaCl buffer, 3 ×15 min, gently stirring.
2. Incubate with 1:75 normal donkey serum in PBS, to reduce non-specific background staining.
3. Incubate with primary (poly- or monoclonal) antibodies raised against the antigens to be localized (the antibodies should be tested to find the best working dilution) diluted in PB-NaCl buffer, overnight at 4°C in a humid chamber.
4. Wash in PB-NaCl buffer, 3×15 min.
5. Apply fluorophore secondary IgG antibodies against the primary antibodies, raised in the same animal as the normal serum in step 2, diluted 1:200 in PB-NaCl buffer, for 30 min at RT.
6. Wash in PB-NaCl buffer, 3×15 min.
7. Add TOTO-3 iodide (1:2000 v/v in PBS) for 20 min at RT to stain DNA and identify cell nuclei.
8. Tip off the TOTO-3 iodide and air-dry sections avoiding direct exposure to light.
9. Coverslip using Vectashield mounting medium.
10. Store in the dark at 4°C until use.

3.2.6   Confocal Microscopy

Immunostained sections are observed under a motorized Leica DM6000 microscope using different magnifications, from a x20 dry objective to a x60 planapochromat oil immersion objective with 1.4 numerical aperture. Fluorescence is detected with a Leica TCS-SL confocal microscope equipped with an Argon and He/Ne mixed gas laser. FITC and TRITC are excited with the 488 nm and 543 nm lines, respectively, and imaged separately in the green and red channels. The nuclear counterstain TOTO-3 iodide is excited with the 633 nm line and visualized in the blue channel.

Before specific staining we usually perform lambda scan records of sections subjected to the procedures described in **Subheading 3.2.5.**, without the primary antibodies (negative controls). The procedure allows us to acquire a series of images within a user-defined wavelength range; each image is detected at a specific emission wavelength. The procedure can be used to measure the emission spectrum of the sample of interest, to determine the emission maximum and optimize parameters during the acquisition process. Images (1024 × 1024) are obtained sequentially from two or three channels using a confocal pinhole of 1.1200, stored as TIFF files and then merged to obtain the final image. Brightness and contrast are adjusted using Photoshop 6 software. Examples of double staining and confocal microscopy analysis on BAT cryosections are given in **Figs. 6.2** and **6.3**.

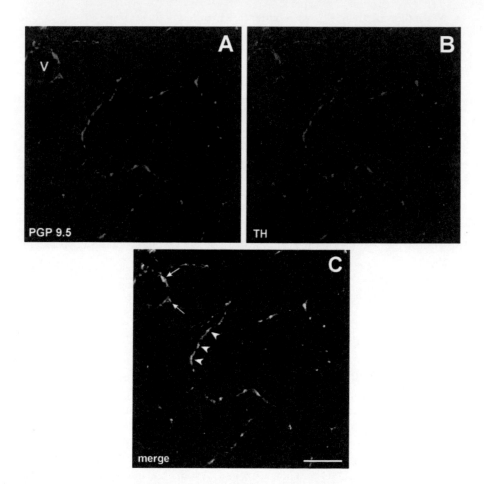

Fig. 6.2 Indirect immunofluorescence in rat brown adipose tissue. A cryosection of the interscapular depot of the animal described in **Fig. 6.1** was double-labeled with PGP 9.5 (**A**, primary antibody: rabbit polyclonal serum from Ultraclone, Isle of Wight, UK, at a dilution of 1:5000 v/v), a molecule present in all peripheral nerves, and tyrosine hydroxylase (**B**), a marker of noradrenergic nerves. Pictures taken at a 0.2-μm thick optical confocal plane. In the merged panel (**C**), all nerves are noradrenergic and are distributed both around blood vessels (arrows) and among the adipocytes (arrowheads). Bar, 20 μm

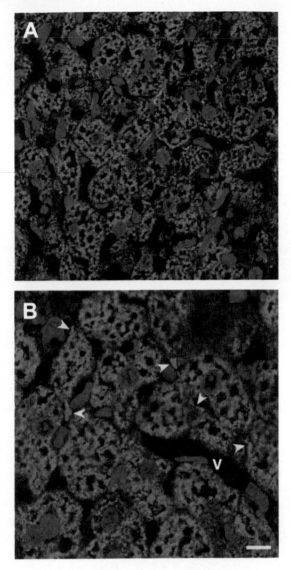

Fig. 6.3 Indirect immunofluorescence in rat brown adipose tissue. A cryosection of the inter-scapular depot of the animal described in **Fig. 6.1** was double-labeled with tyrosine hydroxylase (red), a marker of noradrenergic nerves, and uncoupling protein-1 (green, primary antibody: rabbit polyclonal serum from Chemicon, at a dilution of 1:800 v/v), a mitochondrial molecule that allows brown adipocytes to produce heat. Cell nuclei were stained with TOTO-3 iodide. The merged pictures, taken at a 0.2-µm thick optical confocal plane, show numerous noradrenergic nerves closely apposed to brown adipocytes (arrowheads). Bar, 20 µm (**A**) and 10 µm (**B**)

4 Notes

1. All the fluorescent secondary antibodies by Jackson ImmunoResearch are raised in donkey. This allows us to perform a single blocking step using only normal donkey serum to reduce non-specific background in double and triple staining experiments.

2. Optimum perfusion is demonstrated by muscle tremor, especially in the tail and limbs, and by blood-rich organs such as liver rapidly turning grayish-white. At the end of the perfusion, the body should be stiff. If perfusion is suboptimal, the specimen should be discarded.

3. ABC *Elite* amplifies the immunocomplex much better than ABC. Thus, if the primary antibody is highly specific and the background is low, some nerves expressing very low levels of the antigens, such as the sensory neuropeptides substance P and calcitonin gene-related peptide, are better visualized using ABC *Elite*.

4. Perfusion is a critical step for any type of morphological analysis. Nevertheless, when performing immunoperoxidase staining the presence of a few red blood cells in the capillaries among adipocytes, it may be useful to recognize the nerve fibers running along small blood vessels. Conversely, when performing immunofluorescence experiments it should be borne in mind that erythrocytes are autofluorescent and can considerably hamper detection of parenchymal nerve fibers.

5. Because it is difficult to standardize sectioning of frozen samples, especially section thickness, we usually perform morphometric studies on paraffin sections.

6. This fact is probably closely related to the high lipid content of WAT in normal conditions. In a study of the distribution of nerves in WAT of fasted animals *(13)*, we were able to obtain excellent cryosections of slimmed retroperitoneal and epidydimal adipose depots of fasted animals that had lost more than 20% of body weight at cryostat temperatures ranging from −25°C to −35°C.

7. Superfrost slides work well for both immunoperoxidase staining and immunofluorescence. In contrast, gelatin-coated glasses should not be used for immunofluorescence.

8. Some primary antibodies can detect the antigen on cryosections, but not on paraffin sections. In our experience, a polyclonal serum against the vesicular acetylcholine transporter, a marker of cholinergic nerves, can detect the antigen in cryosections of mediastinal BAT and in positive controls *(12)*, whereas results from paraffin-embedded material are generally poor.

Acknowledgments This work was financed by grants from Marche Polytechnic University (Contributi Ricerca Scientifica) and the Ministry of University (FIRB Internazionalizzazione RBIN047PZY, 2005).

References

1. Cinti S (1999) The adipose organ, Editrice Kurtis, Milan, Italy
2. Cinti S (2005) The adipose organ. Prostaglandins Leukot Essent Fatty Acids 73:9–15

3. Cannon B, Nedergaard J (2004) Brown adipose tissue: function and physiological significance. Physiol Rev 84:277–359
4. Jaworski K, Sarkadi-Nagy E, Duncan R, Ahmadian M, Sul HS (2007) Regulation of triglyceride metabolism. IV. Hormonal regulation in lipolysis in adipose tissue. Am J Physiol Gastrointest Liver Physiol 293:G1–4
5. Bartness TJ, Kay Song C, Shi H, Bowers RR, Foster MT (2005) Brain-adipose tissue cross talk. Proc Nutr Soc 64:53–64
6. Bartness TJ, Bamshad M (1998) Innervation of mammalian white adipose tissue: implications for the regulation of total body fat. Am J Physiol 275:R1399–1411
7. Giordano A, Morroni M, Carle F, Gesuita R, Marchesi GF, Cinti S (1998) Sensory nerves affect the recruitment and differentiation of rat periovarian brown adipocytes during cold acclimation. J Cell Sci 111:2587–2594
8. Slavin BG, Ballard KW (1978) Morphological studies on the adrenergic innervation of white adipose tissue. Anat Rec 191:377–389
9. Giordano A, Morroni M, Santone G, Marchesi GF, Cinti S (1996) Tyrosine hydroxylase, neuropeptide Y, substance P, calcitonin gene-related peptide and vasoactive intestinal peptide in nerves of rat periovarian adipose tissue: an immunohistochemical and ultrastructural investigation. J Neurocytol 25:125–136
10. Norman D, Mukherjee S, Symons D, Jung RT, Lever JD (1988) Neuropeptides in interscapular and perirenal brown adipose tissue in the rat: a plurality of innervation. J. Neurocytol 17:305–311
11. De Matteis R, Ricquier D, Cinti S (1998) TH-, NPY-, SP-, and CGRP-immunoreactive nerves in interscapular brown adipose tissue of adult rats acclimated at different temperatures: an immunohistochemical study. J Neurocytol 27:877–886
12. Giordano A, Frontini A, Castellucci M, Cinti S (2004) Presence and distribution of cholinergic nerves in rat mediastinal brown adipose tissue. J Histochem Cytochem 52:923–930
13. Giordano A, Frontini A, Murano I, Tonello C, Marino MA, Carruba MO, Nisoli E, Cinti S (2005) Regional-dependent increase of sympathetic innervation in rat white adipose tissue during prolonged fasting. J Histochem Cytochem 53:679–687
14. Giordano A, Song CK, Bowers RR, Ehlen JC, Frontini A, Cinti S, Bartness TJ (2006) White adipose tissue lacks significant vagal innervation and immunohistochemical evidence of parasympathetic innervation. Am J Physiol Regul Int Comp Physiol 291:R1243–R1255
15. Hsu SM, Raine L, Fanger H (1981) Use of avidin-biotin-peroxidase complex (ABC) in immunoperoxidase technique: a comparison between ABC and unlabeled antibody (PAP) procedures. J Histochem Cytochem 29:577–580
16. Cinti S, Zingaretti MC, Cancello R, Ceresi E, Ferrara P (2001) Morphologic techniques for the study of brown adipose tissue and white adipose tissue. Methods Mol Biol. 155:21–51
17. Giordano A, Centemeri C, Zingaretti MC, Cinti S (2002) Sibutramine-dependent brown fat activation in rats: an immunohistochemical study. Int J Obesity Relat Metab Disord 26:354–360

Chapter 7
Human Adipose Tissue Blood Flow and Micromanipulation of Human Subcutaneous Blood Flow

Gijs H. Goossens and Fredrik Karpe

Summary Regulation of blood flow in tissues such as skeletal muscle, liver, and adipose tissue is needed to meet the changing local metabolic and physiological demands under varying conditions. In healthy individuals, adipose tissue blood flow (ATBF) is remarkably responsive to meal ingestion, but changes in ATBF in response to other physiological stimuli, such as stress and physical exercise, have also been noted. The ATBF response to nutrient intake may be of particular importance in the regulation of metabolism by facilitating transport of nutrients as well as signaling between adipose tissue and other metabolically active tissues. A reduction in both fasting and postprandial ATBF has been observed in obesity; this impairment is associated with insulin resistance. A better understanding of the physiological basis for (nutritional) regulation of ATBF may therefore give insight to the relationship between disturbances in ATBF and the metabolic disturbances observed in response to insulin resistance. In this chapter, we describe some different approaches to quantify human ATBF, with a particular emphasis on the [133]xenon wash-out technique and a method by which regulatory properties of subcutaneous ATBF can be studied by pharmacological micromanipulation (microinfusion).

Key words Adipose tissue; blood flow; [133]Xenon wash-out; microinfusion; humans.

1 Introduction

Previous reports have demonstrated that adipose tissue blood flow (ATBF) increases after ingestion of carbohydrates or a mixed-meal *(1,2)*. The ATBF response to nutrient intake may be of importance in the regulation of metabolism by facilitating the transport of nutrients as well as conveying signals to and from adipose tissue, skeletal muscle and liver *(3)*.

Reduced fasting *(4–6)* and an impaired increase of postprandial *(4,5)* ATBF are associated with obesity and insulin resistance *(4,7)*. The increase in

From: *Methods in Molecular Biology, Vol. 456:*
Adipose Tissue Protocols, Second Edition. Edited by: Kaiping Yang
© Humana Press, a part of Springer Science+Business Media, Totowa, NJ

postprandial ATBF also appears to be specifically blunted in hypothyroidism *(8)*, but the mechanism is still poorly understood. Adipose tissue extraction of plasma triacylglycerol appears to increase in line with increasing ATBF *(9)*. It can therefore be speculated that an impaired ATBF response after meal intake redirects the storage of lipids derived from postprandial triacylglycerol-rich lipoproteins away from adipose tissue to other organs less well-suited to accommodate triglycerides (liver and skeletal muscle). This may in turn lead to increased production of very low-density lipoprotein from the liver and hypertriglyceridemia, thereby increasing the risk of developing cardiovascular disease. It is also possible that endocrine functions of adipose tissue, either seen in the context of adipose tissue responding to endocrine signals, or as disturbances in the release of endocrine factors from the tissue, can be attenuated as a consequence of impaired ATBF regulation.

Resting adipose tissue blood flow is largely determined by local release of nitric oxide (NO) *(10)*. Inhibition of NO release by local delivery of L-NMMA in the tissue rapidly reduces the ATBF. This implies a continuous tone of NO release in healthy adipose tissue. Another factor contributing to the rate of the resting ATBF is the renin-angiotensin system, within which angiotensin II (Ang II) appears to be responsible for tonic vasoconstriction in the tissue via interaction with the Ang II type 1 receptor *(11)*. Less is known about the mechanisms governing the reduced resting ATBF seen in obesity and insulin resistance as well as the differences in ATBF seen between adipose tissue depots. Gluteal *(12)* and femoral *(13)* adipose tissue have lower ATBF than abdominal adipose tissue. Part of the reduced resting ATBF seen in insulin resistance can be recovered by treatment with insulin sensitizing thiazolidinediones *(14,15)*, but whether that works through NO or an entirely different mechanism is not known. However, the blunted ATBF response after meal intake is not recovered by the thiazolidinedione treatment *(14)*. Overall, in healthy normal adipose tissue of lean subjects the postprandial response is largely regulated by sympathetic activation since it can be blocked by local tissue delivery of beta-adrenergic receptor antagonists *(10)*. Similar mechanisms, often in line with coordinated stress responses to mobilize energy, explain the increase in ATBF in response to physical exercise *(16)* and mental stress *(17)*.

1.1 Methods to Quantify or Estimate Human Adipose Tissue Blood Flow

1.1.1 [133]Xe Washout

The rate of wash-out of ^{133}Xe injected into adipose tissue has been used for four decades to estimate ATBF *(18)*. It is based on the algorithm that ^{133}Xe, which is a lipid-soluble noble gas, has a predictable partition between the water phase and the fat phase in the tissue. Because the lipid phase can be considered constant, any change caused by a difference in the water phase, which in turn is modulated by the blood flow.

Radioactivity can be detected with a skin-mounted detector (CsI (19)) or from a distance by a gamma camera (20). The latter requires the subjects to lie absolutely still whilst the former could allow for continuous monitoring for hours. ^{133}Xe is no longer available as an injectable solution (dissolved in saline), but injection of the gas directly into the tissue works just as well (21). Normally, 1–3 MBq of ^{133}Xe is injected at a standardized depth with the finest possible needle. A volume smaller than 20 µL is difficult to administer correctly whereas a volume greater than 200 µL may take some time to dissolve and a large volume may also alter the architecture of the tissue. After injection a 45–60 min time period is needed for equilibration. The disappearance of radioactivity is monitored continuously thereafter. After collection of the counts log-transformation will give linear slopes against time, which is multiplied by the ^{133}Xe tissue partition coefficient.

ATBF = regression coefficient $\times$ partition coefficient (*see* **Note 1**)

A fixed and theoretical partition coefficient such as the one between oil and water (10/1) can be used or one can make an attempt to extrapolate the conditions in the adipose tissue in question by using related biometric variables (6). The advantage with the ^{133}Xe method is that it approximates absolute flow, it is reproducible, relatively simple to use and, although special equipment is needed, it does not involve a massive investment. The disadvantages are poor resolution within very short time frames (minutes) and the hassle of handling radioactivity including the need to order new substance once or twice a month (half-life is ~5 days).

1.1.2 Positron Emission Scanning (PET)

PET has been used to approximate ATBF (22). It involves intravenous injection of [^{15}O]-labeled water and the scanning of a defined adipose tissue. Simultaneous and continuous withdrawal of arterial blood to monitor the content of [^{15}O]-labeled water is also needed. Although a formal comparison with the ^{133}Xe wash-out technique has not been made, the numerical values for ATBF appear to be similar. A major advantage of this technique is access to deep adipose tissue depot and the possibility of simultaneously monitoring metabolite fluxes, but the need to conduct the studies in a PET scanner is a restriction.

1.1.3 Microdialysis and Ethanol Outflow/Inflow Ratio

This technique has been used in connection with monitoring of interstitial concentrations of metabolites as well as delivering pharmacological substances to the tissue via microdialysis. Ethanol, typically 50 mmol/l, is included in the ingoing fluid (perfusate). To monitor metabolite concentrations in the tissue, a very slow flow rate is needed; otherwise, a near-complete exchange is not achieved. However, to appreciate a change in blood flow, the perfusion rate should be adjusted to a recovery of approximately 50% of the original ethanol concentration. The dialysis probe is sensitive to tissue movements (microtrauma) and the technique will only give an estimation of direction and magnitude of a difference in ATBF (semi-quantitative). In comparison with the ^{133}Xe wash-out method, this technique works well in some hands (23), but less well in others (24).

1.1.4 Laser Doppler (LD)

LD will monitor the movement of red blood cells and thereby give a relative approximation to flow. Used in large vessels where the volume of the vessel can be estimated, LD will help to provide a measurement of absolute flow. Unfortunately, it is our experience that the arteries/arterioles in adipose tissue are too small to accurately quantify the dimension and the flow is too low and variable in direction in the veins, which certainly are big enough to be quantified. Intra-tissue LD appears to be useful to pick up rapid changes in flow in adipose tissue (25). Future developments to pick up LD signals from skin-mounted probes that can monitor deeper structures remain to be tested in adipose tissue.

1.2   Micromanipulation of ATBF Using the ^{133}Xe Washout Technique

The ^{133}Xe wash-out technique can be used to monitor ATBF under fasting conditions, after meal ingestion, during exercise, and during systemic delivery of pharmacological compounds (hormones, drugs). This technique is thus suitable to study ATBF regulation after changes at the whole-body level. However, there is sometimes need to study local blood flow without interference from the systemic circulation, and control experiments within the same individual cannot be performed at the same time in the same subject. Therefore, a new technique was necessary in order to directly examine the underlying mechanisms of ATBF regulation, while avoiding systemic effects. Furthermore, this technique should ideally allow performing a control experiment at the same time in the same subject. Therefore, a novel technique was recently developed, which has been called microinfusion (26).   ·

The microinfusion technique combines local micromanipulation of blood flow using pharmacological agents with the assessment of ATBF using ^{133}Xe wash-out. This method enables quantitative assessment of the local effects of vasoactive agents on adipose tissue blood flow in vivo in humans while avoiding systemic effects of these agents (see **Note 2**). This technique can be applied to study the underlying mechanisms of (dys)regulation of adipose tissue blood flow under different conditions (e.g., fasting, postprandial state) in several (patho)physiological states (obesity, type 2 diabetes, hypertension).

2 Materials

2.1 Catheters and Clinical Disposables

1. Gloves.
2. Alcohol swab.
3. Infusion set (6-mm long, internal diameter 0.38 mm, outer diameter 1.5 mm, dead-space volume 60 μl; Quick-set infusion set, MiniMed, Applied Medical Technology Ltd).
4. Sterile sheet.
5. Fine needle syringe (29 G, Omnican B Braun) (for administration of ^{133}Xe).
6. ^{133}Xe (Isotopen Dienst Benelux, Baarle-Nassau, The Netherlands).
7. 4-cm wide tape.
8. 1-mL syringes (for administration of saline or pharmacological agents).

2.2 Instrumentation

1. Mediscint equipment/software (Oakfield Instruments, Eynsham, UK).
2. γ-counter probe (CsI scintillation detector, Oakfield Instruments).
3. Microinfusion pump (CMA100 pump, CMA microdialysis Ltd, Sunderland, UK).
4. Controlled room temperature.

3 Methods

Conventional approval of using the method should be obtained from the local ethics committee and radioactivity should be handled according to local guidelines. It should be explained to the subject that the best results are obtained if there are as few movement artefacts as possible.

3.1 Inserting the Catheters

Before insertion of the microinfusion catheters, the syringes containing saline and vasoactive agents have to be primed and mounted into the microinfusion pump, with the infusion set connected to each syringe. Catheter insertion points are disinfected with an alcohol swab. The catheters (6-mm long, internal diameter 0.38 mm, outer diameter 1.5 mm, dead-space volume 60 μL) can then be inserted 6–8 cm on either side of the midline of the abdomen into the abdominal subcutaneous adipose tissue (*see* **Notes 3** and **4**). After insertion, the needle is immediately removed, and the head of the infusion line catheter is connected to the infusion set. Thereafter, a saline infusion is started at 2 µL/min (CMA100 pumps, CMA microdialysis Ltd).

3.2   Injecting ^{133}Xe

The volume of ^{133}Xe injected depends on the amount of radioactivity, which can be calculated knowing the specific activity on the day of manufacturing (*see* **Note 5**). After 20 min recovery, gaseous ^{133}Xe (~1.0 MBq per site) can be injected at each site through the port in the hub of the catheter using a fine needle syringe (*see* **Notes 6 and 7**). After injection of ^{133}Xe, the γ-counter probes have to be placed over the infusion device (head of the catheter), so that the probes are placed exactly over the ^{133}Xe-containing adipose tissue. Subsequently, the catheters have to be perfused with saline for 1 min at 60 µL/min to wash the dead space and for a further 40–60 min at 2 µL/min to allow for equilibration of ^{133}Xe. Preferably the collection of cumulative radioactivity data points is made on line and the disappearance rate of ^{133}Xe can be monitored on a computer screen. When a stable decay is obtained (wash-out is mono-exponential) the experiment can start. This normally takes 45–60 min after injection of the ^{133}Xe.

3.3   Measuring ATBF

The principle of ATBF measurement and the equations used to calculate ATBF are explained in **Subheading 1.1.** and **Note 1**, respectively. To obtain accurate measurements of ATBF, it is crucial to firmly tape the detectors over the infusion devices (*see* **Note 8**).

3.4   Infusing Pharmacological Agents or Saline (Control)

After the equilibration period, the saline (control) can be switched from either left or right side (chosen at random), by disconnection at the hub of the infusion set, to a pharmacological agent while the infusion rate is maintained at 2 µL/min. As mentioned earlier (*see* **Subheading 3.1.**), the syringes containing vasoactive agents have to be installed into the microinfusion pump, with the infusion set already connected to each syringe (the vasoactive agent is dripping on a sterile sheet). This ensures that the complete infusion set is flushed with the vasoactive agent before connection to the hub of the catheter and prevents a delay in effect because the vasoactive agent has not yet reached the adipose tissue; the dead volume of the infusion set is 60 µL and without this precaution it would take the vasoactive substance 30 min to reach the tissue at a flow rate of 2 µL/min. The ATBF recording over the next infusion period assesses the effect of the agent, whereas the saline (control) infusion can be continued on the contralateral site to examine whether there are any changes in ATBF during the experiment that can not be attributed to infusion of the pharmacological agent (*see* **Note 9**). An observation time of at least 30 min is needed to establish a clear, new steady state after changing conditions.

3.5   Calculations of Effects on ATBF

Calculation of ATBF is explained in detail in **Note 1**. However, it is important to understand how effects of pharmacological agents can be examined. ATBF is often calculated as the mean of consecutive 10-min time periods *(10,11,26)*. The effect of a pharmacological agent can be analysed by averaging the two or three consecutive time points at the end of each infusion period, depending on the time needed to reach a steady-state of ATBF, after subtraction of baseline blood flow (e.g., t-20 to t0). Responses to pharmacological agents have to be evaluated within individuals by comparison with the saline control site at the same level on the abdomen. Relative changes in ATBF at the experimental site can be presented as changes from baseline values after correction for corresponding changes in ATBF at the control site. When performing an experiment that includes a postprandial period, it is suggested to analyse the effect of the agents on fasting ATBF by averaging the three consecutive time points before meal intake after subtraction of baseline blood flow (e.g., t-20 to t0). Peak ATBF values can be calculated as the mean of three consecutive time points, including the maximum ATBF, which results in the highest mean value within each subject *(11,26)*. The meal effect on ATBF can be analysed using the area under the curve by the trapezoidal rule. Area under the curves can then be divided by the time over which they were calculated.

4   Notes

1. ATBF can be calculated from the semilog plot of disappearance of radioactive counts versus time in 20s intervals. For the calculation of quantitative values of ATBF (per 100 g of tissue), the relative solubility of ^{133}Xe between tissue and blood, i.e., the partition coefficient (λ), is often assumed to be 10 ml/g *(27)*. ATBF can be calculated as follows:

$$\text{ATBF (ml.100 g tissue}^{-1}.\text{min}^{-1}) = \text{slope of semilog plot (ln counts/s)} \times \lambda$$
$$(\text{ml/g}) \times 100 \text{ (g)} \times 60 \text{ (s)} \ \textit{(18)}$$

However, if comparisons are made between people, it is likely that adipose tissues vary and, with that, the estimated partition coefficient. This approach is recommended when comparisons of ATBF are made between subjects with very different body composition (e.g., lean versus obese individuals). For this purpose the lipid fraction of adipose tissue (V) must be known:

Lipid fraction (V):

$$V = 0.327 + (0.0124 \times \% \text{ adiposity}) \ (r = 0.95, p < 0.005) \ \textbf{(28)}$$

Tissue-blood partition coefficient of ^{133}Xe for adipose tissue (λ):

$$\lambda = [V(S_L/S_P - 1) + 1] / [Ht(S_C/S_P - 1) + 1] \ \textbf{(6)},$$

whereby Ht is the hematocrit (expressed as a fraction), S_P (solubility in plasma) is 0.0939 mL/mL, S_L (solubility in lipid) is 1.8276 ml/g, S_C (solubility in red blood cells) is 0.2710 mL/ml. The calculated partition coefficient normally comes to a number in the range of 6.5–8.5 mL/g.

2. Although it is not our experience that infused agents via the microinfusion technique leads to any significant changes in blood pressure or heart rate, it is recommended to monitor blood pressure before the start of infusion and at intervals throughout the study. Knowing the blood pressure and the tissue flow will enable the calculation of vascular resistance.

3. Right and left sides of the abdomen can be studied simultaneously on each subject to allow direct comparison of the effects of the vasoactive compound on one site with the contralateral control site (saline infusion, 9 g/l NaCl). Importantly, the vasoactive compound needs to be administered at the same level on the abdomen as the saline control site, as it has been previously shown that ATBF is greater at the upper level compared to the lower level of the abdomen *(21,29)*, but is not different between the right and the left sides at either level *(11,29)*.

4. It may happen that the catheter is accidentally inserted into or very close to a vessel in the abdominal subcutaneous adipose tissue. This results in a rapid disappearance of the injected ^{133}Xe (injection procedure is explained in **Subheading 3.2.**). To circumvent this problem it may be useful to visualize the vascularisation of the underlying tissue by either using a red cold light source (KL 2500 LCD, Schott, Mainz, Germany) or, alternatively, by ultrasound.

5. When the volume of ^{133}Xe is large, a slight stinging sensation at the injection site can be reported. Therefore, we recommend that the volume injected must not exceed ~100 µL. It is recommended to measure the radioactivity in the syringe immediately before and after injection using scintillation probes in order to calculate the received dose of radioactivity.

6. After withdrawing the ^{133}Xe from it original container, any temperature changes in or around the syringe containing the gas may lead to loss of material. Insure the part of the syringe containing the radioactive gas is not touched by the hand.

7. ^{133}Xe has to be injected through the channel of the insertion needle, which can be retraced. Injecting the ^{133}Xe may take some time, because it is sometimes difficult to find the port in the hub of the catheter. It is strongly recommended to be patient during this procedure; make sure the manipulations with the injection needle are done with gentle movements without any pressure on the connector of the infusion set. Otherwise there is substantial risk of pricking through the

catheter, which will lead to injection of ^{133}Xe into an adipose tissue area different from that which subsequently will receive the infusion. Obviously, potential effects of vasoactive agents are not observed or underestimated in this case. When the injection needle has successfully been inserted through the channel of the insertion needle, ^{133}Xe can be slowly injected over a period of ~60 s.

8. Insufficient taping of the detectors will cause movement artefacts that hamper accurate data collection. Therefore, in addition to taping of the detectors to the skin of the abdominal wall, it is strongly recommended to interconnect all detectors with 4-cm wide tape, resulting in firmly taped detectors that are less prone to movements due to normal respiration. The tape sticks better if the skin is wiped with an alcohol swap before applying the adhesive tape.

9. Regulation of ATBF can be examined both under fasting and postprandial conditions. Thus, local infusion of the pharmacological agents can be started under fasting conditions to assess the effect of the agents on fasting ATBF. Subsequently, a meal (e.g., ingestion of an oral glucose load or mixed-meal) can be provided, while infusion of the agents is continued. In this way, postprandial ATBF regulation can be investigated.

Acknowledgments We are grateful to the colleagues who have worked with us in developing and applying this technique, particularly Prof. Keith Frayn, Dr. Jean-Luc Ardilouze, Dr. Barbara Fielding, Ms Louise Dennis, Prof. Ellen Blaak, and Prof. Marleen van Baak.

References

1. Bulow J, Astrup A, Christensen NJ, Kastrup J (1987) Blood flow in skin, subcutaneous adipose tissue and skeletal muscle in the forearm of normal man during an oral glucose load. Acta Physiol Scand 130:657–661
2. Coppack SW, Evans RD, Fisher RM, Frayn KN, Gibbons GF, Humphreys SM, Kirk ML, Potts JL, Hockaday TD (1992) Adipose tissue metabolism in obesity: lipase action in vivo before and after a mixed meal. Metabolism 41:264–272
3. Frayn KN, Karpe F, Fielding BA, Macdonald IA, Coppack SW (2003) Integrative physiology of human adipose tissue. Int J Obes Relat Metab Disord 27:875–888
4. Jansson PA, Larsson A, Lonnroth PN (1998) Relationship between blood pressure, metabolic variables and blood flow in obese subjects with or without non-insulin-dependent diabetes mellitus. Eur J Clin Invest 28:813–818
5. Summers LK, Samra JS, Humphreys SM, Morris RJ, Frayn KN (1996) Subcutaneous abdominal adipose tissue blood flow: variation within and between subjects and relationship to obesity. Clin Sci (Lond) 91:679–683
6. Blaak EE, van Baak MA, Kemerink GJ, Pakbiers MT, Heidendal GA, Saris WH (1995) Beta-adrenergic stimulation and abdominal subcutaneous fat blood flow in lean, obese, reduced-obese subjects. Metabolism 44:183–187
7. Karpe F, Fielding BA, Ilic V, Macdonald IA, Summers LK, Frayn KN (2002) Impaired postprandial adipose tissue blood flow response is related to aspects of insulin sensitivity. Diabetes 51:2467–2473
8. Dimitriadis G, Mitrou P, Lambadiari V, Boutati E, Maratou E, Panagiotakos DB, Koukkou E, Tzanela M, Thalassinos N, Raptis SA (2006) Insulin action in adipose tissue and muscle in hypothyroidism. J Clin Endocrinol Metab 91:4930–4937

9. Samra JS, Simpson EJ, Clark ML, Forster CD, Humphreys SM, Macdonald IA, Frayn KN (1996) Effects of epinephrine infusion on adipose tissue: interactions between blood flow and lipid metabolism. Am J Physiol 271:E834–E839

10. Ardilouze JL, Fielding BA, Currie JM, Frayn KN, Karpe F (2004) Nitric oxide and beta-adrenergic stimulation are major regulators of preprandial and postprandial subcutaneous adipose tissue blood flow in humans. Circulation 109:47–52

11. Goossens GH, McQuaid SE, Dennis AL, van Baak MA, Blaak EE, Frayn KN, Saris WH, Karpe F (2006) Angiotensin II: a major regulator of subcutaneous adipose tissue blood flow in humans. J Physiol 571:451–460

12. Tan GD, Goossens GH, Humphreys SM, Vidal H, Karpe F (2004) Upper and lower body adipose tissue function: a direct comparison of fat mobilization in humans. Obes Res 12:114–118

13. Engfeldt P, Linde B (1992) Subcutaneous adipose tissue blood flow in the abdominal and femoral regions in obese women: effect of fasting. Int J Obes Relat Metab Disord 16:875–879

14. Tan GD, Fielding BA, Currie JM, Humphreys SM, Desage M, Frayn KN, Laville M, Vidal H, Karpe F (2005) The effects of rosiglitazone on fatty acid and triglyceride metabolism in type 2 diabetes. Diabetologia 48:83–95

15. Viljanen AP, Virtanen KA, Jarvisalo MJ, Hallsten K, Parkkola R, Ronnemaa T, Lonnqvist F, Iozzo P, Ferrannini E, Nuutila P (2005) Rosiglitazone treatment increases subcutaneous adipose tissue glucose uptake in parallel with perfusion in patients with type 2 diabetes: a double-blind, randomized study with metformin. J Clin Endocrinol Metab 90: 6523–6528

16. Simonsen L, Henriksen O, Enevoldsen LH, Bulow J (2004) The effect of exercise on regional adipose tissue and splanchnic lipid metabolism in overweight type 2 diabetic subjects. Diabetologia 47:652–659

17. Linde B, Hjemdahl P, Freyschuss U, Juhlin-Dannfelt A (1989) Adipose tissue and skeletal muscle blood flow during mental stress. Am J Physiol 256:E12–E18

18. Larsen OA, Lassen NA, Quaade F (1966) Blood flow through human adipose tissue determined with radioactive xenon. Acta Physiol Scand 66:337–345

19. Samra JS, Frayn KN, Giddings JA, Clark ML, Macdonald IA (1995) Modification and validation of a commercially available portable detector for measurement of adipose tissue blood flow. Clin Physiol 15:241–248

20. Jansson PA, Larsson A, Smith U, Lonnroth P (1992) Glycerol production in subcutaneous adipose tissue in lean and obese humans. J Clin Invest 89:1610–1617

21. Simonsen L, Enevoldsen LH, Bulow J (2003) Determination of adipose tissue blood flow with local 133Xe clearance. Evaluation of a new labelling technique. Clin Physiol Funct Imaging 23:320–323

22. Virtanen KA, Peltoniemi P, Marjamaki P, Asola M, Strindberg L, Parkkola R, Huupponen R, Knuuti J, Lonnroth P, Nuutila P (2001) Human adipose tissue glucose uptake determined using [(18)F]-fluoro-deoxy-glucose ([(18)F]FDG) and PET in combination with microdialysis. Diabetologia 44:2171–2179

23. Fellander G, Linde B, Bolinder J (1996) Evaluation of the microdialysis ethanol technique for monitoring of subcutaneous adipose tissue blood flow in humans. Int J Obes Relat Metab Disord 20:220–226

24. Karpe F, Fielding BA, Ilic V, Humphreys SM, Frayn KN (2002) Monitoring adipose tissue blood flow in man: a comparison between the (133)xenon washout method and microdialysis. Int J Obes Relat Metab Disord 26:1–5

25. Wellhoner P, Rolle D, Lonnroth P, Strindberg L, Elam M, Dodt C (2006) Laser-Doppler flowmetry reveals rapid perfusion changes in adipose tissue of lean and obese females. Am J Physiol Endocrinol Metab 291:E1025–E1030

26. Karpe F, Fielding BA, Ardilouze JL, Ilic V, Macdonald IA, Frayn KN (2002) Effects of insulin on adipose tissue blood flow in man. J Physiol 540:1087–1093

27. Yeh SY, Peterson RE (1965) Solubility of krypton and xenon in blood, protein solutions, tissue homogenates. J Appl Physiol 20:1041–1047
28. Martin AD, Daniel MZ, Drinkwater DT, Clarys JP (1994) Adipose tissue density, estimated adipose lipid fraction and whole body adiposity in male cadavers. Int J Obes Relat Metab Disord 18:79–83
29. Ardilouze JL, Karpe F, Currie JM, Frayn KN, Fielding BA (2004) Subcutaneous adipose tissue blood flow varies between superior and inferior levels of the anterior abdominal wall. Int J Obes Relat Metab Disord 28:228–233

Chapter 8
Studies of Thermogenesis and Mitochondrial Function in Adipose Tissues

Barbara Cannon and Jan Nedergaard

Summary Brown and white adipose tissues in mammals have a number of similar properties, such as lipid storage and adipokine production, but also distinctive properties. The energy-storing white adipose tissue has few mitochondria and low oxidative capacity. The heat-producing brown adipose tissue has a high density of mitochondria and high oxidative capacity. Mitochondrial function can be investigated in cells and organelles isolated from both brown and white adipose tissues. This chapter describes methods for successful isolation of suitable preparations of adipose tissues and their subsequent use. Questions concerning thermogenic capacity of the tissues, their potential influence on whole body metabolism, and specific properties of the mitochondria and their mode of function may be addressed using these methods.

Key words White adipose tissue; brown adipose tissue; mitochondria; thermogenesis; respiration; membrane potential.

1 Introduction

Adipose tissues in mammals are distinguished as being brown or white. Brown adipose tissue functions to produce heat and, thus, has a high oxidative capacity, evidenced by the extraordinarily high density of mitochondria in the cells. White adipose tissue is primarily an energy-storing tissue with low oxidative capacity. However, studies of metabolic activity are relevant for both tissues.

Because thermogenesis (heat production) is the function of brown adipose tissue, it would be a natural choice to measure this directly. However, it has only been done a few times, mainly because of technical limitations (microcalorimeters are still not widespread equipment in biological laboratories). However, it has been calculated that respiratory determinations are indeed satisfactory measures of heat production (*1*) (this is probably true for most mammalian organs with

From: *Methods in Molecular Biology, Vol. 456:*
Adipose Tissue Protocols, Second Edition. Edited by: Kaiping Yang
© Humana Press, a part of Springer Science + Business Media, Totowa, NJ

good blood supply). It is therefore a routine procedure to perform respiratory measurements and equate the result with that of thermogenesis and metabolic activity in general.

Respiratory measurements can be performed on isolated mitochondria, isolated cells, and on tissue pieces. To obtain sound values, it is essential that the measurements are made under the most optimal conditions possible. This includes provision of an adequate oxygen supply throughout the experiment and also the use of a substrate for respiration that is not limiting.

Because of the requirement for sufficient oxygen, tissue pieces can often be problematic, as oxygen supply may be limited by the diffusion of oxygen through the piece of tissue. Dispersed cells and mitochondria can be more easily oxygenated but are obviously more artificial in other respects.

The supply of a suitable substrate for respiration is often difficult. To estimate maximal capacity, the rate of substrate supply must exceed that of the ongoing respiration. For brown adipose tissue, respiration is uncoupled from phosphorylation under conditions when thermogenesis is activated, and the rate is thus limited by the capacity of the uncoupling protein (UCP1) or by the respiratory chain. In white adipose tissue, respiration is normally coupled to ADP phosphorylation, and the rate is therefore determined by the rate of utilization of ATP. Choice of a nonoptimal substrate, the transport of which is rate-limiting, can provide spurious results, leading to erroneous conclusions.

Isolated mitochondria from brown adipose tissue are relatively easy to study, because the mature cells contain such high mitochondrial density that the mitochondrial population isolated after homogenization of whole tissue is statistically representative for the mature adipocytes. For white adipose tissue, the mitochondrial density in the adipocytes is low; a mitochondrial preparation from total tissue may therefore not be representative for white adipocyte mitochondria and it can therefore be necessary to isolate mitochondria from isolated cells, which leads, however, to very low yields. Mature adipocytes can be conveniently isolated from both tissues based on Rodbell's classical collagenase digestion procedure (2), as the fat-containing cells readily float and can thus be separated from tissue debris in aqueous media.

2 Materials

2.1 Isolation of Mitochondria

1. 0.25 M sucrose (see **Note 1**).
2. 100 mM KCl containing 20 mM K-TES, pH 7.2.
3. Bovine serum albumin, (fraction V), fatty-acid-free. Dissolve in 0.25 M sucrose to a concentration of 0.3%.
4. Glass homogenizer with tight-fitting Teflon pestle.
5. Small glass hand homogenizer.

6. High-speed centrifuge with fixed angle rotor, tube size ≈ 50 mL.
7. Gauze.

2.2 Isolation of Adipocytes

1. Krebs/Ringer phosphate buffer with the following composition (in mM): Na$^+$ 148, K$^+$ 6.9, Ca^{2+} 1.5, Mg^{2+} 1.4, Cl$^-$ 119, SO$_4^{2-}$ 1.4, H$_2$PO$_4^-$ 5.6, HPO$_4^{2-}$ 16.7, glucose 10, fructose 10. Include 4% crude bovine serum albumin. Adjust pH with Tris-OH or HCl to 7.4.
2. Krebs/Ringer bicarbonate buffer with the following composition (in mM): Na$^+$ 145, K$^+$ 6.0, Ca^{2+} 2.5, Mg^{2+} 1.2, Cl$^-$ 128, SO$_4^{2-}$ 1.2, HCO$_3^-$ 25.3, H$_2$PO$_4^-$ 1.2, glucose 10, fructose 10, and fatty-acid-free bovine serum albumin 4%. Bubble the buffer with 5 % CO$_2$ in air at 37°C and adjust the pH with Tris-OH or HCl to 7.4; keep the buffer at 37°C and continuously bubble with a small stream of 5 % CO$_2$ in air until use.
3. Crude and fatty-acid-free bovine serum albumin (Fraction V).
4. Crude collagenase (Type I, Clostridiopeptidase A, EC 3.4.24.3).
5. Water shaker at 37°C.
6. Silk filter cloth (Joymar Scientific, Hicksville, NY).

2.3 Oxygen Electrode

1. Clark type oxygen probe. Available from, e.g., Rank Bros. or Hansatech. Purchased as a complete system with measuring chamber and magnetic stirrer. Alternatively, particularly suitable for small samples, the Oroboros Oxygraph 2k.
2. PowerLab 4/30 data acquisition and analysis system with ChartPro software for Windows or Macintosh.
3. Chart recorder (optional).

3 Methods

3.1 Isolation of Brown Adipose Tissue Mitochondria

For a routine preparation of brown adipose tissue mitochondria, use five mice that have been living at normal animal facility temperatures (*see* **Note 2**). All procedures are conducted at 0–4°C.

1. The animals are anaesthetized for 1–2 min in 79% CO$_2$ and 21% O$_2$ and decapitated. Dissect out the periaortic, cervical, interscapular and axillary brown adipose tissue carefully into a small volume of ice-cold sucrose on a square of parafilm. Rinse in sucrose.

2. Absorb excess sucrose with a medical wipe, mince the tissue with scissors and homogenize the mince in approx. 40 mL of 250 mM sucrose solution (about 5% w/v), in a glass homogenizer with Teflon pestle. Five to six strokes are required.

3. Filter the homogenate through two layers of gauze and centrifuge (all centrifugation steps for 10 min) at 8500 g.

4. Discard the hard-packed fat layer and supernatant by rapidly inverting the tube and wipe the walls of the tube clean with a medical wipe.

5. Resuspend the pellet (containing cell debris, nuclei and mitochondria) in a small volume of sucrose and transfer to a clean tube. Dilute again to about 40 mL sucrose solution and centrifuge at 800 g.

6. Transfer the supernatant carefully to a clean tube. Discard the pellet (that contains debris and nuclei). Centrifuge the supernatant at 8500 g.

7. Resuspend the resulting mitochondrial pellet in 5 mL of sucrose solution with 0.3% fatty-acid-free bovine serum albumin and centrifuge at 8500 g.

8. Further wash the albumin-washed mitochondrial pellet by one of two procedures:

   a. For respiratory studies, resuspend the pellet in about 15 mL of the KCl-TES buffer. (*see* **Note 3**) Centrifuge the suspension at 8500 g.

   b. For other studies, resuspend the pellet in 15 mL of sucrose solution. Centrifuge the suspension at 8500 g.

9. Resuspend the resulting pellet in a minimal volume of the respective medium by hand homogenization.

10. Measure the protein concentration in the final, albumin-washed mitochondrial pellet and dilute the suspension with KCl-TES buffer (*see* **Subheading 2.1.**) or sucrose solution (*see* **Subheading 2.1.**) to a stock concentration of 10 to 20 mg per milliliter for storage on ice.

3.2   Isolation of Brown Adipocytes

For a routine preparation of brown adipocytes, 2 adult (10- to 30-wk-old) Syrian hamsters (*Mesocricetus auratus*) of either sex are used. The hamsters are kept at 20–22°C, one to three per cage, with food and water ad libitum (*see* **Note 4**).

1. The animals are anesthetized by 79% CO_2 and 21% O_2 and decapitated. Dissect out the cervical, interscapular and axillary brown adipose tissue into a small volume of ice-cold Krebs/Ringer phosphate buffer on a square of parafilm and carefully clean from contaminating tissues.

2. Place the brown adipose tissue in a polyethylene vial containing 3 mL Krebs/Ringer phosphate buffer with 4% crude bovine serum albumin and 0.83 mg/mL collagenase.

3. Preincubate the tissue for 5 min in a 1.7-Hz shaking water bath at 37°C. Add 7 mL of the buffer and vortex the vial for 5 s.

4. Filter the contents of the vial onto silk cloth, discard this first filtrate. Transfer the tissue pieces collected on the silk to a small volume of Krebs/Ringer phosphate buffer on a square of parafilm and mince with scissors. Incubate this mince in 3 ml fresh, albumin- and collagenase-containing buffer as above for 25 min, with 5 s vortexing every fifth minute.

5. Add 7 mL of buffer, and vortex the vial for 15 s, and filter the contents as above. Collect the filtrate and centrifuge it (5 min, 65 g). Discard the infranatant by suction with a Pasteur pipet with plastic tubing on the tip connected to a water suction pump (*see* **Fig. 8.1**), add 10 mL of buffer and allow the cells to stand at 4°C.

6. Incubate the tissue pieces remaining on the silk filter as above for 15 min, and collect the cells; the tissue pieces now remaining can be incubated for 10 min and the cells collected, in order to increase the yield.

7. Discard the infranatants in all three tubes and combine the cells. Add 10 mL of buffer, and centrifuge the cells as above for 2 min.

8. Discard the infranatant and count the cells in a Bürker chamber. Store the cells on ice until use, at a concentration of $1–3 \times 10^6$ cells/mL; very little deterioration of cell response is observed during a working day. This is only valid for hamster cells. Cells from rats, and particularly from mice, break easily and should be aliquoted into Eppendorf tubes, kept at room temperature and used immediately (*see* **Notes 5** and **6**).

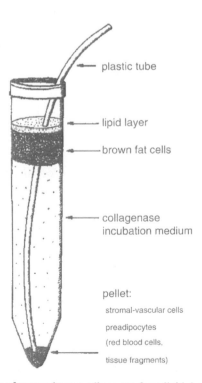

Fig. 8.1 The separation of mature brown adipocytes from lipid droplets and other constituents

9. Hamster cells can be stored overnight at 4°C in 10 ml of KRPB, in which case the cells are washed again the next day by centrifugation (*see* **Notes 5** and **6**).
10. It may be noted that undifferentiated preadipocytes are present in the discarded infranatant. These cells can be collected, cultured in primary culture and differentiated into mature brown adipocytes by the procedures described earlier *(3)*.

3.3   *Measurement of Respiratory Rate*

The rate of oxygen consumption of both isolated brown-fat mitochondria and cells can be readily measured polarographically with a Clark-type oxygen probe *(4)*. Such a probe (obtainable from, e.g., Rank Bros., Hansatech or Oroboros) determines oxygen concentration in aqueous solutions. The current produced by the electrode is proportional to the oxygen tension in the solution. The electrode chamber must be continuously stirred (most practically magnetically). Preferably, the electrode chamber must also be temperature-controlled (e.g., by circulating water from a water bath). Calibrate as follows:

1. Fill the electrode chamber with distilled water at the experimental temperature, do not add any lid (to allow equilibration with atmospheric oxygen), and allow the output to stabilize. At 37°C, this corresponds to 217 nmol O_2/mL (*see* **Note 7**); at 25°C (a traditional but clearly less physiological temperature classically used for mitochondrial experiments), this corresponds to 253 nmol O_2/mL.
2. Add a few crystals of sodium dithionite to the solution; this reduces all oxygen and thus provides for determination of zero oxygen level.
3. Remove the calibration solution, carefully wash the chamber well with distilled water and add the relevant buffer. Note that the output may not fully return to the level it had with distilled water; this is correct, as salt-containing media dissolve less oxygen than distilled water.

   a. For mitochondrial studies, use a medium consisting of 100 mM KCl, 20 mM K-TES, 4 mM KH_2PO_4, 2 mM $MgCl_2$, 1 mM EDTA, final pH 7.2. For introduction of energy conservation and choice of substrate, *see* **Notes 8** and **9**.
   b. For cell studies, use the Krebs/Ringer bicarbonate buffer, bubbled with CO_2, as described previously *(see* **Note 10**).

4. Add 0.2–0.5 mg mitochondrial protein or 50,000–80,000 cells per ml buffer in the electrode chamber. Close the chamber and allow the system to stabilize. Make further additions with a Hamilton syringe through a small hole in the cover of the chamber. Note that an addition artifact may voluntarily be produced by allowing the syringe to momentarily stop the magnetic stirring. Note also that the addition of ethanol (as solvent for some compounds) may lead to a small baseline shift.

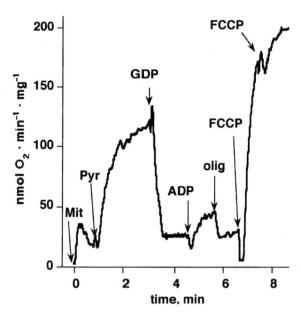

Fig. 8.2 Typical oxygen electrode trace obtained with isolated brown-fat mitochondria. The rate of oxygen consumption is shown. When mitochondria are added, only a very low rate of respiration (thermogenesis) is observed. However, if substrate for combustion, here pyruvate, is added, a high rate of respiration is observed. That this respiration can be inhibited by GDP indicates that it is mediated via the uncoupling protein-1 (UCP1). A low basal rate of respiration is left after addition of GDP. Addition of ADP, which leads to initiation of ATP synthesis, leads, in these mitochondria, to only a very small increase in respiration when contrasted to what is observed in practically all non-brown-fat mitochondria (so-called state-3 rate); the low rate is caused by a very low content of the ATP-synthase enzyme. Addition of oligomycin inhibits the low activity of the ATP-synthase, resulting in so-called state-4 rate. Addition of an artificial uncoupler (here FCCP) reveals the total capacity of the mitochondria for oxidation of the substrate

5. The electrode output can be connected to a computer via an analogue-to-digital converter and the data acquired in a Chart PowerLab application program. (Alternatively, a regular chart recorder can be used.) *See* **Figs. 8.2** and **8.3** for representative results of such determinations.

3.4 Measurement of Mitochondrial Membrane Potential

Mitochondrial membrane potential can be determined by a number of methods that rely upon distribution of a cationic dye within or across the mitochondrial inner membrane in accordance with the membrane potential. Several compounds can be used for this purpose, such as triphenylmethylphosphonium (i.e., TPMP),

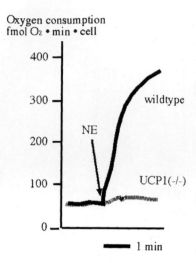

Fig. 8.3 Typical oxygen electrode trace obtained with isolated mature brown-fat cells. Two types of brown-fat cells are compared. In brown-fat cells isolated from mice in which UCP1 has been genetically ablated (UCP1(–/–)), there is a low basal rate of respiration (thermogenesis). In these cells, the addition of norepinephrine (NE) does not lead to an increase in thermogenesis. In brown-fat cells isolated from wild-type mice (WT), the unstimulated rate of respiration is similar to that in brown-fat cells without UCP1, i.e., the mere presence of UCP1 in the mitochondria within the cells does not lead to uncoupling (thermogenesis). However, when the UCP1-containing cells are stimulated with norepinephrine, the high thermogenic potential of the brown-fat cells is demonstrated. Principal sketch based on observations in Matthias et al. *(20)*

tetraphenylphosphonium (TPP, which can also be used in ion-selective electrodes), Rhodamine 123, and safranine O. The method described here is for safranine, because we have found it to function satisfactorily.

The conditions and media should be identical to those described previously for oxygen consumption measurements in mitochondria, so that the results between the two types of determinations can be compared. The changes in absorbance of safranine O (5 μM) at 511 – 533 nm are followed. Increased quenching of the color gives an upward deflection.

1. Add 1 mL of medium (as listed previously for respiratory studies) and 0.5 mg of mitochondrial protein to a suitable cuvette, together with 5 μM safranine.
2. Record the changes in absorbance after the addition of compounds of interest.
3. At the end of the experiment, add 25 μM FCCP, followed by 7 × 10 mM NaOH until the mitochondria are solubilized.
4. For each preparation calibrate the membrane potential as follows. Transfer the mitochondria into a state of energy conservation as in **Note 8**. Subsequently, to the mitochondria in the cuvette, add 9 μM valinomycin followed by KCl, at concentrations between 0.1 mM and 50 mM (choose a range of about 8 concentrations and perform one trace for each concentration). At the end of the experiment, add 25 μM FCCP, followed by 7 × 10 mM NaOH, as above.

5. Plot the change in absorbance against log KCl concentration in order to extrapolate to the internal K$^+$ concentration in the mitochondria and the initial concentration in the medium.
6. Use the Nernst equation and the values obtained in the calibration to calculate the membrane potential under each condition.

4 Notes

1. Some authors use a lightly buffered sucrose solution, containing, e.g., 5 mM K-TES. Our experience is that this has no obvious beneficial effect on the preparation. Similarly, for brown fat mitochondria, (in contrast to mitochondria from skeletal muscle or liver), we have not observed any beneficial effect of the use of a chelator, e.g., EDTA (2 mM) or EGTA (1 mM).
2. Because many studies are now performed on genetically modified mice, the method described here is for such animals. However, mitochondria can also be isolated by the same method from the brown adipose tissue of other mammals (most commonly rats but also Syrian hamsters). They can be isolated from tissue taken from animals kept at colder or warmer environmental temperatures. In these cases, the tissue contains less or more triglyceride and the number of mitochondria per gram tissue is higher in cold-acclimated animals. Triglyceride disturbs the homogenization and a lower relative yield of mitochondria will generally be obtained from animals kept at higher temperatures and similarly higher from animals kept at lower temperatures.
3. When brown adipose tissue mitochondria are isolated, they are uncoupled (*5,6*) and have a collapsed membrane potential. They demonstrate high permeability to many monovalent ions. Presumably as a consequence of this, they have lost the ability to retain osmotic support in the mitochondrial matrix. The matrix is therefore highly condensed after preparation, and oxidation of substrates in the matrix is markedly inhibited (*7–9*). To re-expand the matrix, the mitochondria may be incubated in an iso-osmotic medium of permeant ions (such as KCl). Matrix expansion can also be achieved with low osmolarity sucrose (100 mM), although this probably gives a less controlled expansion.
4. Brown adipocytes can also be prepared by the same method from rats (*10,11*) and mice (*12*). It is our experience that the cells from Syrian hamsters are the most robust and for many studies are therefore very suitable. Their robustness also means that they are an appropriate choice for people learning the technique.
5. The details given here for preparation of brown adipocytes are examples of incubation times with collagenase, collagenase concentrations (and types), centrifugation times (or flotation without centrifugation), which can be applied. Different workers tend to develop personal modifications of these, particularly at times and under circumstances when, for unclear reasons, the preparations are less successful. It is difficult to find convincing evidence that these modifications are of major significance, but this idiosyncrasy also indicates that the details specified here are for guidance and need not be adhered to exactly.

6. The preparation technique is dependent upon the property of the cells to float on top of an aqueous medium. If cells are isolated from animals that are cold-exposed, the triglyceride concentration may be so low that the cells sink in the medium and can therefore not be separated. The yield of cells will thus be lower than normal *(13,14)*. In general, the yield of mature adipocytes as a percentage of total adipocytes in the tissue is not very high. The representativeness of the cell population can perhaps therefore be discussed. Also, if cells are prepared from animals living at thermoneutrality, the cells are very replete with triglycerides and their diameter larger than of cells from animals at room temperature. This large cell size seems to make the cells more fragile and sensitive to mechanical manipulation.

7. This value is, of course, only valid for normal atmospheric pressure. However, the effects of normal fluctuations in atmospheric pressure are normally ignored.

8. To transfer the mitochondria into a state of energy conservation, incubation should be performed in the presence of fatty-acid-free albumin (0.1–0.5 %), to remove fatty acids and related substances, and of purine nucleotides. The nucleotides bind to the brown-fat specific uncoupling protein UCP1 and in so doing close the proton leak through this protein *(9)*. The most commonly used nucleotide is GDP, which is used at a concentration of 0.1 – 1 mM. Other di- and triphosphate purine nucleotides are also more or less efficient. The nucleotide binding site is on the outer side of the inner mitochondrial membrane.

9. When respiratory studies are performed on isolated mitochondria, it is of great importance that a suitable substrate is used. The most relevant is a fatty acid or its derivative, such as long-chain acyl-coenzyme A or acyl-carnitine. In all cases, to permit complete fatty acid oxidation, malate (in 5 mM concentration) must be added to the buffer to replenish citric acid cycle intermediates which have been lost during isolation *(15)*. In some species, the reuptake of malate is low and this may even limit fatty acid combustion. When acyl-carnitine esters are used (50 μM), no further additions (except malate) are required. For acyl-CoA derivatives (similar concentrations), a further addition of L-carnitine (2 mM) guarantees that availability of this compound does not limit oxidation. If free-fatty acids are used, further additions of ATP (100 μM) and coenzyme A (5 μM), in addition to carnitine, allow unlimited fatty acid oxidation. A further NADH-coupled substrate that demonstrates fairly high rates of respiration is pyruvate, used at 5 mM concentration together with malate (5 mM). Glutamate is inappropriate. Brown fat mitochondria also demonstrate a high rate of oxidation of glycerol-3-phosphate (used at mM concentrations), a flavoprotein-coupled substrate which is oxidised on the external face of the inner membrane and thus does not require transport *(16)*. In many species, succinate (which is a classical substrate for studies of liver mitochondria) permeates only poorly into the mitochondria and its use may therefore lead to severe underestimates of oxidative capacity. This is also often the case for other potential substrates, mainly the intermediates of the citric acid cycle, which have a low rate of permeation in certain species *(17)*.

10. Respiration in the brown adipocytes is most notably stimulated by the physiological agent norepinephrine, in which case endogenous lipolysis provides the substrate and also permits uncoupling of respiration from the constraints of a requirement for ATP utilization *(10,18)*. This uncoupling is entirely dependent upon the presence of UCP1 *(19,20)*. Free fatty acids can also be added to the cell suspension and provide an adequate substrate *(18)*. Their combustion is also fully dependent on the presence of UCP1, and this demonstrates that fatty acids can directly or indirectly activate UCP1 *(19,21)*. If other substrates are utilized, there is a transport requirement into the cells, in addition to which rates of respiration are generally low unless respiration is artificially uncoupled with e.g. FCCP $(20\,\mu M)$. The maximum respiratory rates then seen are usually much lower with exogenous substrates such as pyruvate $(5\,mM)$ than with the endogenously generated or exogenously added fatty acids.

White adipose tissue:

Mitochondria from white adipose tissue. To obtain mitochondria representative for the mature adipocytes in the tissue, isolated adipocytes, prepared essentially as described for the brown adipocytes , can be used as the starting material. The cells are homogenized and the mitochondria isolated by routine differential centrifugation, as described. The yields are very low. The mitochondria are well-coupled and can be stimulated to respire on citric acid cycle intermediates, in the presence of ADP *(22)*.

Adipocytes from white adipose tissue: as noted previously, isolated adipocytes can be readily prepared *(2)*. Because of the appearance of the cells (one unilocular fat droplet filling most of the cell volume), intact cells are not easily distinguished from large fat droplets. Few respiratory studies have been performed on such cells. Their rate of respiration is very low and high cells densities must be used. Basal metabolic/respiration rates can be estimated and hormonal stimulation can be performed, but this is generally evaluated in terms of metabolic changes other than respiration. A number of microcalorimetric studies have been performed on isolated white adipocytes from humans. Basal metabolism has been determined *(23)* and comparisons made between tissue taken from obese and lean *(24)* or hypo/euthyroid *(25)* individuals. Effects of hormone stimulation can be determined.

Acknowledgments Our basic research is supported by the Swedish Science Research Council. We thank present and past collaborators for comments on the procedures described here.

References

1. Nedergaard J, Cannon B, Lindberg O (1977) Microcalorimetry of isolated mammalian cells. Nature (Lond) 267:518–520
2. Rodbell M (1964) Metabolism of isolated fat cells. 1. Effects of hormones on glucose metabolism and lipolysis. J Biol Chem 239:375–380

3. Cannon B, Nedergaard J (2001) Cultures of adipose precursor cells from brown adipose tissue and of clonal brown-adipocyte-like cell lines. In: Ailhaud G (ed) Adipose tissue protocols. Humana Press Inc., Totowa NJ., pp 213–224

4. Robinson PK (1994) The Clark oxygen electrode. In: Wilson K, Walker J (eds) Principles and techniques of practical biochemistry. Cambridge University Press, Cambridge, pp 555–562

5. Smith RE, Roberts JC, Hittelman KJ (1966) Nonphosphorylating respiration of mitochondria from brown adipose tissue of rats. Science 154:653–654

6. Lindberg O, DePierre J, Rylander E, Afzelius BA (1967) Studies of the mitochondrial energy-transfer system of brown adipose tissue. J Cell Biol 34:293–310

7. Nicholls DG, Grav HJ, Lindberg O (1972) Mitochondria from hamster brown-adipose tissue. Regulation of respiration in vitro by variations in volume of the matrix compartment. Eur J Biochem 31:526–533

8. Nicholls DG, Lindberg O (1973) Brown-adipose-tissue mitochondria. The influence of albumin and nucleotides on passive ion permeabilities. Eur J Biochem 37:523–530

9. Nicholls DG (1974) Hamster brown-adipose-tissue mitochondria. The control of respiration and the proton electrochemical potential gradient by possible physiological effectors of the proton conductance of the inner membrane. Eur J Biochem 49:573–583

10. Fain JN, Reed N, Saperstein R (1967) The isolation and metabolism of brown fat cells. J Biol Chem 242:1887–1894

11. Zhao J, Cannon B, Nedergaard J (1998) Thermogenesis is β_3- but not β_1-adrenergically mediated in rat brown fat cells, even after cold acclimation. Am J Physiol 275: R2002–R2011

12. Zhao J, Cannon B, Nedergaard J (1998) Carteolol is a weak partial agonist on β_3-adrenergic receptors in brown adipocytes. Can J Physiol Pharmacol 76:428–433

13. Nedergaard J (1982) Catecholamine sensitivity in brown fat cells from cold-acclimated hamsters and rats. Am J Physiol 242:C250–C257

14. Svartengren J, Svoboda P, Cannon B (1982) Desensitisation of β-adrenergic responsiveness in-vivo. Decreased coupling between receptors and adenylate cyclase in isolated brown fat cells. Eur J Biochem 128:481–488

15. Cannon B (1971) Control of fatty-acid oxidation in brown-adipose-tissue mitochondria. Eur J Biochem 23:125–135

16. Bukowiecki L, Lindberg O (1974) Control of sn-glycerol 3-phosphate oxidation in brown adipose tissue mitochondria by calcium and acyl-CoA. Biochim Biophys Acta 348: 115–125

17. Cannon B, Bernson VMS, Nedergaard J (1984) Metabolic consequences of limited substrate anion permeability in brown fat mitochondria from a hibernator, the golden hamster. Biochim Biophys Acta 766:483–491

18. Prusiner SB, Cannon B, Lindberg O (1968) Oxidative metabolism in cells isolated from brown adipose tissue. I. Catecholamine and fatty acid stimulation of respiration. Eur J Biochem 6:15–22

19. Matthias A, Jacobsson A, Cannon B, Nedergaard J (1999) The bioenergetics of brown fat mitochondria from UCP1-ablated mice. UCP1 is not involved in fatty acid-induced de-energization. J Biol Chem 274:28150–28160

20. Matthias A, Ohlson KEB, Fredriksson JM, Jacobsson A, Nedergaard J, Cannon B (2000) Thermogenic responses in brown-fat cells are fully UCP1-dependent: UCP2 or UCP3 do not substitute for UCP1 in adrenergically or fatty-acid induced thermogenesis. J Biol Chem 275:25073–25081

21. Shabalina IG, Jacobsson A, Cannon B, Nedergaard J (2004) Native UCP1 displays simple competitive kinetics between the regulators purine nucleotides and fatty acids. J Biol Chem 279:38236–38248

22. Marshall SE, McCormack JG, Denton RM (1984) Role of Ca^{2+} ions in the regulation of intramitochondrial metabolism in rat epididymal adipose tissue. Evidence against a role for Ca^{2+} in the activation of pyruvate dehydrogenase by insulin. Biochem J 218:249–260

23. Monti M, Nilsson-Ehle P, Sörbis R, Wadsö I (1980) Microcalorimetric measurement of heat production in isolated human adipocytes. Scand J Clin Lab Invest 40:581–587
24. Olsson SA, Monti M, Sörbis R, Nilsson-Ehle P (1986) Adipocyte heat production before and after weight reduction by gastroplasty. Int J Obes 10:99–105
25. Valdemarsson S, Fagher B, Hedner P, Monti M, Nilsson-Ehle P (1985) Platelet and adipocyte thermogenesis in hypothyroid patients: a microcalorimetric study. Acta Endocrinol (Copenh) 108:361–366

Chapter 9
Application of Lipidomics and Metabolomics to the Study of Adipose Tissue

Ismo Mattila, Tuulikki Seppänen-Laakso, Tapani Suortti, and Matej Orešič

Summary Role of specific reactive lipids as well as amino acids in control of insulin signalling in adipose tissue is well recognized. Since it is practically impossible to measure the levels of all metabolites in the biological sample simultaneously with a single analytical platform, we utilize multiple platforms to study the lipids and metabolites of relevance to adipose tissue metabolism and insulin signalling. Two screening platforms cover a broad range of lipid molecular species (UPLC/MS based lipidomics platform) as well as organic acids and sterols (GCxGC-TOF platform). A targeted platform for amino acids (UPLC) is also applied.

Key words Metabolomics; lipidomics; branched chain amino acids; liquid chromatography; gas chromatography; mass spectrometry; insulin resistance.

1 Introduction

Metabolomics is a discipline dedicated to the global study of metabolites, their dynamics, composition, interactions, and responses to interventions or to changes in their environment, in cells, tissues, and biofluids (1). Metabolites are known to be involved as key regulators of systems homeostasis. Concentration changes of specific groups of metabolites may be descriptive of system responses to environmental or genetic interventions, and their study may therefore be a powerful tool for characterization of complex phenotypes as well as for development of biomarkers for specific physiological responses.

The role of specific reactive lipids as well as amino acids in control of insulin signaling in adipose tissue is well recognized. For example, two mechanisms have been proposed to explain how expansion of the adipose tissue stores affects insulin sensitivity. One mechanism suggests that increased adiposity induces a chronic inflammatory state characterized by increased cytokine production by adipocytes

From: *Methods in Molecular Biology, Vol. 456:*
Adipose Tissue Protocols, Second Edition. Edited by: Kaiping Yang

and/or the macrophages infiltrating adipose tissue. Cytokines produced by these adipocytes or macrophages may directly antagonize insulin signalling. In this context, the role of eicosanoids as key mediators of inflammatory signalling is well recognized (2). A second nonexclusive mechanism is the lipotoxic hypothesis. The lipotoxic hypothesis states that if the amount of fuel entering a tissue exceeds its oxidative or storage capacity, toxic metabolites that inhibit insulin action are formed. For example, lipid metabolites, such as ceramides and diacylglycerol, or reactive oxygen species generated from hyperactive oxidative pathways, have been shown to inhibit insulin signaling (3,4). There is also increasing evidence that branched chain amino acids affect the insulin signalling in adipose tissue (5–7). The increased amino acid levels in circulation as a result of nutrient overload may therefore lead to insulin resistance in peripheral tissues (7).

It is practically impossible to measure the levels of all metabolites in the biological sample simultaneously with a single analytical platform. The reason for this is that metabolites are (bio)chemically diverse and can cover a dynamic range of over 10 orders of magnitude in concentration, therefore a single extraction and detection method for all metabolites from biological matrices is unfeasible (1). Additionally, complexity of biological samples may also affect the efficiency and reliability of detection, for example due to ion suppression effects in mass spectrometry based approaches (8). Multiple analytical platforms are commonly applied in parallel to cover the broad range of metabolites and typically include different extraction methods for specific groups of metabolites (1). Analytical technologies based on gas chromatography coupled to mass spectrometry, liquid chromatography–mass spectrometry, capillary electrophoresis coupled to mass spectrometry, as well as nuclear magnetic resonance have most commonly been applied (9,10).

To cover the metabolites most relevant to adipose tissue metabolism, we apply four different platforms. Two screening platforms cover a broad range of lipid molecular species (UPLC/MS based lipidomics platform) as well as organic acids and sterols (GCxGC-TOF platform). A targeted platform for amino acids (UPLC) is also applied.

2   Materials

2.1   Lipidomics Profiling Platform (UPLC/MS)

1. Tissue samples: 20-mg aliquots are weighed for extraction.
2. Internal standard mixture: GPCho(17:0/0:0), GPCho(17:0/17:0), GPEtn (17:0/17:0), GPGro(17:0/17:0)[rac], Cer(d18:1/17:0), GPSer(17:0/17:0), GPA (17:0/17:0) and D-erythro-Sphingosine-1-Phosphate (C17 Base) from Avanti Polar Lipids and MG(17:0/0:0/0:0)[rac], DG(17:0/17:0/0:0)[rac] and TG (17:0/17:0/17:0) from Larodan Fine Chemicals.
3. Solvent: 10 μL of 0.15 M (0.9%) sodium chloride.

4. Extraction buffer: mixture of chloroform and methanol (2:1; 100 μL, HPLC-grade, Rathburn).
5. Standard mixture added after extraction contains 3 labeled lipid compounds: GPCho (16:0/0:0-D$_3$), GPCho(16:0/16:0-D$_6$) and TG(16:0/16:0/16:0–^{13}C3) from Larodan Fine Chemicals.

2.2 Metabolomics Platform (GCxGC-TOF)

1. Tissue samples: 10- to 20-mg aliquots are weighed for extraction.
2. Extraction solvent: methanol from Rathburn (HPLC-grade, Walkerburn, Scotland).
3. Derivatization reagents: 2% MOX (methoxyamine hydrochloride) in pyridine and MSTFA (*N*-Methyl-*N*-(trimethylsilyl)trifluoroacetamide) from Pierce (Rockford, IL).
4. Internal standard: Palmitic-16,16,16-d3 acid from Isotec (Miamisburg, OH) 1 g/L in methanol.

2.3 Amino Acid Platform (UPLC)

1. External standard mixture of D-Leucine, D-Isoleucine, D-Valine in borate buffer. Concentrations ranging from 12 μ*M* to 0.3 μ*M* and containing 10 μ*M* α-aminobutyric acid as internal standard and dissolved in water.
2. Derivatization reagent kit. AccQ˙ Tag reagent kit from Waters Inc (Milford, MA; WAT052880).
3. Centrifugal filters Nanosept MF GHP 0.45 μm (PALL Life Sciences).
4. Acquite UPLC BEHC18 column 1.0 × 50 mm 1.7-mm particles from Waters Inc. (WAT0186002344).
5. Acetonitrile (HPLC quality, Rathburn).
6. Solvent A for chromatography: 1.2% formic acid (V/V) (Riedel-Haen) adjusted to pH 5.3–5.7 with 5 N ammonia (Sigma).
7. Solvent B for chromatography: 2% Formic acid (V/V) in acetonitrile.
8. Methanol (HPLC, Rathburn).
9. Water (>18 MΩ quality).

3 Methods

3.1 Lipidomics Profiling Platform (UPLC/MS)

1. The UPLC/MS system consists of an Acquity Ultra Performance LC™ (UPLC) combined with a Waters Q-Tof Premier mass spectrometer. A sample organizer is used for the automatic sampling (*see* **Note 1**).

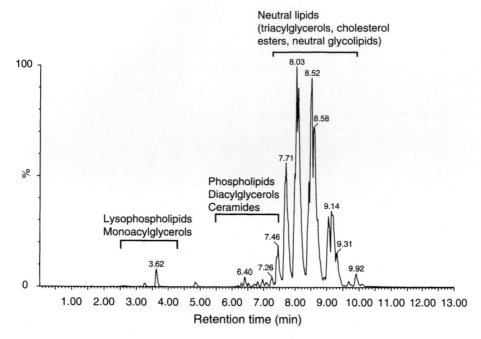

Fig. 9.1 Total ion chromatogram (positive ion mode) from UPLC/MS lipidomics platform of white adipose tissue lipid extract

2. The column used is an Acquite UPLC BEH C_{18} column $1.0 \times 50\,\text{mm}$ $1.7\,\mu\text{m}$ particles from Waters Inc. The temperature of the column is 50°C.
3. The solvent system includes A. water (1% 1 M NH_4Ac, 0.1% HCOOH) and B. AcCN / 2-Propanol (5:2; 1% 1 M NH_4Ac, 0.1% HCOOH).
4. Start gradient at 65% A/35% B, set to reach 100% B in 6 min and remain there for 7 min. The total run time including a 5 min re-equilibration step is 18 min. The flow rate is 0.200 mL/min and the injected amount of lipid extract is 0.75 µL. The temperature of the sample organizer is 10°C.
5. Lipid compounds are detected by using electrospray ionization in positive ion mode. The data is collected in continuum mode using extended dynamic range at mass range of m/z 300–1200 with a scan duration of 0.2 s.
6. Data processing using the MZmine software (*see* **Note 2** (*11,12*)). An example chromatogram is shown in **Fig. 9.1**.

3.2 Metabolomics Platform (GCxGC-TOF)

1. Weigh tissue samples (10–20 mg each) into Eppendorf-tubes.
2. Add 10 µL of 500 ppm labeled palmitic-16,16,16-d3 acid as internal standard.

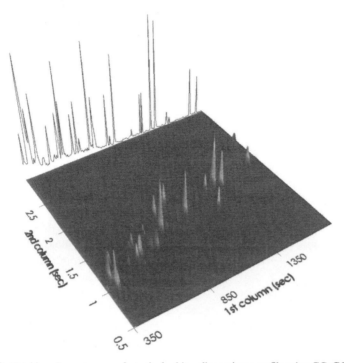

Fig. 9.2 Total ion chromatogram of a typical white adipose tissue profile using GCxGC-TOF metabolomics platform. The horizontal axes are the retention times from the two successive GC columns

3. Extract samples with 500 μL of methanol (2 min vortexing, 30 min extraction time) and centrifuge at 10,000 g for 3 min.
4. Evaporate the separated supernatants to dryness under a stream of nitrogen.
5. Derivatize the residues as follows; add 25 μL of 2% methoxyamine HCl in pyridine and heat at 30°C for 90 min followed by addition of 50 μL of N-Methyl-N-(trimethylsilyl)trifluoroacetamide and heating at 37°C for 30 min.
6. Run the samples on a GCxGC-TOF instrument (Agilent 6890N gas chromatograph and LECO Pegasus 4D mass spectrometer). For instrument parameters, *see* **Note 3**. Example chromatogram is shown in **Fig. 9.2**.

3.3 Amino Acid Platform (UPLC)

1. Weight 10 mg of adiopose tissue, add 250 μL of 1:10 diluted borate buffer supplied in derivatization kit and 50 μL of 10 μM α-aminobuturic acid in water.
2. Add 50 mg of glass beads (diameter 0.45–0.75 mm).
3. Heat sample in waterbath at 45°C for 5 min and homogenize in Sartorius Dismembrator at 3000 cps for 3 min.

Table 9.1 Chromatographic conditions for the amino acid platform

Flow	Time	%A	%B	Gradient
0.2	0.00	99.9	0.1	6
0.2	0.47	99.9	0.1	6
0.2	5.02	90.9	9.1	7
0.2	13.77	78.8	21.2	6
0.2	14.04	40.4	59.6	6
0.2	14.56	40.4	59.6	6
0.2	14.64	99.9	0.1	6
0.2	15.31	99.9	0.1	6

4. Cool in ice-bath for 10 min and centrifuge for three minutes at 10,000 g.
5. Take solution and wash precipitate with 200 µL of 1:10 diluted borate buffer supplied in AccQ˙ Tag reagent kit and centrifuge it again. Combine original and wash solution.
6. Dry under nitrogen and dissolve in 25 µL of borate buffer supplied in AccQ˙Tag reagent kit (*see* **Materials 2.3.2**) and add 5 µL of AccQ˙ Tag reagent.
7. Inject into UPLC instrument; the chromatographic solvents A (*see* **Subheading 2.3, step 6**) and solvents B (*see* **Subheading 2.3, step 7**). Acquire UPLC BEHC18 column (*see* **Subheading 2.3, step 4**) at 55°C and injection volume 4 µL. Acquity 2996 detector at 260 nm. The chromatographic conditions are described in **Table 9.1**. Example chromatograms are shown in **Fig. 9.3**.

4   Notes

1. The method parameters are optimized for the specific analytical system used. The primary aim of the described method is rapid screening of lipid molecular species across broad range of lipid classes. Although use of positive ion mode is described, the same platform can also be applied using electrospray ionization in negative ion mode, which will lead to better sensitivity for specific phospholipid classes such as phosphatidylinositols, phosphatidylserines, and phosphatidic acids. The method can be transferred to a different LC/MS system. Ideally such system should have MS/MS and/or accurate mass capabilities.
2. MZmine data processing parameters need to be optimized for a specific analytical system used in order to account for different peak shapes and lengths as well as different mass spectrometer resolution. Differential profiling of multiple samples requres steps such as peak detection, alignment (matching of peaks across multiple samples), and normalization using internal standards.
3. The gas chromatograph is operated in split mode (1:20) using helium as carrier gas in constant pressure mode. The injection volume is 1 µl. The first column is relatively non-polar RTX-5, 10 m × 180 µm × 0.20 µm (Restek) and the second

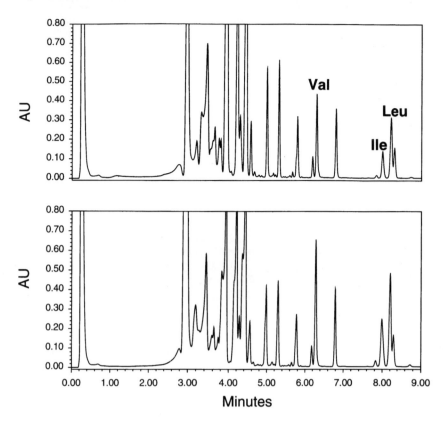

Fig. 9.3 Chromtogram of adipose tissue sample (upper trace) and sample where 5 μM of leucine (Rt = 8.2 min), Isoleucine (Rt = 8 min), and valine (Rt = 6.3) has been added

column is polar BPX-50, 1.10 m × 100 μm × 0.10 μm (SGE, Australia). The temperature programme is as follows; Primary oven: initial 50°C, 1 min. ~ 280°C, 7°C/min, 5 min. Secondary oven: initial 60°C, 1 min. ~ 290°C, 7°C/min, 5 min. Modulator is kept at 25°C above the primary oven temperature. The second dimension separation time is set to 4 s. Data are acquired at 100 spectra s^{-1} from m/z 40 to 700.

References

1. Oresic M, Vidal-Puig A, Hanninen V (2006) Metabolomic approaches to phenotype characterization and applications to complex diseases. Expert Rev Mol Diagn 6:575–585
2. Funk CD (2001) Prostaglandins and leukotrienes: advances in eicosanoid biology. Science 294:1871–75

3. Summers SA (2006) Ceramides in insulin resistance and lipotoxicity. Prog Lipid Res 45:42–72
4. Medina-Gomez G, Virtue S, Lelliott C, Boiani R, Campbell M, Christodoulides C, Perrin C, Jimenez-Linan M, Blount M, Dixon J, Zahn D, Thresher RR, Aparicio S, Carlton M, Colledge WH, Kettunen MI, Seppanen-Laakso T, Sethi JK, O'Rahilly S, Brindle K, Cinti S, Oresic M, Burcelin R, Vidal-Puig A (2005) The link between nutritional status and insulin sensitivity is dependent on the adipocyte-specific Peroxisome Proliferator-Activated Receptor-{gamma}2 isoform. Diabetes 54:1706–1716
5. Lynch CJ, Patson BJ, Anthony J, Vaval A, Jefferson LS, and Vary TC (2002) Leucine is a direct-acting nutrient signal that regulates protein synthesis in adipose tissue. Am J Physiol Endocrinol Metab 283:E503–E513
6. Hinault C, Van Obberghen E, Mothe-Satney I (2006) Role of amino acids in insulin signaling in adipocytes and their potential to decrease insulin resistance of adipose tissue. J Nutr Biochem 17:374–378
7. Um SH, D'Alessio D, Thomas G (2006) Nutrient overload, insulin resistance, and ribosomal protein S6 kinase 1, S6K1. Cell Metab 3:393–402
8. de Hoffmann E, Stroobant V (2001) Mass spectrometry: principles and applications, John Wiley & Sons, New York
9. van der Greef J, Stroobant P, Heijden R v. d. (2004) The role of analytical sciences in medical systems biology. Curr Opin Chem Biol 8:559–565
10. Lindon JC, Holmes E, Nicholson JK (2004) Metabonomics and its role in drug development and disease diagnosis. Expert Rev Mol Diag 4:189–199
11. Katajamaa M, Oresic M (2005) Processing methods for differential analysis of LC/MS profile data. BMC bioinformatics [electronic resource] 6:179
12. Katajamaa M, Miettinen J, Oresic M (2006) MZmine: toolbox for processing and visualization of mass spectrometry based molecular profile data. Bioinformatics 22:634–636

Chapter 10
Applications of Proteomics to the Study of Adipose Tissue

Sonja Hess and Xiaoli Chen

Summary Determination of the complex secretory proteome of adipocytes and its metabolic changes induced by drug treatment such as insulin or rosiglitazone is possible with the advanced proteomics technologies described herein. To study the secreted proteins of adipocytes, a 2D- liquid chromatography/mass spectrometry/mass spectrometry protocol has been established. With the use of reversed-phase high-performance liquid chromatography, intact proteins were separated in the first dimension into eight fractions, then digested with Lys-C and trypsin. Comparative differences after drug treatment were assessed using ^{18}O proteolytic labeling strategies. With the advent of more and more sophisticated instrumentation and data analysis tools, protocols like this one will likely become standard tools for scientists in the research fields of endocrinology, obesity, and diabetes. These protocols enable researchers to study the dynamic drug-induced changes in a comprehensive and systematic manner that was inconceivable just a few years ago.

Key words Adipose tissue; quantitative proteomics; ^{18}O-labeling; diabetes; mass spectrometry; electrospray.

1 Introduction

Advanced mass spectrometry-based proteomics technologies have revolutionized the way we have conducted biological studies during the last decade *(1–4)*. Although it is still a major challenge to determine global changes in protein expression, focussed approaches on organelles and selected tissues are now feasible *(1–5)*. Proteomics technologies have also enabled scientists in the research fields of endocrinology, obesity, and diabetes to study complex changes, for instance, in the secretory proteome of adipose cells through liquid chromatography-tandem mass spectrometry (2D-LC-MS/MS *(5–11)*). This concept is particularly interesting because adipose tissue plays a pivotal role as an endocrine organ in the regulation of energy metabolism and glucose homeostasis and because dysfunction of adipose

From: *Methods in Molecular Biology, Vol. 456:*
Adipose Tissue Protocols, Second Edition. Edited by: Kaiping Yang
© Humana Press, a part of Springer Science+Business Media, Totowa, NJ

tissue secretion is associated with obesity and its linked metabolic syndrome *(12–15)*. Because cultured adipose cells secrete their proteins directly into the medium, they are thus amenable to extensive 2D-LC-MS/MS studies.

To retain as much information in the intact proteins as possible, the intact proteins are separated in the first dimensional reversed phase chromatography into eight fractions. Adipokines of the individual fractions are then digested with Lys-C in the presence of urea and trypsin prior to LC-MS/MS analysis. Using ^{18}O proteolytic labeling strategies, comparative differences after drug treatment such as insulin or rosiglitazone can additionally be assessed by measuring the isotopic ratios of the labeled and unlabeled peptides. With the use of this technique, autocrine and endocrine effects of drug treatment of adipose cells can systematically be studied. To eliminate false-positive proteins that may have leaked out of the damaged tissue, all identified proteins are checked for signal peptides, common to secreted proteins using Signal P predictions. Studies like this give significant insight into the complexity of signaling cascades and the multiple effects of a drug treatment on the global changes in the secretory proteome of adipose cells in healthy and diseased states.

2 Materials

2.1 Rat Adipose Tissue and Primary Cell Culture

1. Dulbecco's modified Eagle's medium (DMEM; Invitrogen), high glucose (4.5 g/L D-glucose), no sodium pyruvate.
2. Krebs Ringer bicarbonate HEPES buffer (KRBH buffer): 120 mmol/L NaCl, 4 mmol/L KH_2PO_4, 1 mmol/L $MgSO_4$, 1 mmol/L $CaCl_2$, 10 mmol/L $NaHCO_3$, 200 nmol/L adenosine, and 30 mmol/L HEPES, pH 7.4.
3. Collagenase type 1 (Worthington Biochemical).
4. Bovine serum albumin (BSA; Intergen).
5. Adenosine (Sigma).
6. 173 µmol insulin (Sigma), prepare in 0.1 *M* HCl.
7. 2 mg/ml Rosiglitazone (GSK): dissolve 8 mg pill in water, prepare fresh.

2.2 High-Performance Liquid Chromatography (HPLC) Separation of Intact Proteins

1. Hewlett Packard HP1100 system consisting of a qarternary pump, degasser, autosampler, and UV detector (Hewlett Packard, now Agilent Technologies).
2. HP Chemstation for data acquisition.
3. Zorbax 300SB-C3 reversed phase column (150 mm × 4.6 mm ID, 5 µm) equipped with a guard column (C3 reversed phase).

4. Water (Chromasolv, LC-MS quality, Riedel de Haen).
5. Solvent A: 0.1% (v/v) Trifluoroacetic acid (99+%, glacial acetic acid 99%; Sigma).
6. Solvent B: acetonitrile (Chromasolv, LC-MS quality, Riedel de Haen).

2.3 Digestion With Lys-C and Trypsin

1. 5 mM dithiothreitol (DTT; Sigma), prepare in 5 M urea.
2. 50 mmol iodoacetamide (Sigma), prepare in water.
3. Endoproteinase Lys-C (Roche).
4. Trypsin, modified sequencing grade (Roche).

2.4 ^{18}O Labeling Procedure

1. ^{18}O-water (Sigma).
2. Immobilized trypsin (Applied Biosystems).

2.5 Electrospray Mass Spectrometry Instrumentation

1. Hybrid LTQFT (Thermo Fisher, San Jose, CA) mass spectrometer, equipped with a Waters CapLC (Waters, Milford, MA) for all liquid chromatography-electrospray ionization mass spectrometry experiments (*see* **Note 1**).
2. Solvent A: 0.2% formic acid, 1% acetonitrile, and 98.8% water.
3. Solvent B: 0.2% formic acid, 1% water, and 98.8% acetonitrile.

3 Methods

The following methods are described below in detail: (1) rat adipose tissue and primary adipose cell culture (*16*); (2) liquid chromatography of secreted intact adipokines; (3) tryptic digestion of the fractionated adipokines; (4) ^{18}O-Water labeling of digested peptides (*5,17–21*); and (5) bioinformatic tools: database searching and parsing, SignalP prediction and quantification.

3.1 Rat Adipose Tissue and Primary Adipose Cell Culture

1. Remove adipose tissue from the epididymal fat pads of 150- to 200-g male Zucker *fa/fa* rats (Charles River Laboratories) treated with or without Rosiglitazone (3 mg/kg body weight) via gavage for 12 d.

2. Finely mince the tissue, then wash by centrifugation at 1200×g for 2 min, twice with KRBH buffer containing 0.1% BSA, and twice with serum-free, BSA-free DMEM medium.
3. Proceed to tissue culture (**Subheading 3.1.1.**) or adipose cell culture (**Subheading 3.1.2.**)

3.1.1 Rat Adipose Tissue Culture

1. Culture tissue (~150 mg/mL) in serum-free, BSA-free DMEM at 37°C for 48 h in 5% CO_2.
2. Collect conditioned medium and centrifuge at 1200×g for 10 min to separate cell debris.
3. Filter supernatant medium using a 0.45-μm syringe-driven filter to remove remaining possible cell debris.
4. Desalt conditioned medium and concentrate using a Macrosep centrifugal device with molecular weight cutoff of 1 KDa (Pall Life Sciences, MI).
5. Determine protein concentration using Pierce BCA Protein Assay (*see* **Note 2**)

3.1.2 Primary Adipose Cell Culture

1. Digest minced tissue for 2 h with collagenase (2 mg/mL solution) in digestion vials containing KRBH buffer, pH 7.4, and 0.1% BSA.
2. Separate adipocytes and stromal-vascular (SV) cells by centrifugation at 1200 g for 10 min.
3. Wash isolated adipocytes by centrifugation at 1200 g for 2 min, twice with KRBH buffer containing 0.1% BSA, and twice with serum-free, BSA-free DMEM medium.
4. Culture cells at a density of 0.5 × 10⁵/mL in serum-free, BSA-free DMEM at 37°C for 48 h in 5% CO_2.
5. Collect conditioned medium and centrifuge at 1200 g for 10 min to separate cell debris.
6. Filter supernatant medium using a 0.45-μm syringe-driven filter to remove the remaining adipocytes.
7. Desalt conditioned medium and concentrate using a Macrosep centrifugal device with molecular weight cutoff of 1 KDa (Pall Life Sciences, Ann Arbor, MI).
8. Determine protein concentration using Pierce BCA Protein Assay (*see* **Note 2**).

3.2 HPLC Separation of Intact Proteins

1. Inject 100 μL of concentrated extracts of the culture medium.

2. Perform HPLC using a gradient of 5 to 40% solvent B within 55 min, flow rate of 700 μL/min, UV detector set at 254 nm, and fraction collection to give 8 fractions in total.
3. Remove excess solvents using speedvac (*see* **Note 2**).

3.3 Digestion with Lys-C and Trypsin

1. Dissolve and reduce fraction in 20 μL of 5 *M* urea containing 5 m*M* DTT.
2. Akylate with 10 mmol iodoacetamide.
3. Digest with 0.5 μg of endoproteinase Lys-C.
4. Dilute with water to a final concentration of 2 *M* urea.
5. Digest with 1 μg of trypsin.

3.4 ^{18}O Labeling Procedure

1. Evaporate aliquots of the tryptic peptides from basal and drug-treated samples to dryness. (*see* **Note 3**).
2. Redissolve in $H_2^{16}O$ and $H_2^{18}O$, respectively.
3. Add immobilized trypsin to basal ($H_2^{16}O$) and drug-treated ($H_2^{18}O$) samples.
4. Incubate overnight in the presence of 0.1 M ammonium carbonate.
5. Repeat **steps 1** through **4** three times (*see* **Note 4**).
6. Stop labeling procedure after 16 h.
7. Mix aliquots of untreated, unlabeled control peptides and ^{18}O-labeled treated peptides at a 1:1 ratio.
8. Analyze by LC-MS.

3.5 Electrospray Mass Spectrometry Instrumentation

1. Introduce sample with an autosampler.
2. Concentrate on a Waters Symmetry300 C18 5 μm trap column.
3. Direct flow onto a Microtech Scientific C18 column (100 mm × 150 μm ID, 5 μm) after a 5-min delay.
4. Prepare solvent A and solvent B (*N.B.* solvent composition is that in **Subheading 2.5.**)
5. Use gradient, ramped from 5% solvent B to 95% solvent B within 175 min at a flow rate of 6 μL/min and a split ratio of 1:10.
6. For protein identification, use MS/MS data in a data-dependent mode to switch automatically between MS and MS/MS mode. Full spectra (m/z 400–20000

are acquired in the FTICR cell with R=25,000. Using collisionally induced dissociation with Helium gas, the ten most abundant ions are subjected to MS/ MS fragmentation (*see* **Fig. 10.1**)

7. For protein quantification, use MS data (*see* **Fig. 10.2**).

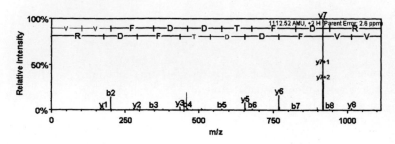

Fig. 10.1 MS/MS spectrum of the doubly charged 557.27 ion, corresponding to the amino acid sequence VVFDDTFDR of carbonic anhydrase; The b-ion series is shown in light grey and the y-ion series is shown in dark grey.

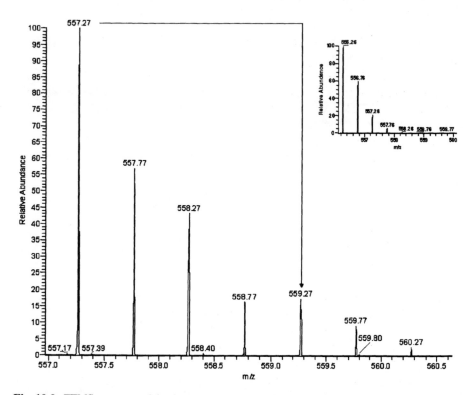

Fig. 10.2 FTMS spectrum of the doubly charged 557.27 ion of carbonic anhydrase together with the isotopically ^{18}O-labeled 559.27. The inset shows the theoretical isotopic distribution for the 557.27 ion. Quantification can either be done manually by determining the peak height or with advance software (*see* **Note 5**)

3.6 Bioinformatics Tools

3.6.1 Database Searching

1. Process data using Bioworks to create .dta files (data files) (*see* **Note 5**).
2. Merge .dta files to merged .mrg files.
3. Use .mrg files for subsequent database searching using a MASCOT search engine *(25)*.
4. Use the following parameters: MSDB; 1 missed cuts; iodoacetamidation of cysteines; deamidation; and charge states +2, +3, and +4.
5. Choose a window of 10 ppm mass accuracy for precursor ions and 0.6-Da mass accuracy for MS/MS data.
6. Consider probability-based MASCOT scores as significant when greater than the cutoff score that indicates either identity or homology ($p < 0.05$) for individual ions.
7. Confirm, by de novo sequencing, one peptide hit sequence assignment.
8. To increase confidence in the assignment as secretory proteins and reduce the risk of contamination from cellular material, check all proteins meet the criteria set by SignalP predictions.
9. Use SignalP 3.0 server (www.cbs.dtu.dk/services/SignalP) to predict the presence and location of signal peptide cleavage sites in amino acid sequences of proteins. The S-score, C-score, Y-max, S-mean, and D-score for the signal peptide prediction, the hidden Markov model calculation for signal peptide probability, and the eukaryotic HMM model calculation for signal anchor probability are all evaluated for amino acid sequences. The proteins with all five high scores and high signal peptide probability are considered secreted proteins.
10. Use Scaffold (www.proteomesoftware.com) to parse data from several runs (*see* **Table 10.1**).
11. Use either manual interpretation or software to quantify individual peptides (*see* **Note 6**).

4 Notes

1. Other general settings: spray voltage is 3.0 kV, capillary temperature 100°C, normalize collision energy using wide-band activation mode, 35% for MS2. Ion selection thresholds are 1000 counts for MS2. An activation $q = 0.25$ and activation time of 30 ms are applied in MS2 acquisitions.
2. Should be approximately 650 ng/μL.
3. It is important not to precipitate the proteins during the fractionation and evaporation of the solvent. Avoid full evaporation to dryness.
4. To force the equilibrium towards the double label, two strategies are used: removing an excess of ^{16}O-H_2O and replacing it with ^{18}O-H_2O and the use of immobilized trypsin.

Table 10.1 Representative data collected from one untreated and one rosiglitazone treated fraction. M_R is molecular weight

No.	Identified proteins	M_R	Untreated	Treated
1	B-factor, properdin.		100% (7)	100% (6)
2	Carbonic anhydrase 3 (EC 4.2.1.1)	29 kDa	100% (2)	100% (1)
3	Contrapsin-like protease inhibitor 6 precursor	47 kDa	100% (8)	100% (10)
4	SPARC precursor (Osteonectin)	34 kDa	100% (5)	100% (4)
5	Vimentin	54 kDa	100% (3)	99% (1)
6	Fatty acid-binding protein, adipocyte (AFABP)	15 kDa	100% (3)	100% (1)
8	Cathepsin B, preproprotein.		100% (2)	100% (1)
9	Serum albumin precursor	69 kDa	100% (3)	100% (1)
10	Gelsolin precursor (Actin-depolymerizing factor)	86 kDa	100% (2)	100% (2)
12	Gamma-synuclein (Persyn)	13 kDa	100% (2)	100% (2)
13	Clusterin precursor (Sulfated glycoprotein 2)	51 kDa	100% (1)	100% (1)
15	Complement component 4, gene 2.		100% (2)	100% (3)
16	Thrombospondin 1.		100% (2)	100% (1)
17	Galectin-1 (Beta-galactoside-binding lectin L-14-I)	15 kDa	100% (2)	100% (2)
18	Angiotensinogen precursor	52 kDa	100% (2)	100% (3)
19	Carboxylesterase 3 precursor (EC 3.1.1.1)	62 kDa	100% (3)	100% (2)

The calculated percentages of certainty are given followed by the number of unique peptides in parentheses.

Table 10.2 Differential data collected from one untreated and one rosiglitazone treated fraction

Number	Protein	Avg XPRESS XPRESS (H:L)
1	PAI1_RAT (P20961) Plasminogen activator inhibitor-1 precursor (PAI-1)	1:0.95
2	HSP7C_RAT (P63018) Heat shock cognate 70 kDa protein (HSP70)	1:0.78
3	CATL_RAT (P07154) Cathepsin L precursor (EC 3.4.22.15) (Major excreted protein)	1:1.92
4	SPTA2_RAT (P16086) Spectrin alpha chain	1:2.88
5	CPI6_RAT (P09006) Contrapsin-like protease inhibitor 6 precursor (CPI-26)	1:2.30

Avg XPRESS gives the average ratio of the light (L) to heavy (H) isotope ratio.

5. The program Bioworks comes with a version of XPress that can be used for automatic interpretation of the data (*see* **Table 10.2** *(22)*). More sophisticated software such as MSQuantification *(23)* or ProRata *(24)* can be used by more advanced users.

6. Use tryptic peptides as internal standards. Their ratio must be 1:1. It is also recommended to use only doubly charged ions for the manual interpretation. Due to the overlap of the isotopes from triply charged ions, interpretation may otherwise lead to misinterpretation.

Acknowledgments The authors thank NIDDK for internal funding.

References

1. Anderson JS, Mann M (2006) Organellar proteomics: turning inventories into insights. EMBO Rep 7:874–879
2. Ong SE, Mann M (2005) Mass spectrometry-based proteomics turns quantitative. Nat Chem Biol 1:252–262
3. Domon B, Aebersold R (2006) Challenges and opportunities in proteomics data analysis. Mol Cell Proteomics 5:1921–1926
4. Domon B, Aebersold R (2006) Mass spectrometry and protein analysis Science 312: 212–217
5. Chen X, Pannell LK, Cushman S, Hess S (2005) Quantitative proteomic analysis of the secretory proteins from rat adipose cells using a 2D liquid chromatography-MS/MS approach. J Proteome Res 4:570–577
6. Kratchmarova I, Kalume DE, Blagoev B, Scherer PE, Podtelejnikov AV, Molina H, Bickel PE, Andersen JS, Fernandez MM, Bunkenborg J, Roepstorff P, Kristiansen K, Lodish HF, Mann M, Pandey A(2002) A proteomic approach for identification of secreted proteins during the differentiation of 3T3-L1 preadipocytes to adipocytes. Mol Cell Proteomics 1:213–222
7. Sidhu RS (1979) Two-dimensional electrophoretic analyses of proteins synthesized during differentiation of 3T3-L1 preadipocytes. J Biol Chem 254:11111–11118
8. Spiegelman BM, Green H (1980) Control of specific protein biosynthesis during the adipose conversion of 3T3 cells. J Biol Chem 255:8811–8818
9. Wilson-Fritch L, Burkart A, Bell G, Mendelson K, Leszyk J, Nicoloro S, Czech M, Corvera S (2003) Mitochondrial biogenesis and remodeling during adipogenesis and in response to the insulin sensitizer rosiglitazone. Mol Cell Biol 23:1985–1094
10. Tsuruga H, Kumagai H, Kojima T, Kitamura T (2000) Identification of novel membrane and secreted proteins upregulated during adipocyte differentiation. Biochem Biophys Res Commun 272:293–297
11. Wang P, Mariman E, Keijer J, Bouwman F, Noben J-P, Robben J, Renes J (2004) Profiling of the secreted proteins during 3T3-L1 adipocyte differentiation leads to the identification of novel adipokines. Cell Mol Life Sci 61:2405–2417
12. Ahima RS, Flier JS (2000) Adipose tissue as an endocrine organ. Trends Endocrinol Metab 11:327–332
13. Tilg H, Moschen AR (2006) Adipocytokines: mediators linking adipose tissue, inflammation and immunity. Nat Rev Immunol 6:772–783
14. Rosen ED, Spiegelman BM (2006) Adipocytes as regulators of energy balance and glucose homeostasis. Nature 444:847–853
15. Van Gaal LF, Mertens IL, De Block CE. (2006) Mechanisms linking obesity with cardiovascular disease. Nature 444:875–880
16. Weber TM, Joost HG, Simpson IA, Cushman SW (1998) The insulin receptor. A. R. Liss, Inc., New York, pp 171–187
17. Stewart II, Thomson T, Figeys D (2001) ^{18}O labeling: a tool for proteomics. Rapid Commun Mass Spectrom 15:2456–2465
18. Mirgorodskaya OA, Kozmin YP, Titov MI, Korner R, Sonksen CP, Roepstorff P (2000) Quantitation of peptides and proteins by matrix-assisted laser desorption/ionization mass spectrometry using ^{18}O-labeled internal standards. Rapid Commun Mass Spectrom 14:1226–1232
19. Yao X, Freas A, Ramirez J, Demirev PA, Fenselau C (2001) Proteolytic ^{18}O labeling for comparative proteomics: model studies with two serotypes of adenovirus. Anal Chem 73: 2836–2842
20. Reynolds KJ, Yao X, Fenselau C (2002) Proteolytic ^{18}O labeling for comparative proteomics: evaluation of endoprotease Glu-C as the catalytic agent. J Proteome Res 1:27–33
21. Krijgsveld J, Ketting RF, Mahmoudi T, Johansen J, Artal-Sanz M, Verrijzer CP, Plasterk RH, Heck AJ (2003) Metabolic labeling of C. elegans and D. melanogaster for quantitative proteomics. Nat Biotechnol 21:927–931

22. Gygi SP, Rist B, Gerber SA, Turecek F, Gelb MH, Aebersold R (1999) Quantitative analysis of complex protein mixtures using isotope–coded affinity tags. Nat Biotechnol 17:994–999
23. Schulze WX, Mann M (2004) A novel proteomic screen for peptide-protein interactions. J Biol Chem 279:10756–10764
24. Pan C, Kora G, McDonald WH, Tabb DL, VerBerkmoes NC, Hurst GB, Pelletier DA, Samatova NF, Hettich RL (2006) ProRata: A quantitative proteomics program for accurate protein abundance ratio estimation with confidence interval evaluation. Anal Chem 78:7121– 7131
25. www.matrixscience.com

Chapter 11
Application of DNA Microarray to the Study of Human Adipose Tissue/Cells

Paska A. Permana, Saraswathy Nair, and Yong-Ho Lee

Summary Adipose tissue is increasingly recognized as a metabolically active endocrine organ with multiple functions beyond its lipid storage capability. Various constituents of the tissue, such as mature adipocytes and stromal vascular cells, have distinct functions. For example, they express and secrete different kinds of bioactive molecules collectively called adipokines. Altered adipokine secretion patterns characterize obesity and insulin resistance, which are major risk factors for type 2 diabetes mellitus. The contribution of dysregulated adipokine expression to these diseases may be assembled from transcriptomic profiles of the tissue and/or its cellular constituents. The gene expression profiles may also complement genetic approaches to identify disease susceptibility genes. Here, we describe an application of gene expression profiling using DNA microarrays to study human adipose tissue, adipocytes, and stromal vascular cells.

Key words Microarray; adipose tissue; adipocyte; stromal vascular cells; preadipocyte; gene expression.

1 Introduction

Adipose tissue plays a major role in obesity and metabolic disorders, as discussed in a separate chapter of this volume. The main functions of adipose tissue include: secretion of adipokines (a diverse set of bioactive molecules produced by adipose tissue), energy dissipation/thermogenesis, and energy storage. Different cell types of adipose tissue serve different functions. For example, adipocytes store energy in the form of lipids *(1)* as well as secrete adiponectin, an insulin-sensitizing hormone *(2)*, and leptin, a signal of energy balance *(2)*. However, nonadipocyte cells, which comprise the stromal vascular fraction, provide structural support *(3)* and secrete angiogenesis-related cytokines *(4)*. Dysregulation of these cellular functions may be reflected in altered expression levels of the genes

From: *Methods in Molecular Biology, Vol. 456:*
Adipose Tissue Protocols, Second Edition. Edited by: Kaiping Yang
© Humana Press, a part of Springer Science + Business Media, Totowa, NJ

involved in pertinent pathways. Defining signature gene expression profiles of adipose tissue or its cellular constituents in disease states may also complement genetic approaches to identify disease susceptibility genes *(5)*. The process of isolating the cellular components of adipose tissue may alter some gene expression profiles. For example, the transcription of many pro-inflammatory genes are increased after standard adipocyte isolation *(6)*. Regardless of this difficulty, the pursuit of better understanding of the different roles each cellular component of adipose tissue plays is warranted.

The use of DNA microarrays to investigate normal and dysregulated adipose tissue gene activity has grown exponentially in the past decade. This technology provides comprehensive expression profiles of multitudes of genes simultaneously. Microarrays commonly use cDNA or oligonucleotide probes. cDNA microarrays involve spotting 3' expressed sequence tags or known genes on glass slides *(7)*, whereas oligonucleotide microarrays contain combinatorially synthesized short oligonucleotides complementary to expressed genes as well as expressed sequence tags and all identified exons *(8)*. The amount of labeled complementary deoxyribonucleic acid (cDNA) from experimental samples that hybridize to these probes correlates with the amount of messenger ribonucleic acid (mRNA) in the original samples. It is generally expected that the concentrations of specific mRNA in the samples reflect the concentrations and activity of protein products of the genes, although this is not always the case *(9)*.

Here, we describe the utility of DNA microarrays to study gene expression profiles of human adipose tissue, adipocytes, and stromal vascular cells. Although different kinds of microarrays, analytical methods, and data mining methods exist, we will specifically describe sample preparation and hybridization methods for human oligonucleotide microarrays synthesized by Affymetrix (Santa Clara, CA) and the resulting data analyzed by Affymetrix software. The data can be further mined using the continually updated databases on genomic networks as described below. We have reported an application of these methods to investigate gene expression profiles in adipocytes and cultured stromal vascular cells of adipose tissue from obese compared to non-obese subjects *(10,11)*.

2 Materials

2.1 Adipose Tissue Biopsy

For needle biopsy:

1. Reusable Hypodermic Needle with Luer-Lok, 13-gauge × 2 inches, short bevel (BD Sciences, San Jose, CA).
2. Several (up to 5 per incision site) 60-mL syringe, each containing ~10 mL of sterile 0.9% NaCl solution.

2.2 Adipose Tissue Digestion and Separation

1. Nylon mesh (pore size 250 µm), autoclaved.
2. Fatty acid-free bovine serum albumin (BSA).
3. Type I collagenase.
4. Sterile 230-µm and 25-µm stainless-steel tissue sieve.
5. M199: Medium 199 supplemented with 1 µg/mL of amphotericin B, 100 units/ mL penicillin G sodium, 100 µg/mL streptomycin sulfate, and 2 mM Glutamax-1 (Invitrogen).
6. Heat-inactivated fetal bovine serum (FBS) (Invitrogen).

2.3 cRNA Preparation

All the reagents to synthesize labeled complementary (c)RNA are from Affymetrix as the supplier of the oligonucleotide arrays.

1. One-Cycle Target Labeling and Control Reagents.
2. IVT Labeling Kit (*see* **Note 1**).
3. One-Cycle cDNA Synthesis Kit.

2.4 Microarray Hybridization and Scanning

All of the reagents and equipment for hybridization are from Affymetrix unless otherwise mentioned.

1. GeneChip Eukaryotic Hybridization Control Kit. Completely resuspend the cRNA frozen stock by heating to 65°C for 5 min before aliquoting.
2. Herring Sperm DNA (Promega).
3. BSA solution (50 mg/mL; Invitrogen).
4. 12x MES stock buffer: 1.22 M MES, 0.89 M [Na$^+$].
5. 2x Hybridization buffer: 200 mM MES, 2 M [Na$^+$], 40 mM EDTA, 0.02% Tween-20.
6. Oligonucleotide microarrays (e.g., Human Genome U-133).
7. Hybridization Oven 640.
8. Affymetrix® Microarray Suite 5.0 (MAS 5.0) or GeneChip Operating Software (GCOS) on PC-compatible workstation (*see* **Note 2**).
9. Fluidics Station 400, or 450/250.
10. Wash Buffer A (1 L): Non-Stringent Wash Buffer: 6x SSPE, 0.01% Tween-20
11. Wash Buffer B (1 L): Stringent Wash Buffer: 100 mM MES, 0.1 M [Na$^+$], 0.01% Tween-20
12. 2X Stain Buffer: 200 mM MES, 2 M[Na$^+$], 0.1% Tween-20.
13. R-Phycoerythrin Streptavidin (Invitrogen).

14. Goat IgG Stock: resuspend 50 mg of Goat IgG (Sigma) in 5 mL of 150 m*M* NaCl. Store at 4°C.
15. Anti-streptavidin antibody (goat), biotinylated (Vector Laboratories).
16. Affymetrix GeneChip® Scanner 3000 or Agilent GeneArray® Scanner.
17. Agilent 2100 Bioanalyzer or agarose gel electrophoresis system.

2.5   *Microarray Data Analysis*

1. PC-compatible workstation.
2. Microarray Suite 5.0 (MAS 5.0) or GeneChip Operating Software (GCOS).
3. The following software has the capacity to manage, integrate, quality check, analyze, and annotate microarray data from various sources: http://www.insightful. com/products/s-plus_arrayanalyzer/keyfeatures.asp.

3   Methods

3.1   *Adipose Tissue Biopsy*

Adipose tissue (typically from abdominal subcutaneous or omental depot) can be obtained surgically *(12)* or using needle biopsy/aspiration *(13)* after the administration of local anesthesia. The latter procedure is only applicable to subcutaneous adipose tissue.

3.1.1   Surgical Samples

1. Put tissue samples (approximately 1–3 g) in a sterile plastic tube containing Hank's Balanced Salt Solution (HBSS) or M199 (supplemented as described above; *see* **Note 3**).
2. Mince tissue sample coarsely (into ~25 mg pieces) using sterile scalpels and forceps in a laminar flow hood. Clean the tissue pieces from visible connective scar tissue, blood clots, and blood vessels in HBSS supplemented with 5.5 m*M* glucose.

3.1.2   Needle Biopsy Samples

1. Aspirate adipose tissue into a 60-mL syringe containing 0.9% NaCl solution. Place tissue on a sterile nylon mesh and rinse with more NaCl solution (*see* **Note 4**).
2. In a laminar flow hood, clean the already fragmented tissue sample from visible connective tissue and blood vessels in HBSS supplemented with 5.5 m*M* glucose.

3.1.3 Preparation for Subsequent Processing

1. For RNA extraction: Place the tissue samples in 5–10 volume of RNALater (Ambion) solution to preserve for later extraction (stored at 4°C overnight then transferred to −80°C for longer term storage) or proceed with RNA extraction immediately. Tissue samples can also be frozen quickly in liquid nitrogen and stored at −80°C for RNA extraction (covered in a different chapter of this issue) at a later time.
2. For digestion: Transfer the tissue samples into a sterile 125-mL Erlenmeyer flask and proceed with digestion protocol below.

3.2 Adipose Tissue Digestion and Separation of Adipocytes and Stromal Vascular Cells

1. Digest the tissue in HBSS buffer containing 5.5 mM glucose, 5% fatty-acid free BSA, and 3.3 mg/mL type I collagenase for 30–60 min in a shaking 37°C water bath at 120–150 strokes/min.
2. Pass the digestion mixture through a sterile 230-μm stainless-steel tissue sieve into a sterile 50-mL tube.
3. After allowing the adipocytes to float by buoyancy, pipet the adipocyte layer carefully into a separate tube. Using this floatation method, rinse the adipocytes with M199 media containing 3% FBS and 5% fatty-acid free BSA twice. After the final wash, the adipocytes can be lysed for RNA extraction (covered in a different chapter of this issue).
4. Add equal volume of M199 to the infranatant containing stromal vascular cells. Centrifuge this mixture at 585 g for 5 min. Discard supernatant carefully, leaving the pellet in ~1–2 mL of media.
5. Using the centrifugation method, rinse the stromal vascular pellet with M199 and subsequently with HBSS. At the end of the last rinse, discard as much of the supernatant as possible without disturbing the pellet. The stromal vascular cells can be lysed immediately or expanded in culture (*see* **Note 5**) for RNA extraction, similar to the procedure for adipocytes.

3.3 cRNA Preparation

The following procedures (recommended by Affymetrix) are used to prepare biotinylated cRNA target using 5 μg of purified total RNA as the starting sample, which will provide sufficient amount of cRNA target for hybridization of one or both subarrays of the HG-U133 set. High-quality (1.9–2.1, A_{260}/A_{280}) total RNA is required for DNA microarray analysis.

3.3.1   First-Strand cDNA Synthesis (One-Cycle cDNA Synthesis Kit)

1. Mix 5 μg of RNA sample, 2 μL of diluted poly-A RNA controls, and 2 μL of T7-oligo(dT) primer, adding RNase-free water to a final volume of 12 μL in a 0.2-mL PCR tube.
2. Flick the tube to mix and spin the tube.
3. Incubate the reaction for 10 min at 70°C using a PCR machine, quickly chill the tube (by putting on ice), and keep at 4°C at least 2 min.
4. Spin the tube briefly.
5. Prepare sufficient First-Strand Master Mix by adding 4 μL of 5x First Strand Reaction Buffer, 2 μL of DTT (0.1 M), and 1 μL of dNTP (10 mM) for each sample.
6. Mix well, spin down and transfer 7 μL of First-Strand Master Mix to each denatured sample tube.
7. Incubate for 2 min at 42°C.
8. Add 1 μL of SuperScript II to each RNA sample, mix by flicking, spin down, and immediately incubate for 1 h at 42°C, then cool the sample for at least 2 min at 4°C.
9. Spin down the tube and immediately proceed to Second-Strand cDNA Synthesis.

3.3.2   Second-Strand cDNA Synthesis (One-Cycle cDNA Synthesis Kit)

1. Prepare sufficient Second-Strand Master Mix in a separate tube with 91 μL of RNase-free water, 30 μL of 5x Second Strand Reaction Mix, 3 μL of dNTP, 1 μL of *E. coli* DNA ligase, 4 μL of *E. coli* DNA polymerase I, and 1 μL of RNase H for each sample.
2. Add 130 μL of Second-Strand Master Mix to each first-strand synthesis sample and incubate for 2 hrs at 16°C.
3. Add 2 μL of T4 DNA Polymerase, incubate for 5 min at 16°C, then hold at 4°C.
4. Add 10 μL of 0.5 M EDTA and proceed to Clean-up of Double-Stranded cDNA step (see below).

3.3.3   Clean-up of Double-Stranded cDNA (Sample Clean-up Module at Room Temperature [RT])

1. Transfer second-strand cDNA mixture into a 1.5-mL microfuge tube and add 600 μL of cDNA Binding Buffer. Vortex the tube for 3 sec to mix and spin briefly.
2. Check that the color of the mixture is yellow. If the color is orange or violet, add 10 μL of 3 M sodium acetate, pH 5.0, with which the mixture will turn yellow.
3. Pipet 500 μL of the sample into the cDNA Clean-up Spin Column in a 2-mL collection tube, and centrifuge for 1 min at ≥9000 g. Discard the flow-through.
4. Repeat the step using the remainder of the sample. Discard the collection tube with the flow-through.
5. Transfer the column into a new 2-mL collection tube. Pipet 750 μL of the cDNA Wash Buffer onto the spin column and centrifuge for 1 min at ≥9000 g. Discard flow-through.

6. Open the cap of the column and centrifuge for 5 min at maximum speed. Discard the collection tube with the flow-through.
7. Transfer the column into a 1.5-mL collection tube, and pipette 14 μL of cDNA Elution Buffer directly onto the membrane. Incubate for 1 min at RT and centrifuge for 1 min at maximum speed. The average volume of eluate is 12 μL from 14 μL of Elution Buffer.

3.3.4 Synthesis of Biotin-Labeled cRNA by In Vitro Transcription (IVT) Reaction (IVT Labeling Kit)

1. Transfer all of the eluate from the clean-up procedure into a 0.2-mL PCR tube and add 8 μL of RNase-free water, 4 μL of 10x IVT Labeling Buffer, 12 μL of NTP Mix, and 4 μL of Enzyme Mix.
2. Carefully mix by tapping the tube and spin briefly.
3. Incubate at 37°C for 16 hrs in a thermal cycler.
4. Biotin-labeled cRNA can be stored at −20°C or −70°C, if not used for clean-up immediately.

3.3.5 Clean-up and Quantification of Biotin-Labeled cRNA (Sample Clean-up Module at RT)

1. Save an aliquot (0.5 μL) of the unpurified IVT product for analysis by gel electrophoresis or Agilent 2100 Bioanalyzer to estimate the yield and size distribution of labeled transcripts.
2. Transfer cRNA sample to a 1.5-mL tube, add 60 μL of RNase-free water, and vortex for 3 sec.
3. Add 350 μL of IVT cRNA Binding Buffer and vortex.
4. Add 250 μL of ethanol (96–100%) and mix well by pipetting. Do not centrifuge.
5. Apply sample to the IVT cRNA Clean-up Spin Column in a 2-mL collection tube and centrifuge for 15 sec at $\geq 9000\,g$.
6. Transfer the column into a new 2-mL collection tube, pipette 500 μL of IVT cRNA Wash Buffer, and centrifuge for 15 s at $\geq 9000\,g$. Discard flow-through.
7. Pipette 500 μL 80 % ethanol onto the column and centrifuge for 15 sec at $\geq 9000\,g$. Discard flow-through.
8. Open the cap of the column and centrifuge for 5 min at maximum speed. Discard the collection tube with the flow-through.
9. Transfer the column into a 1.5-mL collection tube, and pipet 11 μL of RNase-free water directly onto the membrane. Incubate for 1 min at RT and centrifuge for 1 min at maximum speed.
10. Pipet another 10 μL of RNase-free water onto the membrane and elute as above. Combined eluate (21 μL) can be stored at −20°C or −70°C if not used for quantification immediately.

11. To determine cRNA concentration and purity, check the absorbance at 260 nm and 280 nm of 1:100 dilution of 1 µL cRNA sample. Ratios between 1.9 and 2.1 of A^{260}/A^{280} are acceptable.

12. Calculate adjusted cRNA yield to reflect carryover of unlabeled total RNA by subtracting 5 µg of starting total RNA from the cRNA yield. More than 50 µg of adjusted cRNA yield can be expected by starting amount of 5 µg of total RNA. Use adjusted cRNA concentration in fragmentation procedure.

13. Check the quality of unfragmented samples by gel electrophoresis or Agilent 2100 Bioanalyzer.

3.3.6 Fragmentation of cRNA (Sample Clean-up Module)

1. Transfer 40 µg of cRNA of adjusted concentration into a 0.2-mL PCR tube and add 16 µL of 5x Fragmentation Buffer and RNase-free water to a final volume of 80 µL to prepare fragmented cRNA for two sub-arrays of the HG-U133 set.

2. Incubate at 94°C for 35 min in a thermal cycler. Put on ice following the incubation.

3. Save an aliquot (1 µL) for analysis on the Bioanalyzer or gel electrophoresis to confirm the fragmentation.

4. Store the undiluted fragmented cRNA at −70°C until ready to perform the hybridization.

3.4 Microarray Hybridization and Scanning

Target cRNA hybridization step should be followed by washing, staining, and scanning without delay for the best results.

3.4.1 Target Hybridization

1. Prepare 600 µL of hybridization cocktail for two subarrays of the HG-U133 set by mixing 30 µg of fragmented cRNA, 10 µL of control oligonucleotide b2, 30 µL of eukaryotic hybridization controls, 6 µL of herring sperm DNA, 6 µL of BSA, 300 µL of 2x hybridization buffer, 60 µL of DMSO, and nuclease-free water to final volume.

2. Heat the hybridization cocktail for 5 min at 95°C and then incubate the tube in a 45°C heat block for 5 min. Centrifuge the hybridization cocktail tube at maximum speed for 5 min.

3. Equilibrate probe array to RT and fill it with 1X Hybridization Buffer through one of the septa using a micro-pipettor with another tip for venting through the other septum. Incubate array at 45°C for 10 min.

4. Remove the buffer from the array and cartridge, fill it with 200 µL (for each array) of hybridization cocktail, and hybridize for 16 hr in a Hybridization Oven with 60 rpm at 45°C.

3.4.2 Fluidics Station Setup

1. Define file location if MAS is used by selecting Tools → Defaults → File → Locations from the menu bar and verify that all three file locations (files for probe information, fluidics protocols, and experiment data) are set correctly. (If GCOS is used, this step is not necessary.)
2. Launch MAS 5.0 or GCOS and enter experiment name, probe array type, and the additional information of sample name, sample type, and project if using GCOS.
3. Turn on the Fluidics Station and select Run → Fluidics from the GCOS menu bar.
4. Change the intake buffer reservoir A to Wash Buffer A and intake buffer reservoir B to Wash Buffer B and prime the Fluidics Station by selecting Protocol, Prime, All Modules, then clicking Run.

3.4.3 Probe Array Washing, Staining, and Scanning

1. Prepare 1200 µL of fresh streptavidin phycoerythrin (SAPE) stain solution for each array by mixing 600 µL of 2x stain buffer, 48 µL of BSA, 12 µL of SAPE, and 540 µL of nuclease-free water and divide into two aliquots of 600 µL each for staining **steps 1** and **3**.
2. Prepare 600 µL of fresh Antibody Solution Mix with 300 µL of 2x Stain Buffer, 24 µL of BSA, 6 µL of Goat IgG Stock, 3.6 µL of biotinylated antibody, and 266.4 µL of nuclease-free water.
3. Remove hybridization cocktail from the probe array immediately after 16 hr hybridization and completely fill with ~250 µL of Wash Buffer A. This probe array can be stored at 4°C for up to 3 h if washing and staining is not available immediately after hybridization.

   Perform washing and staining of the probe array using one of either Fluidics Station 450/250 or Fluidics Station 400 following the automated protocol (EukGE-WS2v4) for the probe array format of HG-U133.
4. Remove the probe arrays from the Fluidics Station at the end of washing and staining and check the probe array glass window for bubbles or air pockets. If there are large bubbles, fill the probe array without making large bubbles using Fluidics Station or manually. Keep the probe array in the dark at 4°C until ready for scanning.
5. Warm up Affymetrix GeneChip Scanner 3000 for at least 10 min or Agilent GeneArray Scanner for at least 15 min before scanning.
6. Clean the excess fluid from the probe array and apply one Tough-Spot to each of the two septa, pressing to ensure that the spots remain flat. Clean the glass window of probe arrays. Insert the cartridge into the scanner and test the auto-focus to ensure that the Tough-Spots do not interfere.
7. Click the Option button to check for the correct pixel value and wavelength of the laser beam in case GeneArray Scanner is used (Pixel value, 3 µm; Wavelength, 570 nm).
8. Select Run and Scanner from the the menu bar or click the Start Scan in the tool bar in MAS 5.0 or GCOS. Select experiment name and click Start button.

9. Open the sample door on the scanner and apply the probe array without forcing the probe array into the holder. Close the door.
10. Click OK in the Start Scanner dialog box.
11. Each complete probe array image is stored in a separate data file identified by the experiment name and is saved with a data image file (.dat) extension.

3.5   Microarray Data Analysis

This is a simplified version of analysis options using Affymetrix microarray data analysis software. The technical manuals of the analysis software have extensive descriptions for choosing the various options and we refer the reader to those websites. Describing them here is beyond the scope of this chapter. The general workflow is presented in **Fig. 11.1**.

3.5.1   Checking Scanned Image and Converting Fluorescence Intensities to Numerical Values

1. Double-click on each *.dat file to open it.
2. Under the tab for "Image settings." choose autoscale and pseudocolor, then click "OK."
3. Click on "**Grid**" tab, press "G" and select View→ Grid from menu bar to superimpose grid lines on the scanned image.
4. Make sure that the corners of the scanned image fit within the grid (*see* **Note 6**).
5. Manually align the grid by click-dragging the double arrows on the grid perimeters.
6. Select Run→Analysis from menu bar.
7. At the end of each analysis of a *.dat file (~2 min), cell intensity data is computed and a *.cel file is created (single intensity value is computed for each probe cell).
8. To analyze absolute expression values for an array, go to **Tools→Analysis settings→Expression**.

   a. Make sure that in the Expression Analysis Settings dialog box, the "Use baseline comparison file" box is unselected.
   b. Click **Scaling** tab; choose "**All probe sets**"; choose default or user defined scaling value (*see* **Note 7**).
   c. Choose default values for normalization, probe mask, baseline and parameter.
   d. Choose Probe array type (eg. HG-U133A).

9. Select Run→Analysis from menu bar. The cell intensity data is analyzed and, in the Expression Analysis Window (EAW), the analysis output file (*.chp file) is displayed. For each transcript on the probe array a probeset identifier, signal, detection call, detection p-value, and description are generated (*see* **Note 8**).
10. "**Save results as**" dialog box opens and the *.chp files can be saved at desired loacation with any name.

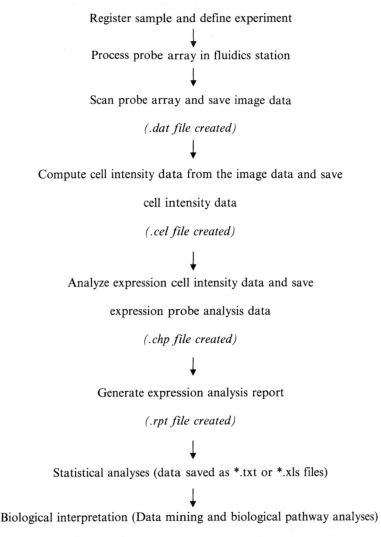

Register sample and define experiment

↓

Process probe array in fluidics station

↓

Scan probe array and save image data

(.dat file created)

↓

Compute cell intensity data from the image data and save

cell intensity data

(.cel file created)

↓

Analyze expression cell intensity data and save

expression probe analysis data

(.chp file created)

↓

Generate expression analysis report

(.rpt file created)

↓

Statistical analyses (data saved as *.txt or *.xls files)

↓

Biological interpretation (Data mining and biological pathway analyses)

Fig. 11.1 The flowchart outlines the sequence of steps involved in collecting, processing, analyzing, and interpreting microarray data

11. Double click on any *.chp file to open. In the EAW, click **Pivot** tab and view absolute analysis results in the pivot table. [A *.rpt file (expression analysis report file that has quality control metrics) can be generated from each *.chp file by **File→Report**].

12. There are several pivot table display options. The pivot table can be exported as a text file (*.txt) or excel file (*.xls) by clicking on **File→Save As.**

13. The *.chp files can also be published into Publish databases (eg. GCOS server). Published data can be queried using data mining software as described below.

3.5.2 Statistical Analyses, Data Mining, and Pathway Analyses

Microarray data often are log-transformed to approximate normal distribution and to reduce the influence of single measurements. Standard statistical analyses such as the parametric Student's t-test and Mann-Whitney tests (when the data are not normally distributed) can be used to assess the significance of the data. Depending on the experiment or researcher's needs, other statistical tests such as multivariate tests (e.g., analysis of variance) and multiple comparison corrections (e.g., Bonferroni correction or False Discovery Rate *(10)*) can be applied.

Biological interpretation is the next step after lists of genes with statistically significant expression patterns are generated. Databases with functional gene information, such as Ensembl (http://www.ensembl.org/index.html), Entrez, Locus link, RefSeq, and the Gene Ontology database (http://amigo.geneontology.org/cgi-bin/amigo/go.cgi), may provide useful information. The following databases contain annotated biological information and serve as integrated single interfaces through which one can link to multiple databases to mine data about any given gene or sequence.

1. Affymetrix (NetAffx™ Analysis Center): http://www.affymetrix.com/analysis/index.affx
2. UCSC genomic browser website: http://genome.ucsc.edu/
3. NCBI Entrez website: http://www.ncbi.nlm.nih.gov/gquery/gquery.fcgi

To biologically interpret expression from sets of genes and identify relevant biological pathways with co-regulated genes, one can utilize various biological pathway analysis software, some of which are listed below. Most of these data mining software can upload *.chp or *.xls files after data analyses. These software incorporate graphics to summarize networks of biological interactions. One can place global gene expression datasets in the context of these pathway images to identify existing or unique pathways that may be significant in the tested hypothesis.

1. Genesifter: http://www.genesifter.net/web/dataCenter.html
2. Metacore: http://www.genego.com/metacore.php
3. Affymetrix portal: http://www.affymetrix.com/products/software/compatible/pathway.affx#GeneGo.
4. Ingenuity: http://www.ingenuity.com/.

4 Notes

1. If IVT cRNA Binding Buffer forms a precipitate, redissolve in a 30°C water bath and then place the buffer at RT. Before using for the first time, add 24 mL and 20 mL of ethanol (96–100%) to cDNA Wash Buffer and IVT cRNA Wash Buffer, respectively.
2. MAS 5.0 software has been widely used but has recently been upgraded into GCOS by Affymetrix. Both software programs share basic similarities in

data collection, management and analysis. Product and technical support including manuals for both may be obtained at the following link: http://www.affymetrix.com/support/technical/byproduct.affx?cat=software. Also refer to the following site for the most recent Affymetrix software and manuals/technical sheets: http://www.affymetrix.com/products/software/index.affx.

3. Visceral adipose tissue samples are usually taken from patients undergoing abdominal surgery. If adipose tissue samples need to be transferred to another place for processing, they should be carried in tubes containing saline solution at RT, not in a cold carrier or on ice, to prevent changes in gene expression profiles caused by exposure to cold temperature.

4. If needle biopsy samples need to be transferred to another place for subsequent processing, they need to be diluted with more saline solution with gentle agitation to prevent blood clotting inside the syringes.

5. If the number of stromal vascular cells needs to be expanded to obtain adequate RNA, the cells can be cultured for several population doublings. Stromal vascular cells are usually composed of a mixture of preadipocytes, macrophages, endothelial cells and fibroblasts. We have shown that under 2-wk culture conditions, preadipocytes predominate in the culture and contamination from other cell types is minimal *(11,13)*. To culture the stromal vascular cells, resuspend the pellet in 10 ml of M199 containing 10% FBS and transfer the resuspended cells to a culture flask/dish. Culture in a 37°C incubator for a day before washing the attached stromal vascular cells with PBS twice to get rid of red blood cells. Continue culturing the cells in fresh M199 media containing 10% FBS until there are approximately 1.5×10^6 cells (e.g., the cells can be propagated in a 15-cm dish, with media change every 2–3 d, and harvested for RNA when they are about 90% confluent). RNA can then be extracted from these cells.

6. A picture of the image data is displayed in an image window when *.dat is opened. The software represents the fluorescence intensity values from each pixel on the array in a grayscale or pseudocolor mode and superimposes a grid on the image to delineate the probe cells (Ref: Technical manual). This is important because each grid cell represents a probe cell and the software calculates the average of intensities of the middle squares of each probe cell to generate a *.cel file during analyses. It is rare to have to adjust the grid.

7. Absolute analysis results from several experiments (using the same type of probe arrays) can be directly compared if a user specified target signal value is chosen consistently for all arrays using the "All probe sets" scaling option.

8. The far left column is the unique Affymetrix probeset identifier and the right most column provides a brief description of the sequence that the probeset represents. The "signal" denotes intensity. A "detection call" may be "Present," "Absent," or "Marginal" for each probeset, and the "detection p-value" provides an assessment of the statistical significance of each call.

References

1. Large V, Peroni O, Letexier D, Ray H, Beylot M (2004) Metabolism of lipids in human white adipocyte. Diabetes Metab 30:294–309
2. Havel PJ (2004) Update on adipocyte hormones: regulation of energy balance and carbohydrate/lipid metabolism. Diabetes 53 Suppl 1:S143–151
3. Aoki S, Toda S, Ando T, Sugihara H (2004) Bone marrow stromal cells, preadipocytes, and dermal fibroblasts promote epidermal regeneration in their distinctive fashions. Mol Biol Cell 15:4647–4657
4. Nakagami H, Morishita R, Maeda K, Kikuchi Y, Ogihara T, Kaneda Y (2006) Adipose tissue-derived stromal cells as a novel option for regenerative cell therapy. J Atheroscler Thromb 13:77–81
5. Permana PA, Del Parigi A, Tataranni PA (2004) Microarray gene expression profiling in obesity and insulin resistance. Nutrition 20:134–138
6. Ruan H, Zarnowski MJ, Cushman SW, Lodish HF (2003) Standard isolation of primary adipose cells from mouse epididymal fat pads induces inflammatory mediators and down-regulates adipocyte genes. J Biol Chem 278:47585–47593
7. Duggan DJ, Bittner M, Chen Y, Meltzer P, Trent JM (1999) Expression profiling using cDNA microarrays. Nat Genet 21:10–14
8. Lipshutz RJ, Fodor SP, Gingeras TR, Lockhart DJ (1999) High density synthetic oligonucleotide arrays. Nat Genet 21:20–24
9. Cagney G, Park S, Chung C, Tong B, O'Dushlaine C, Shields DC, Emili A (2005) Human tissue profiling with multidimensional protein identification technology. J Proteome Res 4:1757–1767
10. Lee YH, Nair S, Rousseau E, Allison DB, Page GP, Tataranni PA, Bogardus C, Permana PA (2005) Microarray profiling of isolated abdominal subcutaneous adipocytes from obese vs non-obese Pima Indians: increased expression of inflammation-related genes. Diabetologia 48:1776–1783
11. Nair S, Lee YH, Rousseau E, Cam M, Tataranni PA, Baier LJ, Bogardus C, Permana PA (2005) Increased expression of inflammation-related genes in cultured preadipocytes/stromal vascular cells from obese compared with non-obese Pima Indians. Diabetologia 48:1784–1788
12. Kern PA, Di Gregorio GB, Lu T, Rassouli N, Ranganathan G (2003) Adiponectin expression from human adipose tissue: relation to obesity, insulin resistance, and tumor necrosis factor-alpha expression. Diabetes 52:1779–1785
13. Permana PA, Nair S, Lee YH, Luczy-Bachman G, Vozarova De Courten B, Tataranni PA (2004) Subcutaneous abdominal preadipocyte differentiation in vitro inversely correlates with central obesity. Am J Physiol Endocrinol Metab 286:E958–E962

Chapter 12
Differentiation of Adipose Stem Cells

Bruce A. Bunnell, Bradley T. Estes, Farshid Guilak, and Jeffrey M. Gimble

Summary The broad definition of a stem cell is a cell that has the ability to self-renew and differentiate into one or more specialized terminally differentiated cell types. It has become evident that stem cells persist in, and can be isolated from, many adult tissues. Adipose tissue has been shown to contain a population of cells that retain a high proliferation capacity in vitro and the ability to undergo extensive differentiation into multiple cell lineages. These cells are referred to as adipose stem cells and are biologically similar, although not identical, to mesenchymal stem cells derived from the bone marrow. Differentiation causes stem cells to adopt the phenotypic, biochemical, and functional properties of more terminally differentiated cells. This chapter will provide investigators with some background on stem cells derived from adipose tissue and then provide details on adipose stem cell multilineage differentiation along osteogenic, adipogenic, chondrogenic, and neurogenic lineages.

Key words Adipose stem cells; differentiation; osteogenesis; adipogenesis; chondrogenesis; neurogenesis.

1 Introduction

Adult mesenchymal stem cells can give rise to diverse cell types in response to appropriate extrinsic or intrinsic developmental cues *(1,2)*. Although originally identified within the bone marrow stroma, there is growing evidence that such mesenchymal stem cells exist in many adult tissues, including adipose tissue. These cells, referred to as adipose stem cells (ASCs), adipose-derived adult stem cells, and other similar terms, have been shown to be biologically similar, but not identical, to bone marrow mesenchymal stem cells. The multilineage differentiation potential of adipose stem cells offers significant potential for the application of ASCs for the treatment of human degenerative diseases. ASCs have been demonstrated to differentiate into a broad spectrum of different cell lineages, such as bone,

From: *Methods in Molecular Biology, Vol. 456:*
Adipose Tissue Protocols, Second Edition. Edited by: Kaiping Yang
© Humana Press, a part of Springer Science+Business Media, Totowa, NJ

cartilage, endothelium (cells that line blood vessels and other body cavities), liver, and neural cells *(3,4)*.

The transition from stem cell to predetermined mature cell type is called "differentiation." Many studies have evaluated mechanisms that mediate stem cell differentiation along various lineages and tissues. For example, precursor cells derived from muscle were reported to form blood cells *(5)*. Stem cells derived from bone marrow have also been reported to generate neurons and glial cells *(6–8)*, hepatic oval cells *(9,10)*, and muscle cells *(11,12)*. Stem cells isolated from other tissues such as brain have been found to be capable of giving rise to blood elements *(13,14)*, and muscle cells *(15–17)*. In addition, stem cells derived from liver have been reported to form pancreatic islet cells *(18)*. The data from these studies have led to a newly proposed theory of stem cell differentiation, which involves the genetic reprogramming of stem cells to acquire the biologic genotype and phenotype of another committed cell lineage. The resulting terminally differentiated cells retain the identical DNA composition; however, novel patterns of gene expression are induced as a result of differentiation. Thus, a potent genetic regulatory system has evolved to coordinate and maintain tissue-specific patterns of gene expression. As stem cells differentiate, specific subsets and families of genes are activated at the level of transcription, whereas others are transcriptionally silenced. The activation and silencing events are exquisitely coordinated as cells differentiate along specific lineages. During the process of differentiation, the alterations in gene function result in the expression of genes that give rise to a terminally differentiated phenotype. As the stem cells become terminally differentiated, characteristic markers of the differentiated cells such as proteins or secreted factors (e.g., cytokines, growth factors) are detectable.

The differentiation process is tightly regulated by specific families of elements such as inducers, or processes such as cell–to-cell and cell-matrix interactions. For example, there is growing evidence for a role of the extracellular matrix in regulating cell phenotype and differentiation, both through direct interactions with cell surface receptors and through regulation of cell shape. Physiologic inducers of differentiation are primarily biological substances that drive cells towards specific lineages, among them are hormones (e.g. hydrocortisone, glucagon, thyroxin), vitamins (vitamin D), ions (Ca^{2+}), and oxygen tension. In addition, there are chemical inducers that can be used as supplements in differentiation medium, such as DMSO and cytotoxic drugs (e.g., 5′ azacytidine, methotrexate, mitomycin C).

The differentiation of ASCs in vitro is governed by the following criteria:

- Presence of a correct stem cell population maintained in appropriate conditions with proper media.
- Proper density of cells in culture on or within an appropriate extracellular matrix.
- A medium formulation that promotes differentiation rather than proliferation.
- Use of relevant differentiation-inducing agents.
- Use of proper timing (some inducers may be needed at specific differentiation stages for specific periods of time).

The procedures described in this chapter detail the conditions necessary to induce ASCs to differentiate along osteogenic, adipogenic, chondrogenic and neurogenic lineages.

1.1 Osteogenic Differentiation

ASCs can differentiate into cells of osteogenic lineage. In the presence of ascorbate, β-glycerophosphate, and dexamethasone, rat, nonhuman primate and human ASCs differentiate into osteoblast-like cells. During osteogenesis, the composition of the ASC culture changes from the characteristic fibroblast monolayer to a multilayer fibroblast culture. As the cells differentiate, the organization of cytoskeletal elements leads to these changes in cell morphology. A marked change in organization of the cytoskeleton is observed in the ASCs during osteogenic differentiation. These changes in the assembly and disassembly kinetics of actin microfilaments may be critical in supporting the osteogenic differentiation of ASCs *(19)*. ASCs also acquire the expression of osteogenic specific markers such as alkaline phosphatase activity and mineralization (calcium deposition) *(20)*.

1.2 Adipogenic Differentiation

Adipogenesis is the formation of adipose tissue (fat cells). Several groups have shown that human ASCs can differentiate into adipocytes in vitro using induction cocktails containing insulin, methylisobutylxanthine (a phosphodiesterase inhibitor resulting in elevated cyclic AMP levels), hydrocortisone or dexamethasone (a glucocorticoid receptor agonist), and indomethacin or thiazolidinedione (a peroxisome proliferator activated receptor γ ligand). Under adipogenic differentiation conditions, proliferating ASCs differentiate into preadipocytes, which then proliferate at sites of adipogenesis. Preadipocytes undergo a second differentiation step and begin to accumulate lipid in the form of small vacuoles or droplets. The adipocytes can continue to enlarge by accumulating additional lipid during the differentiation process. Adipogenesis in vitro follows a highly ordered and well-characterized temporal sequence. One week to 10 d after exposure, the ASCs increase their intracellular stores of neutral lipid as detected by Oil Red O or Nile Red staining, increase their secretion of leptin protein, and express adipogenic mRNAs, including the fatty acid binding protein, aP2 *(21–25)*. Compared with the differentiation of many other cellular lineages, adipogenic differentiation in vitro is rather authentic, recapitulating most of the key features of adipogenesis in vivo. This includes morphological changes, cessation of cell growth, expression of many lipogenic enzymes, extensive lipid accumulation, and the establishment of sensitivity to most or all of the key hormones that have an impact on this cell type, including insulin *(26)*.

1.3 Chondrogenic Differentiation

ASCs can be induced to undergo chondrogenesis and become chondrocytes. ASCs require appropriate signals to differentiate specifically to cartilage. Chondrogenesis can be promoted by several different factors but appears to require the maintenance of cells in a three-dimensional, rounded morphology, similar to that required for primary culture of chondrocytes. Cells grown in a micromass pellet culture or within a hydrogel that maintains a rounded shape (e.g., alginate, agarose) will stop dividing and *(27,28)* begin to express extracellular matrix proteins that are characteristic of hyaline articular cartilage, such as type II collagen and the aggregating proteoglycan, aggrecan *(29)*. In the presence of transforming growth factor beta (TGF- β), ascorbate, and dexamethasone, the expression of these biochemical markers associated with mature chondrocytes is greatly increased *(3,25,29–32)*. Furthermore, other studies have shown that various growth factors such as bone morphogenic protein 6 (BMP6), alone or in combination with insulin-like growth factor (IGF), transforming growth factor- beta 1 and 3 (TGF-β1 and TGF-β3) can induce ASCs to differentiate into chondrocytes under specific culture conditions *(27,28)*. The extracellular matrix can also have significant effects on the phenotypic expression of chondrogenically induce ASCs *(33)*. Human ASCs have been shown to maintain their chondrogenic phenotype in vivo for up to 12 wk based on the analysis of alginate matrices implanted subcutaneously in immunodeficient mice *(29)*.

1.4 Neurogenic Differentiation

ASCs may also serve as a viable source of stem cells capable of undergoing neural differentiation. ASCs isolated from rats, nonhuman primates, and humans have been differentiated into neuron-like cells that express many neuronal markers *(34–36)*. Under high-density culture in neurogenic differentiation medium, the ASCs generate compact spheroid bodies that become neurospheres. Cells from these neurospheres undergo extensive morphologic differentiation and begin to express a multitude of neural genes and proteins. In the presence of antioxidants and the absence of serum, murine and human ASCs display a bipolar morphology, similar to neuronal cell lineages *(25,36)*. Under neural induction conditions, ASCs express neural-associated proteins, including nestin, intermediate filament M, glial fibrillary acidic protein (GFAP), and Neu N *(25,35–37)*. After induction of ASCs with indomethacin, insulin, and isobutylmethylxanthine, murine, nonhuman primate, and human cells have also been found to express additional neuronal markers: S-100, MAP2, and β-III tubulin, and the glutamate receptor subunits NR-1 and NR-2 *(35,38)*.

2 Materials

2.1 Osteogenic Differentiation

1. Osteogenic induction medium: BGJ$_b$ medium (Fitton-Jackson Modification; Invitrogen), 1 *M* HEPES, pH 7.4, (Biologos), 1 *M* β-glycerophosphate (Sigma), 50 mg/mL ascorbate-2-phosphate (Sigma), 10 m*M* 1,25 (OH)$_2$ vitamin D3

(BioMOL), 10 mM dexamethasone (Sigma), 10,000 U of Penicillin / 10,000 mg of Streptomycin (Invitrogen), fetal bovine serum 10% (Hyclone).

2. 40 mM Alizarin red, pH 4.1 (Sigma).
3. 0.9% NaCl.
4. 0.1 N HCl.
5. 3 N NaOH.
6. 1 mM MgCl$_2$.
7. 50 mM sodium bicarbonate (Sigma).
8. P-nitrophenyl phosphate (Sigma).
9. Alkaline phosphatase enzyme activity kit (Sigma).
10. 10% cetylpyridinium chloride.
11. Total calcium kit.
12. Mouse anti-human osteonectin (Abcam).

2.2 Adipogenic Differentiation

1. Adipogenic induction medium (*see* **Note 1**): Dulbecco's Modified Eagle's Medium/ F-12 Ham's Medium, High glucose (DMEM/F-12), 0.5 mM isobutyl-methylxanthine (IBMX) (Sigma), 50 μM Indomethacin (Sigma) or 5 μM rosiglitazone (AK scientific), 0.5 μg/mL dexamethasone (Sigma), 10 μL/mL Insulin (Sigma), fetal bovine serum 20% (Hyclone).
2. Oil red O (Sigma; *see* **Note 2**).
3. Triethanolamine/HCl, pH 7.5 (Sigma).
4. EDTA.
5. NADH.
6. Dihydroxyacetone phosphate (Sigma).
7. β-mercaptoethanole (Sigma).
8. Mouse anti-human leptin (Abcam).

2.3 Chondrogenic Differentiation

1. Chondrogenic induction medium (*see* **Note 3**): Several different media formulations have been used to induce chondrogenesis. Consistent with bone marrow derived MSCs, ASCs have shown the ability to be induced toward chondrogeneisis with the following medium components: DMEM high-glucose (Invitrogen), 10 μg/mL TGF-β1 (R&D Systems), 50 mg/mL ascorbic acid 2-phosphate (Sigma), 50 μM dexamethasone (Sigma), 5 mg/mL insulin and transferrin, 5 μg/ mL selenious acid, 1.25 μg/mL bovine serum albumin, insulin-transferrin-selenium plus ITS+ (Collaborative Biomedical), 10,000 U penicillin/10,000 mg streptomycin (Invitrogen), fetal bovine serum (Hyclone) 1% (3; 25) or 10% (4; 29; 30; 33; 39; 40). More recently, in lieu of TGF-β1 and dexamethasone, 500 ng/mL bone

morphogenetic protein 6 (BMP-6, R&D Systems) has been used to induce chondrogenesis *(27,28)*.
2. 1.2% Alginate solution: dissolve 1.2 g of alginic acid sodium salt in 100 mL of 150 mM NaCl. Heat on a hot plate and stir thoroughly. Filter (0.22 μm pore size) to sterilize, and store at 4°C.
3. 102 mM $CaCl_2$.
4. 50 mM sodium chloride and 55 mM sodium citrate buffer.
5. Papain (125 μg/mL papain, 0.1 M $NaH_2PO_4H_2O$, 5 mM EDTA, 5 mM cysteine-HCl).
6. DMB dye solution: 1,9-dimethtlemethylene blue is dissolved in 21 mg of 5 mL of absolute ethanol with 2.0 g of sodium formate in 800 mL of distilled water. The dye solution is then titrated with formic acid solution to a pH of 3.0, and the final volume adjusted to 1000 mL using distilled water.
7. 0 to 30 mg/mL whale chondroitin-4-sulfate standard curve.
8. Acetate citrus buffer: 5 g/L citric acid (monohydrate), 12 g/L sodium acetate (trihydrate), 3.4 g/L sodium hydroxide, and 1.2 mL/L glacial acetic acid in distilled water, pH 6.0.

2.4   Neurogenic Differentiation

1. Neurogenic induction medium (*see* **Note 4**): neurobasal medium with 1x B27 serum-free supplements (Invitrogen), 1 mM retinoic acid (Sigma), 10 ng/mL neural growth factor (NGF, Promega), 10 ng/mL brain-derived neurotrophic factor (BDNF, Promega), 10 ng/mL neurotrophin-3 (NT-3, Promega), 10,000 U of penicillin/ 10,000 μg of streptomycin (Invitrogen), 20% fetal bovine serum (Hyclone).

3   Methods

3.1   Osteogenic Differentiation

1. ASCs are plated at a density of 25,000 cell/cm² and expanded in standard culture medium to reach the confluency of at least 90% (*see* **Note 5**).
2. To induce osteogenesis, replace the Stromal Medium with osteogenic induction medium and continue to grow cells for 21–28 d.
3. The induction medium is replaced every third day. Cells that have undergone in vitro osteogenic differentiation proliferate rapidly and form tightly packed colonies. In culture, dense, granular areas appear within individual colonies, and multiple layers of cells often form toward the endpoint of the assay. In some cases, these colonies give rise to dense nodules from which radiated, highly elongated, spindle-shaped cells with large nuclei are formed.

3.1.1   Evaluation of Osteogenesis

There are several methods to assess osteogenic differentiation, such as demonstration of mineralization by staining with Alizarin red, quantitation of alkaline phosphatase (AP) activity, analysis of the intracellular levels of calcium, and detection of the induction of lineage specific gene and protein expression.

3.1.1.1   Alizarin Red Staining

1. Remove differentiation media from culture plate and wash cells two times with 0.9% sodium chloride.
2. Fix cells with 1% paraformaldehyde for 10 min at room temperature.
3. Stain cells with 2.0% alizarin red solution (pH adjusted to 4.1–4.3) for 45 min at room temperature.
4. Wash the cells two times with 0.9% NaCl.
5. Determine the amount of mineralization using light microscopy.
6. Elute the stain by adding 50 mL of 10% cetylpyridinium chloride for 30 min at room temperature.
7. Determine optical density at 540 nm using a plate reader.

3.1.1.2   AP Activity

The AP activity assay described below was originally described by Martinez and colleagues *(41)*.

1. Rinse differentiated cells with Tris-buffered saline.
2. Fix cells in 3.7% formaldehyde/90% ethanol for 30 s at room temperature.
3. Incubate fixed cells with 1 mL of alkaline phosphatase substrate P-nitrophenyl phosphate (1 mg/mL) in 50 mM bicarbonate buffer (pH 9.6) containing 1 mM MgCl$_2$ for 20 min at 37°C.
4. Stop the reaction with 0.5 mL of 3 N NaOH.
5. Photograph the cells for evidence of staining.

3.1.1.3   Calcium Production

Calcium accumulation can be assessed using colorimetric system as described by Zuk and colleagues *(25)*.

1. Aspirate media from differentiated cells and wash 3–4 times with Ca^{2+} and Mg^{2+} free PBS.
2. Solubilize cells with 0.6 N HCl.

3. React o-cresol-phthalein-complex to cell lysate (Test Combination Calcium; Boehringer Mannheim).
4. Determine the absorbance of the reaction using colorimetric methods at 570 nm.

3.1.1.4 Lineage-Specific Gene Expression

Up-regulation of osteocyte specific genes are detectable in differentiated cells such as osteonectin, osteopontin, and osteocalcin using qRT-PCR on total RNA; oligo primer sequences for the osteocyte genes are listed in **Table 12.1**.

Table 12.1 Some common human osteocyte, adipocyte, and chondrocytes lineage specific genes and primer sequences

	Gene	Forward	Reverse	Product size
Osteogenesis	Osteonectin	5′-TGTGGGAGCT AATCCTGTCC-3′	5′-T CAGGACGTT CTTGAGCCAGT-3′	400 bp
	Osteopontin	5′-GCTCTAGAATGA GAATTGCACTG-3′	5′-GTCAATGGAG TCCTGGCTGT-3′	270 bp
	Osteocalcin	5′-GCTCTAGAATGG CCCTCACACTC-3′	5′-GCGATATCCTA GACCGGGCCGTAG-3′	302 bp
	CBFA-1[a]	5′-CTCACTACCACA CCTACCTG-3′	5′-TCAATATGGTCGC CAAACAGATTC-3′	320 bp
	Collagen I	5′-GAGAGAGAGGC TTCCCTGGT-3′	5′-CACCACGATC ACCACTCTTG-3′	300 bp
	Alkaline phosphatase	5′-TGAAATATG CCCTGGAGC-3′	5′-TCACGTTGTT CCTGTTTAG-3′	475 bp
	Retinoid X receptor α	5′-ACATGGCTT CCTTCACCAAG-3′	5′-CAGCTCAGCC TCCAGGATCC-3′	300 bp
	Vitamin D receptor	5′-CTCGTCCAG CTTCTCCAATC-3′	5′-GCTCCTCCTC ATGCAAGTTC-3′	400 bp
	c-fos	5′-CCTGTCAAG AGCATCAGCAG-3′	5′-GTCAGAGGAA GGCTCATTGC-3′	348 bp
	MSX2	5′-TTACCACAT CCCAGCTCCTC-3′	5′-GCATAGGTTT TGCAGCCATT-3′	201 bp
	Distal-less 5	5′-TTGCCCGAGT CTTCAGCTAC-3′	5′-TCTTTCTCT GGCTGGTTGGT-3′	254 bp
Adipogenic	aP2	5′-TGGTTGATTTT CCATCCCAT-3′	5′-TACTGGGCC AGGAATTTGAT-3′	150 bp
	LPL	5′-GAGATTTCTCT GTATGGCACC-3′	5′-CTGCAAATG AGACACTTTCTC-3′	276 bp
	GLUTA	5′-AGCAGCTCTCT GGCATCAAT-3′	5′-CAATGGAG ACGTAGCACATG-3′	275 bp
	Leptin	5′-GGCTTTGGCCC TATCTTTTC-3′	5′-GCTCTTAGA GAAGGCCAGCA-3′	325 bp
	PPAR γ1	5′-GCTCTAGAATG ACCATGGTTGAC-3′	5′-ATAAGGTGGA GATGCAGGCTC-3′	250 bp
	PPAR γ2	5′-GCTGTTATGGG TGAAACTCTG-3′	5′-ATAAGGTGGAGAT GCAGGTTC-3′	325 bp

(continued)

Table 12.1 (continued)

	Gene	Forward	Reverse	Product size
Chondrogenesis	COL2A1	5'-TTCAGCTATGGA GATGACAATC-3'	5'-AGAGTCCTAGA GTGACTGAG-3'	472 bp
	COL10A1	5'-CACCAGGCATT CCAGGATTCC-3'	5'-AGGTTTGTTGGT CTGATAGCTC-3'	825 bp
	COPM[b]	5'-TGGGCCCGCA GATGCTTC-3'	5'-AGGTTTGTTGGT CTGATAGCTC-3'	474 bp
	Link protein	5'-CCTATGATGAA GCGGTGC-3'	5'-TTGTGCTTGT GGAACCTG-3'	618 bp
	SOX-4	5'-CAAACCAACA ATGCCGAGAAC-3'	5'-CTCTTTTTCTGC GCCGGTTTG-3'	584 bp
	SOX-5	5'-AGCCAGAGTTA GCACAATAGG-3'	5'-CATGATTGCCT TGTATTC-3'	619 bp
	SOX-9	5'-GAACGCACATC AAGACGGAG-3'	5'-TCTCGTTGATTT CGCTGCTC-3'	631 bp
	PTHrP[c]	5'-CTCGGAGCGTGT GAACATTCC-3'	5'-CTTCCGGAAAGT TGATTCCAC-3'	216 bp
	PTHrP-R	5'-AGGAACAGATCT TCCTGCTGCA-3'	5'-TGCATGTGGATGT AGTTGCGCGT-3'	571 bp
	BMP-2	5'-CAGAGACCCA CCCCCAGCA-3'	5'-CTGTTTGTGTTTG GCTTGAC-3'	688 bp
	BMP-6	5'-CTCGGGGTTCA TAAGGTGAA-3'	5'-ACAGCATAACATG GGGCTTC-3'	412 bp

[a] Core binding factor α-1.
[b] Cartilage oligomeric protein.
[c] Parathyroid hormone-related peptide.

3.1.1.5 Lineage-Specific Protein Expression

Lineage specific proteins such as osteonectin are detectable using immuno-histochemistry.

1. Grow and differentiate ASCs on slide chambers.
2. Wash differentiated cells two to three times with PBS
3. Fix cells by incubating them with 1% paraformaldehyde in PBS for 3–5 min.
4. Permeabilize cells with 0.5% Triton X-100 in PBS for 15 min (*see* **Note 6**).
5. Post-fix cells for 10 min in 4% paraformaldehyde in PBS.
6. Block nonspecific binding sites with blocking buffer for 30 min room temperature (*see* **Note 7**).
7. Make a proper dilution of primary antibody (osteonectin, Molecular Probes) in blocking reagent, and apply on the cells for 1 h room temperature.
8. Wash three to five times with PBS at 5-min intervals.
9. Label bound antibody-antigen using conjugated secondary antibody in appropriate dilution for 30 min at room temperature.

10. Wash the cells three to five times with PBS.
11. Detect the intracellular distribution of osteonectin using fluorescent microscopy.

3.2   Adipogenic Differentiation

1. Grow ASCs in maintenance media up to 70–100% confluency (*see* **Note 8**).
2. Induce adipogenesis with differentiation medium containing induction agents such as isobutyl-methylxanthine, indomethacine, insulin, and dexamethasone, for 3 d.
3. Replace induction medium after 3 days with adipocyte maintenance medium (identical to induction medium except for the omission of isobutylmethylxanthine and the PPARγ ligand). ASCs treated with induction medium develop lipid containing intracellular vacuoles. Vacuoles are visible in differentiated cells as early as 5–6 d after induction. Feed the ASCs a minimum of every 3 days with adipocyte maintenance medium for upto 15 days.

3.2.1   Evaluation of Adipogenesis

There are several methods for evaluating the adipogenic differentiation, such as staining of lipid vesicles, lipoxegenase enzyme activity, and adipocyte specific gene and protein regulation.

3.2.1.1   Staining

The most direct method is detection of the intracellular lipid deposition, which can be demonstrated by staining the differentiated ASC monolayers with Oil Red O or Nile Red.

1. Wash differentiated cells 2–3 times with PBS.
2. Fix cells in 1% paraformaldehyde for 10 min at room temperature.
3. Stain cells with 30% Oil Red O in isopropanol for 60 min.
4. Visualize lipid deposition in differentiated cells using light microscopy.

3.2.1.2   Lipogenic Enzyme Activity

Lipogenic enzyme glycerophosphate dehydrogenase activity is assessed according to Wise and Green *(42)*.

1. Grow and differentiate ASCs.
2. Mix up to 100 μg of protein of differentiated and control undifferentiated cells with buffer containing triethanolamine/HCl, EDTA, NADH, dihydroxyacetone phosphate, and β-mercaptoethanol.
3. Determine the absorbance of the mixture with a spectrophotometer at 340 nm.

3.2.1.3   Lineage-Specific Gene and Protein Expression

1. Adipogenic induction also results in lineage specific gene and protein expression. To confirm adipogenesis, up regulation of adipogenic specific genes is examined by qRT-PCR for the expression of aP2, LPL, leptin etc. (list of the adipocyte specific genes and their primer sequences are available in **Table 12.1**).
2. The expression of lineage specific proteins, such as LPL and leptin could be assessed by immunocytochemical staining of committed cells with specific anti-bodies (*see* **Subheading 3.1.1.5.**).

3.3   *Chondrogenic Differentiation*

3.3.1   Preparation of (Adipose-Derived) Cell-Seeded Alginate Beads

1. Aspirate the media off the culture plates.
2. Wash the plate (or flask) twice with D-PBS.
3. Add sufficient trypsin-EDTA to coat the cells and incubate for a maximum of 10 minutes at room temperature.
4. Dilute cell suspension in DMEM containing 10% FBS to inactivate the trypsin.
5. Collect the cell suspension into a centrifuge flask. Centrifuge at $450\,g$ for 10 min at room temperature.
6. Remove the supernatant.
7. Resuspend the cell pellet in 10 mL of DMEM.
8. Count the cells on a hemocytometer and determine their viability using the Trypan blue exclusion method 300g for 5 minutes.
9. Spin the cell suspension at $450\,g$ for 10 min at room temperature.
10. Remove the supernatant and resuspend the cell pellet in 1.2% alginate solution at 4×10^6 cells/ml. Pipette up and down without creating bubbles to mix thoroughly.
11. Draw the cell-seeded alginate suspension into a sterile 10-mL syringe using a 22-gauge needle.
12. Add 3 mL of $CaCl_2$ (102 mM) into one well of a six-well culture plate.
13. Carefully and slowly dispense equal-sized droplets of the cell-seeded alginate solution into the $CaCl_2$ solution (dispense 10–30 beads per well taking care to avoid clumping the beads). Cure the beads in the $CaCl_2$ solution for 10 min at room temperature. Alternatively, the cell suspension may be drawn into a 1ml pipette tip, and larger ~4mm in diameter beads may be formed by pipetting large drops into $CaCl_2$.
14. Aspirate the $CaCl_2$ solution off. Be careful not to suck the beads into your glass pipette.
15. Wash the cell-seeded alginate beads 3 times with NaCl and then one more time with DMEM.

16. Add 3 mL of chondrogenic differentiation medium.
17. Incubate at 37°C, 5.0% CO_2, and 95% relative humidity for the duration of the experiment, changing the feeding media every third day.

3.3.2 Releasing the ASCs From the Alginate Beads

1. Aspirate the media from the plate leaving the cell-seeded alginate beads in the well.
2. Add about 3 mL of sodium chloride (50 mM)/sodium citrate (55 mM) buffer solution.
3. Incubate at room temperature for 20 min. The Ca^{2+} is chelated leading to the gel breakdown thus releasing the cells and their pericellular matrix.
4. Collect the cell suspension into a centrifuge tube. Centrifuge at 450 g for 10 min at room temperature.
5. Remove the supernatant with care to avoid sucking away the beads.
6. Resuspend the cell pellet in the appropriate media for the next step.

3.3.3 Evaluation of Chondrogenesis

3.3.3.1 Quantitation of Sulfated Glycosaminoglycans

After being digested with papain, the sulfated glycosaminoglycan (S-GAG) content of construct digests can be measured using a dimethyl methylene blue (DMB) assay as we have reported previously *(30,33)*. In brief, 125 µL of DMB dye solution is added to 40 µL of sample, and the optical density of the solution is read at 595 nm (GENios microplate reader). The concentration (0–30 µg/mL) of S-GAG is calculated from a standard curve generated with whate chondroitin 4-sulfate (Calbiochem).

3.3.3.2 Quantitation of Hydroxyproline (HOP) Content (A Measure of Total Collagen Production)

OHP content is measured with Ehrlich's reaction assay.

1. Hydrolyze aliquots (100 µL) of construct digests (using papain) in 6N HCl at 110°C for 18 h and then lyophilized.
2. Reconstitute in 200 µL of acetate citrus buffer, then filter through activated charcoal.
3. Add an equal volume of 62 mM chloramine-T to the filtered samples and allow completion of the oxidation reaction at room temperature (15 min).
4. Aliquot the samples in triplicate (× 50 µL) in a 96 well plate with 50 µL of 0.94 M dimethylaminobenzaldehyde (p-DMBA, Sigma) and incubate for 30 min at 37°C to allow for color development.

5. Measure the optical densities of the assayed samples using a plate reader at 540 nm, and calculate the concentration of OHP (0–300 μg/mL) from a standard curve generated with *trans*-4-hydroxy-*L*-proline (Aldrich Chemicals).

3.3.3.3   Radiolabel Incorporation

1. Add [³H]-proline (10 μCi/mL) and of [³⁵S]-sulfate (5 μCi/mL) to the culture medium in order to quantify total protein and S-GAG synthesis, respectively (27; 29; 30; 33; 43).
2. After 24 h of labeling, wash the constructs 5 × 15 min at 4°C in PBS containing non-radioactive proline (1 m*M*) and nonradioactive Na_2SO_4 (0.8 m*M*). Subsequently, digest the constructs in 125 μg/mL papain for 18 h at 60°C.
3. Quantify radiolabel incorporation using a scintillation counter (Packard Instrument) after the addition of 5 mL of Bio-Safe II scintillation mixture (Research Products International), and normalize the resulting data to DNA content.

3.3.3.4   Lineage-Specific Gene and Protein Expression

1. The expression of chondrocyte lineage specific genes such as COL2A1, SOX-4, and SOX-6 is detectable using qRT-PCR on total RNA obtained from differentiated cells (*see* **Table 12.1**).
2. Chondrocyte specific proteins are detectable by immunohistochemistry (IHC); aggrecan and collagen II are indicators of chondrocytes (*see* **Subheading 3.1.1.5** for cells grown directly on slides).
3. For IHC methods involving whole constructs (e.g., ASCs embedded in alginate or agarose), the following method for immunohistochemistry of paraffin embedded constructs may be used:

   a. Fix overnight in a solution of 4% paraformaldehyde, 100 m*M* sodium cacodylate (pH 7.4) at 4°C.
   b. Dehydrate constructs with a series of increasing ethanol concentrations (30 minutes per EtOH concentration to 100%; 100% for 1 h to ensure complete dehydration).
   c. Clear with xylene: 50% xylene, 50% EtOH for 30 min. 100% xylene for 30 min × 2.
   d. Embed in paraffin wax: 50% paraffin (molten), 50% xylene for 60 minutes at 60° C; 100% paraffin for 60 min at 60°C; and replace paraffin completely and incubate for another 60 min at 60°C.
   e. Place in embedding tray and allow paraffin to solidify.
   f. Cut 5- to 10-μm sections using a microtome (Reichert-Jung rotary microtome) and mount on glass slides.
   g. Perform immunohistochemistry on 5-μm sections using monoclonal antibodies to the following proteins (as examples for detecting cartilaginous protein deposition): type I collagen (Sigma), type II collagen (II-II6B3 AB, Developmental

Studies Hybridoma Bank, The University of Iowa, Iowa City, IA), type X collagen (Sigma), and chondroitin 6-sulfate (3B3+ antibody). Use Digest-All (Zymed Laboratories, San Francisco, CA) for pepsin digestion on all sections except those labeled for chondroitin 6-sulfate. Treat sections to be labeled for chondroitin 6-sulfate with trypsin (Sigma), then with soybean trypsin inhibitor (Sigma), and then with chondroitinase (Sigma) to allow the antibody to interact with a chondroitin 6-sulfate epitope (digestion protocols are specific to each antibody). Block sections using undiluted goat serum for 1 hour at room temperature. Apply the primary antibody at its proper dilution and incubate the slides overnight at 4°C. Link the primary antibody to the appropriate secondary antibody for 30 minutes at room temperature. Apply aminoethyl carbazole for 10 min, which effectively binds to the horseradish peroxidase and acts at the enzyme substrate/chromogen. Prepare the appropriate positive controls for each antibody and examine to ensure antibody specificity: cartilage for type II collagen and chondroitin 6-sulfate, deep layer and calcified zone of cartilage for type X collagen, and ligamentous/fibrous tissues for type I collagen. Negative controls utilizing secondary antibodies can also be prepared to rule out non-specific labeling. Finally, apply GVA mounting medium (Zymed) and place coverslips.

3.4 Neurogenic Differentiation

1. Grow ASCs in Stromal Medium up to 70% confluency (*see* **Note 8**).
2. Induce neurogenesis with differentiation media containing induction agents, including retinoic acid, neural growth factor *(44)*, brain-derived neurotrophic factor, and neurotrophin-3 for 10–12 d.
3. Replace induction medium every 2–3 d. ASCs treated with induction medium develop a bipolar morphology, similar to neurons. The cells develop extended cell processes that begin to exhibit branching of smaller processes if cultured for sufficient time.

3.4.1 Evaluation of Neurogenesis

There are several methods for the evaluation of the neurogenic differentiation, such as qRT-PCR for the induction of neural genes and immunohistochemical staining for neural-specific protein expression.

1. Neurogenic induction results in lineage specific gene and protein expression. To confirm neurogenesis: examine induced gene expression for neural-specific genes by qRT-PCR for the expression of Nestin, Neu-N, intermediate filament M, GFAP, S-100, MAP2, and β-III tubulin.
2. The expression of lineage specific proteins such as Nestin, Neu-N, intermediate filament M, GFAP, S-100, MAP2, and β-III tubulin can be assessed by immunocytochemical staining of differentiated cells with specific antibodies (*see* **Subheading 3.1.1.5**).

4 Notes

1. This medium is stable at 4°C for up to 2 wk.
2. Use 37% Oil red O in 70% isopropanol. Oil red O is difficult to dissolve in 70% isopropanol, mix for several hours, and filter it with 0.45-μm filter before use.
3. This medium can be stored for 6 mos at 2–8°C.
4. Make this medium fresh
5. Confluency of cells in culture is important; induction should start when cells reach 50–60% confluency. Since cell replication continues during osteogenesis induction, it is recommended to start the induction with cultures at 50–60% confluency to avoid any over accumulation of cells in culture dishes.
6. Triton X-100 permeabilizes cells and permits the antibody to detect the intracellular antigens.
7. Blocking reagent is 5% goat serum in PBS.
8. Because growth in adipogenic differentiated cells arrests upon commitment of the ASCs, it is important to start the differentiation with large numbers of cells.

References

1. Caplan AI (2005) Review: mesenchymal stem cells: cell-based reconstructive therapy in orthopedics. Tissue Eng 11:1198–1211
2. Pittenger MF, Mackay AM, Beck SC, Jaiswal RK, Douglas R, Mosca JD, Moorman MA, Simonetti DW, Craig S, Marshak DR (1999) Multilineage potential of adult human mesenchymal stem cells. Science 284:143–147
3. Zuk PA, Zhu M, Mizuno H, Huang J, Futrell JW, Katz AJ, Benhaim P, Lorenz HP, Hedrick MH (2001) Multilineage cells from human adipose tissue: implications for cell-based therapies. Tissue Eng 7:211–228
4. Guilak F, Lott KE, Awad HA, Cao Q, Hicok KC, Fermor B, Gimble JM (2006) Clonal analysis of the differentiation potential of human adipose-derived adult stem cells. J Cell Physiol 206:229–237
5. Jackson KA, Mi T, Goodell MA (1999) Hematopoietic potential of stem cells isolated from murine skeletal muscle. Proc Natl Acad Sci U S A 96:14482–14486
6. Eglitis MA, Mezey E (1997) Hematopoietic cells differentiate into both microglia and macroglia in the brains of adult mice. Proc Natl Acad Sci U S A 94:4080–4085
7. Brazelton TR, Rossi FM, Keshet GI, Blau HM (2000) From marrow to brain: expression of neuronal phenotypes in adult mice. Science 290:1775–1779
8. Woodbury D, Schwarz EJ, Prockop DJ, Black IB (2000) Adult rat and human bone marrow stromal cells differentiate into neurons. J Neurosci Res 61:364–370
9. Petersen BE, Bowen WC, Patrene KD, Mars WM, Sullivan AK, Murase N, Boggs SS, Greenberger JS, Goff JP (1999) Bone marrow as a potential source of hepatic oval cells. Science 284:1168–1170
10. Lagasse E, Connors H, Al-Dhalimy M, Reitsma M, Dohse M, Osborne L, Wang X, Finegold M, Weissman IL, Grompe M (2000) Purified hematopoietic stem cells can differentiate into hepatocytes in vivo. Nat Med 6:1229–1234
11. Ferrari G, Cusella-De Angelis G, Coletta M, Paolucci E, Stornaiuolo A, Cossu G, Mavilio F (1998) Muscle regeneration by bone marrow-derived myogenic progenitors. Science 279:1528–1530

12. Gussoni E, Soneoka Y, Strickland CD, Buzney EA, Khan MK, Flint AF, Kunkel LM, Mulligan RC (1999) Dystrophin expression in the mdx mouse restored by stem cell transplantation. Nature 401:390–394
13. Bjornson CR, Rietze RL, Reynolds BA, Magli MC, Vescovi AL (1999) Turning brain into blood: a hematopoietic fate adopted by adult neural stem cells in vivo. Science 283:534–537
14. Vescovi AL, Galli R, Gritti A (2001) The neural stem cells and their transdifferentiation capacity. Biomed Pharmacother 55:201–205
15. Clarke DL, Johansson CB, Wilbertz J, Veress B, Nilsson E, Karlstrom H, Lendahl U, Frisen J (2000) Generalized potential of adult neural stem cells. Science 288:1660–1663
16. Galli R, Borello U, Gritti A, Minasi MG, Bjornson C, Coletta M, Mora M, De Angelis MG, Fiocco R, Cossu G, Vescovi AL (2000) Skeletal myogenic potential of human and mouse neural stem cells. Nat Neurosci 3:986–991
17. Tsai RY, Kittappa R, McKay RD (2002) Plasticity, niches, and the use of stem cells. Dev Cell 2:707–712
18. Yang L, Li S, Hatch H, Ahrens K, Cornelius JG, Petersen BE, Peck AB (2002) In vitro transdifferentiation of adult hepatic stem cells into pancreatic endocrine hormone-producing cells. Proc Natl Acad Sci U S A 99:8078–8083
19. Rodriguez JP, Gonzalez M, Rios S, Cambiazo V (2004) Cytoskeletal organization of human mesenchymal stem cells (MSC) changes during their osteogenic differentiation. J Cell Biochem 93:721–731
20. Heng BC, Cao T, Stanton LW, Robson P, Olsen B (2004) Strategies for directing the differentiation of stem cells into the osteogenic lineage in vitro. J Bone Miner Res 19:1379–1394
21. Halvorsen YD, Franklin D, Bond AL, Hitt DC, Auchter C, Boskey AL, Paschalis EP, Wilkison WO, Gimble JM (2001) Extracellular matrix mineralization and osteoblast gene expression by human adipose tissue-derived stromal cells. Tissue Eng 7:729–741
22. Hauner H, Entenmann G, Wabitsch M, Gaillard D, Ailhaud G, Negrel R, Pfeiffer EF (1989) Promoting effect of glucocorticoids on the differentiation of human adipocyte precursor cells cultured in a chemically defined medium. J Clin Invest 84:1663–1670
23. Izadpanah R, Joswig T, Tsien F, Dufour J, Kirijan JC, Bunnell BA (2005) Characterization of multipotent mesenchymal stem cells from the bone marrow of rhesus macaques. Stem Cells Dev 14:440–451
24. Sen A, Lea-Currie YR, Sujkowska D, Franklin DM, Wilkison WO, Halvorsen YD, Gimble JM (2001) Adipogenic potential of human adipose derived stromal cells from multiple donors is heterogeneous. J Cell Biochem 81:312–319
25. Zuk PA, Zhu M, Ashjian P, De Ugarte DA, Huang JI, Mizuno H, Alfonso ZC, Fraser JK, Benhaim P, Hedrick MH (2002) Human adipose tissue is a source of multipotent stem cells. Mol Biol Cell 13:4279–4295
26. Rosen ED, Spiegelman BM (2000) Molecular regulation of adipogenesis. Annual Review of Cell and Developmental Biology 16:145–171
27. Estes BT, Wu AW, Guilak F (2006) Potent induction of chondrocytic differentiation of human adipose-derived adult stem cells by bone morphogenetic protein 6. Arthritis Rheum 54:1222–1232
28. Estes BT, Wu AW, Storms RW, Guilak F (2006) Extended passaging, but not aldehyde dehydrogenase activity, increases the chondrogenic potential of human adipose-derived adult stem cells. J Cell Physiol 209:987–995
29. Erickson GR, Gimble JM, Franklin DM, Rice HE, Awad H, Guilak F (2002) Chondrogenic potential of adipose tissue-derived stromal cells in vitro and in vivo. Biochem Biophys Res Commun 290:763–769
30. Awad HA, Halvorsen YD, Gimble JM, Guilak F (2003) Effects of transforming growth factor beta1 and dexamethasone on the growth and chondrogenic differentiation of adipose-derived stromal cells. Tissue Eng 9:1301–1312
31. Wickham MQ, Erickson GR, Gimble JM, Vail TP, Guilak F (2003) Multipotent stromal cells derived from the infrapatellar fat pad of the knee. Clin Orthop:196–212

32. Abe T, Miyatake T, Norton WT, Suzuki K (1979) Activities of glycolipid hydrolases in neurons and astroglia from rat and calf brains and in oligodendroglia from calf brain. Brain Res 161:179–182
33. Awad HA, Wickham MQ, Leddy HA, Gimble JM, Guilak F (2004) Chondrogenic differentiation of adipose-derived adult stem cells in agarose, alginate, and gelatin scaffolds. Biomaterials 25:3211–3222
34. Kang SK, Lee DH, Bae YC, Kim HK, Baik SY, Jung JS (2003) Improvement of neurological deficits by intracerebral transplantation of human adipose tissue-derived stromal cells after cerebral ischemia in rats. Exp Neurol 183:355–366
35. Kang SK, Putnam LA, Ylostalo J, Popescu IR, Dufour J, Belousov A, Bunnell BA (2004) Neurogenesis of Rhesus adipose stromal cells. J Cell Sci:jcs.01264
36. Safford KM, Hicok KC, Safford SD, Halvorsen YD, Wilkison WO, Gimble JM, Rice HE (2002) Neurogenic differentiation of murine and human adipose-derived stromal cells. Biochem Biophys Res Commun 294:371–379
37. Ashjian PH, Elbarbary AS, Edmonds B, DeUgarte D, Zhu M, Zuk PA, Lorenz HP, Benhaim P, Hedrick MH (2003) In vitro differentiation of human processed lipoaspirate cells into early neural progenitors. Plast Reconstr Surg 111:1922–1931
38. Safford KM, Safford SD, Gimble JM, Shetty AK, Rice HE (2004) Characterization of neuronal/glial differentiation of murine adipose-derived adult stromal cells. Exp Neurol 187:319–328
39. Erickson GR, Franklin D, Gimble JM, Guilak F (2001) Adipose tissue-derived stromal cells display a chondrogenic phenotype in culture. In 47th Annual Meeting of the Orthopaedic Research Society San Francisco, CA
40. Wickham MQ, Erickson GR, Gimble JM, Vail TP, Guilak F (2003) Multipotent stromal cells derived from the infrapatellar fat pad of the knee. Clin Orthop Relat Res:196–212
41. Martinez J, Silva S, Santibanez JF (1996) Prostate-derived soluble factors block osteoblast differentiation in culture. J Cell Biochem 61:18–25
42. Wise LS, Green H (1979) Participation of one isozyme of cytosolic glycerophosphate dehydrogenase in the adipose conversion of 3T3 cells. J Biol Chem 254:273–275
43. Wang DW, Fermor B, Gimble JM, Awad HA, Guilak F (2005) Influence of oxygen on the proliferation and metabolism of adipose derived adult stem cells. J Cell Physiol 204:184–191
44. Cullingford TE, Bhakoo K, Peuchen S, Dolphin CT, Patel R, Clark JB (1998) Distribution of mRNAs encoding the peroxisome proliferator-activated receptor alpha, beta, and gamma and the retinoid X receptor alpha, beta, and gamma in rat central nervous system. J Neurochem 70:1366–1375

Chapter 13
Methods That Resolve Different Contributions of Clonal Expansion to Adipogenesis in 3T3-L1 and C3H10T1/2 Cells

Colin R. Jefcoate, Suqing Wang, and Xueqing Liu

Summary The mouse embryo fibroblast cell lines 3T3-L1 and C3H10T1/2 differentiate to adipocytes that exhibit similar insulin regulation of lipogenesis. These cell lines, however, differ appreciably in the processes that produce the major regulator PPARγ. Each line is stimulated by a mixture of insulin, dexamethasone, and methylisobutylxanthine (IDM). In the first 24 h, IDM activates each cell type to produce similar regulatory changes and cell contraction. However, the increase in PPARγ is delayed by 24 h in typical 3T3-L1 cells compared with C3H10T1/2 cells. This delay is caused by the need for one or two rounds of cell division (clonal expansion) for PPARγ synthesis in 3T3-L1 cells. This expansion also occurs in C3H10T1/2 cells, but is not needed for PPARγ synthesis and differentiation. Other 3T3-L1 sublines have been described that follow the C3H10T1/2 pattern of differentiation. Culture conditions and inhibitors are described here that remove clonal expansion in C3H10T1/2 cells. With these constraints the cells retain full commitment to differentiation. This distinction is significant because many agents suppress differentiation in 3T3-L1 cells through inhibition of clonal expansion. Other effects on differentiation may be seen in C3H10T1/2 cells that are obscured in 3T3-L1 cells due to this inhibition of proliferation. Human preadipocytes do not need clonal expansion for adipogenesis, thus paralleling C3H10T1/2 cells.

Key words Culture conditions; mouse adipogenesis; clonal expansion; PPARγ; cell contraction; C3H10T1/2; 3T3L1 cell lines.

1 Introduction

1.1 Adipogenesis in Mouse Cells Lines

This chapter focuses on the early stages of adipogenesis in two cell culture models, C3H10T1/2 and 3T3-L1 cells. Each line has been used to dissect the differentiation process but the respective processes differ appreciably (*(1,2)*; *see* **Note 1**). We will

From: *Methods in Molecular Biology, Vol. 456:*
Adipose Tissue Protocols, Second Edition. Edited by: Kaiping Yang
© Humana Press, a part of Springer Science + Business Media, Totowa, NJ

concentrate on the conversion of these cells from the preadipocyte stage to the point where they are committed to insulin-dependent conversion of glucose and fatty acids to triglycerides. The generation of fat cells (mature adipocytes) in vivo starts from mesenchymal stem cells in a fat pad and progresses through generation of preadipocytes to early adipocytes (small lipid droplets) and then to mature adipocytes (enlarged cells with giant lipid droplets). This process may be different in different fat pads (3). Mature adipocytes then undergo apoptosis with subsequent removal by macrophages that are abundant in fat tissue.

In these two cell culture models, most work has been carried out using a hormonal cocktail (IDM) consisting of insulin, the glucocorticoid dexamethasone, and the phosphodiesterase inhibitor methylisobutylxanthine (MIX) (1,2). This hormonal cocktail is likely to produce substantial differences in regulation compared with physiological stimuli. Also of critical importance is the local environment provided by blood vessels and endothelia (4) and by extracellular matrix proteins (5,6).

Specific deletion and overexpression of PPARγ isoforms demonstrate that this nuclear receptor is essential for adipogenesis both in the fat pad and in these cell culture models (7). Two forms, PPARγ1 and PPARγ2, are derived by alternative splicing from the same gene, but differ in their expression through their alternative promoters (8). There is a short additional sequence at the N-terminus of PPARγ2, which provides some differences in co-activator recruitment and gene stimulation. The γ2 isoform is more widely expressed, for example in muscle and cells of the immune system, whereas the γ1 isoform is more selective to fat cells.

Most work on adipogenesis has been carried out with the committed 3T3-L1 preadipocyte line. The C3H10T1/2 (10T1/2) cell line however can differentiate into multiple lineages. This was first recognized after chronic treatment with the hypomethylation agent azacytosine when adipocytes, myocytes, and chondrocytes each appeared in the cultures ((9); see Note 2). More recent work has refined the selective differentiation conditions (10). Many laboratories have now exploited 10T1/2 cells for studies of these various differentiation processes. 10T1/2 cells, like NIH3T3 cells, arise from C3H mouse embryo fibroblasts but are selected through passaging to an immortalized state at higher densities. The more proliferative NIH3T3 cells do not differentiate unless transfected with nuclear factors (C/EBPβ, PPARγ), which are essential for differentiation (11). These transfection methods, particularly when linked to inducible promoters, have become a powerful tool to identify specific contributions of individual factors.

1.2   Sequence of Adipogenesis In Vitro

The standard IDM cocktail that is used for initiation of adipogenesis is usually administered with a change in the serum-containing medium. The same protocol provides effective differentiation of both 3T3-L1 and 10T1/2 cells, although they undergo differentiation to adipocytes at different rates and with different efficiencies.

PPARγ activators such as BRL49653 (BRL, rosiglitazone) or troglitazone enhance the process in 10T1/2 cells and are typically included for these cells. After each further 48 h, insulin is added alone, again with renewal of the serum medium. Adipogenesis is most obviously recognized by the increasing presence of triglyceride lipid droplets. In 10T1/2 cells these appear after 24 h of insulin treatment (*see* **Note 3**). Lipid droplets are seen by staining fixed cells with Oil Red O or, at later stages, directly through Phase contrast light reflectance from the larger lipid droplets.

Despite broad similarities, the processes in these cells are substantially different. We will also draw attention to problems arising from heterogeneity of these cultures. In many respects 10T1/2 cells are closer to primary mouse embryo fibroblasts. The 10T1/2 cells are tetraploid and exhibit chromosomal instability and multiple variant sub-lines. However, even when clonally isolated, the most adipogenic sub-lines commit less to adipogenesis than 3T3-L1 cells and show more dependence on a PPARγ ligand *(12)*. Increased passaging and particularly maintenance of the cells at confluence for over 48 h can cause a progressive suppression of differentiation. We will describe how to follow the progress of differentiation by simultaneously measuring protein and mRNA.

After the addition of IDM, differentiation in each cell type progresses through increases in first C/EBPβ (and c/EBPδ), PPARγ, and then C/EBPα. Several suppression factors decline in this period. Suppression of cyclin D1 may be important since this G1 cyclin attenuates PPARγ activity *(13)*. Suppression of constitutive wnt family glycoproteins is critical since elevation of β-catenin blocks both PPARγ expression and activity *(14,15)*. An increase in chibby, an antagonist to β-catenin also facilitates differentiation in vitro and in vivo *(16)*. These effects are demonstrated in vivo through wnt10b suppression of adipogenesis in vivo. Pref-1 is another early suppressor, which is lowered by dexamethasone *(17)*.

There are appreciable differences between 3T3-L1 and 10T1/2 cells in these responses. In 10T1/2 cells, PPARγ increases more rapidly between 12 and 24 h and peaks at about 36 h. C/EBPβ becomes activated through phosphorylation but expression increases only slightly. C/EBPα rises after PPARγ between 36 and 48 h and is further enhanced by PPARγ ligands (BRL). C/EBPα thus functions in 10T1/2 cells predominantly as a transcription target of PPARγ1 (*see* **Fig. 13.1**). Removal of insulin from the cocktail makes very little difference to these early increases in 10T1/2 cells. The response in 3T3-L1 cells is much slower. PPARγ expression only starts to rise after 48 hours and reaches a peak close to 96 hours (*see* **Fig. 13.2**). C/EBPβ typically increases appreciably and may be responsible for the rise in C/EBPα that precedes the PPARγ increase.

Dexamethasone alone produces an early increase in PPARγ1 in each cell line (Zheng and Jefcoate, unpublished *(18)*) without increasing C/EBPβ, C/EBPα, or adipogenesis. Dexamethasone stimulates C/EBPδ but also removes negative effects of Pref-1 and HDAC-1 *(17,19)*. The addition of the PPARγ ligand troglitazone restores the increase of C/EBPα and lipogenesis. MIX, however, stimulates only CREB and C/EBPβ. These experiments together suggest that C/EBPβ functions primarily to replace troglitazone or BRL. C/EBPβ may therefore stimulate PPARγ1

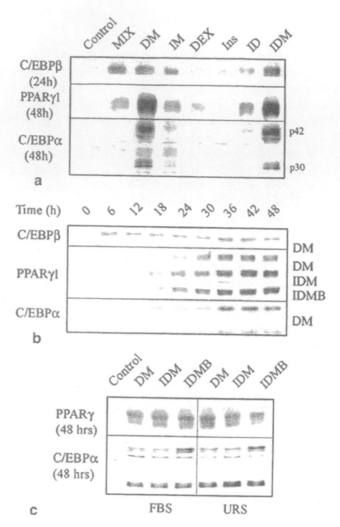

Fig. 13.1 (A) Stimulation of C/EBPβ, C/EBPα and PPARγ in C3H10T1/2 cells by components of the differentiation mixture. (B) Time course, BRL is also added. (C) Stimulation in 10T1/2 cells with serum renewal (FBS) and without renewal (URS)

activity by enhancing either endogenous ligand synthesis or PPARγ co-activators. Recent work has demonstrated that as yet unidentified ligands are generated via stimulation of xanthine oxidase and superoxide radicals (20). The insulin-dependent lipogenesis is mediated by the combined interactive action of PPARγ isoforms and C/EBPα (1). Early PPARγ responses include stimulation of the Glut-4 glucose transporter, CD36 (an essential component of the fatty acid uptake process), and perilipin (a structural component of lipid droplets). Other key markers include the fatty acid binding protein FABP4 (aP2) and various enzymes involved in the synthesis of triglycerides. We have used qRT-PCR to measure these markers (see **Table 13.1**).

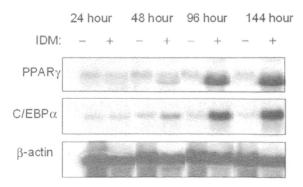

Fig. 13.2 Time course for stimulation of PPARγ and C/EBPα in 3T3L1 cells after the addition of IDM

Table 13.1 qRT-PCR primers for markers of adipogenesis

Gene name	Primer set
PPARγ	Forward ACC CCC TGC TCC AGG AGA T
	Reverse TGC AAT CAA TAG AAG GAA CAC GTT
C/EBPα	Forward CCC CAG TCA GAC CAG AAA GC
	Reverse CCA CAA AGC CCA GAA AC
C/EBPβ	Forward TCG GGA CTT GAT GCA ATC C
	Reverse CGC AGG AAC ATC TTT AAG GTG AT
FABP4	Forward AAG TGG GAG TGG GCT TTG C
	Reverse CCC CAT TTA CGC TGA TGA TCA
CD36	Forward TCC TCT GAC ATT TGC AGG TCT ATC
	Reverse AAA GGC ATT GGC TGG AAG AA
Perilipin	Forward GCT TGA CCA TCA GAA CCA ATT TT
	Reverse GAA TCT GCC CAC GAG AAA GG
ADRP	Forward AGG CCA AAC AAA AGA GCC AGG AGA CCA
	Reverse ACC CTG AAT TTT CTG GTT GGC ACT GTG CAT
Cyclophilin	Forward AGC GTT TTG GGT CCA GGA AT
	Reverse AAA TGC CCG CAA GTC AAA AG

In 10T1/2 cells, addition of a thiazolidenedione ligand for PPARγ increases the percent of cells containing lipid droplets accompanied by increases in C/EBPα and downstream lipogenic markers (*see* **Fig. 13.1 *(12)*)**. In 3T3-L1 cells, this additional stimulation has less affect on these markers, suggesting that endogenous generation of PPARγ ligands is higher.

1.3 Clonal Expansion

Adipogenesis is preceded by a distinctive cell cycle arrest that occurs despite high expression of many proliferative markers *(21)*. The much slower differentiation of 3T3-L1 cells probably arises from the need for an intervening cell division prior to this arrest and the rise in PPARγ synthesis (*see* **Note 4** *(2)*). This mitosis can occur in 10T1/2 cells but is not needed for the stimulation of PPARγ *(22)*. A second type of 3T3-L1 line has been identified that differentiates much faster than these standard 3T3-L1 lines at a similar rate to 10T1/2 cells and without the need for cell division *(23)*. Here we will refer to these standard/conventional 3T3-L1 cells as Type A, and the fast differentiating 3T3-L1 cells as Type B. Typically, publications do not distinguish these types and investigators are recommended to examine the rate of appearance of PPARγ as a guide to this important distinction.

The serum change associated with the IDM addition produces a large transient activation of the MEK/Erk MAP kinase pathway. This is needed for the increases in PPARγ and C/EBPα in 3T3-L1 cells *(24)*. 10T1/2 cells do not require this mitogenic stimulus and, indeed, when this is omitted, PPARγ stimulation occurs even earlier. We refer to this modification as the unrenewed serum protocol (URS). **Figure 13.3** shows the effects of various media on the increase in cell numbers in relation to differentiation. Inhibitors of DNA synthesis completely remove PPARγ expression and adipogenesis in 10T1/2 cells whereas inhibitors of mitosis have no

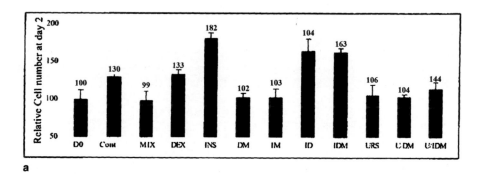

a

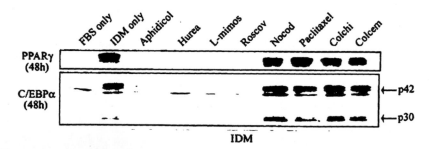

b

Fig. 13.3 Differentiation in C3H10T1/2 cells does not depend on clonal expansion. (**A**) Changes in cell numbers after 48 hrs following various treatments as designated in **Fig. 13.1**. (**B**) Effects of DNA synthesis inhibitors and mitosis inhibitors on stimulations of PPARγ and C/EBPα by IDM

effect *(21,22)*. URS differentiation conditions, even with insulin omitted, modestly stimulate DNA synthesis. The increase in PPARγ is insensitive to MEK inhibitors in 10T1/2 cells, but is inhibited in 3T3-L1 cells. Applications of these various manipulations in 10T1/2 cells provide the means to separate the early steps in differentiation from proliferation.

The linkage between clonal expansion and adipogenesis has led to the hypothesis that mitogenesis is necessary for a reorganization of chromatin prior to activation of PPARγ synthesis by C/EBPβ *(2)*. However, these mitosis-induced changes that precede PPARγ1 activation in type A 3T3-L1 cells seem to be more readily achieved in 10T1/2 cells or in the type B 3T3-L1 sub-lines. A more modest reorganization associated with the need for DNA synthesis and cell cycle advance may be sufficient in 10T1/2 cells. There may also be differences in the activation of C/EBPβ through sequential phosphorylation by Erk and GSK3β that are required in 3T3-L1 cells. The lack of dependence on Erk in 10T1/2 cells suggests less dependence on this activation of C/EBPβ. Human pre-adipocytes appear to mimic 10T1/2 cells in undergoing adipogenesis without the need of mitosis *(19)*.

1.4 Cell Morphology Changes

The expression of genes associated with cell structure, cell adhesion, extracellular matrix, and extracellular proteases changes dramatically in the first 24 h after IDM stimulation of 10T1/2 cells in parallel with the cell rounding *(1)*. By using URS conditions (*see* **Subheading 3.3**) we circumvented many irrelevant serum induced changes. Similar adhesion associated expression changes at this time are seen in 3T3-L1 cells *(25,26)*. We also showed that selective reversal of these adhesion changes reverses not only the PPARγ1 increase but also a set of differentiation linked gene expression changes. Adhesion-associated signaling therefore probably plays a key role in the early PPARγ stimulation. This dependence on decreased adhesion is also seen in 3T3-L1 differentiation. Thus, addition of either fibronectin or type 1 collagen increase adhesion and prevent differentiation *(6,27)*. Cell–cell interactions and cytoskeletal regulation are further suggested by the role of Enc1 in differentiation *(28)*. Metalloproteases, such as ADAM12 and MT-MMP1, play essential roles by modifying the extracellular matrix environment *(29)*. There are large decreases in the TIMP metalloprotease inhibitors during the early stage of adipogenesis and addition of TIMP1 prevents adipogenesis in 3T3-L1 cells. A more complete analysis of adhesion changes has shown that integrin-mediated adhesion changes as adipogenesis progresses *(6)*. **Figure 13.4** shows morphology changes induced by DM in 10T1/2 cells after 24 h. Phalloidin staining reveals the contraction of actin stress fibers whereas paxillin antibodies show the focal adhesion complexes as points of intense staining. These protein clusters are formed by interactions between αβ integrins, focal adhesion kinase, paxillin, and actin *(30)*.

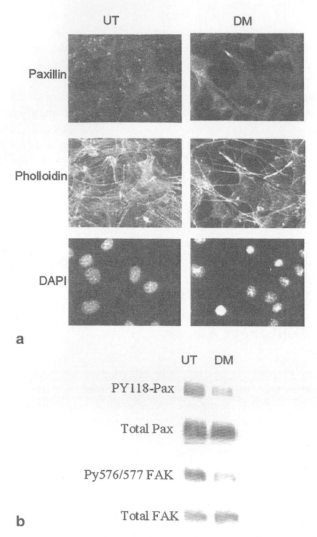

Fig. 13.4 Morphology changes in C3H10T1/2 cells induced after 24 hrs by a Dex/MIX combination. Cells are treated on coverslips, fixed and stained. (**A**) DM removes paxillin foci (white dots) recognized by anti-paxillin antibody and decreases intracellular actin stress fibers displayed by phaloidin. (**B**) Immunoblots show parallel decreases in phosphorylation of paxillin and focal adhesion kinase

1.5   Summary of Changes

In **Fig. 13.5** we summarize the differences between the progression of differentiation in 10T1/2 cells and in 3T3-L1 cells. The two cell lines exhibit similar cell contraction and proliferation responses, but the latter is not necessary for differentiation in 10T1/2 cells. The delay in the expression of PPARγ in 3T3L1 cells may

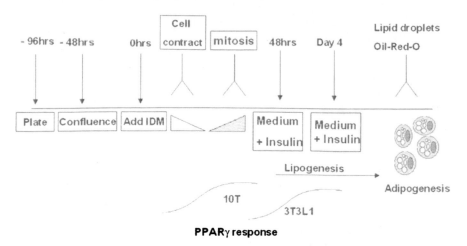

Fig. 13.5 Diagram to represent the sequence of changes produced by adipogenic (IDM) stimulation of pre-adipocytes, Differences in PPAR expression between C3H10T1/2 cells and 3T3L1 cells are highlighted

reflect the requirement for cell proliferation to reorganize chromatin around the PPARγ promoter such that regulatory factors have access. By contrast, 10T1/2 cells only need advance into S phase. The methods described here address ways to resolve these different responses.

2   Materials

2.1   Cells

1. C3H10T1/2 (ATCC).
2. Type A (*see* **Introduction 1.3**) 3T3-L1 cells (ATCC).
3. Type B (*see* **Introduction 1.3**) 3T3-L1 cells (from specific laboratories).

2.2   Cell Culture Medium

1. DMEM-F12 FBS Medium for C3H10T1/2 cells: dissolve Dulbecco Modified Eagle Media/F-12 nutrient mix (DMEM-F12) (Invitrogen) in 800 mL of doubly distilled sterile filtered water (dd H_2O). Add 1.2 g of $NaHCO_3$, mix well, adjust the pH to 7.2–7.3 and bring the volume to 1000 ml with ddH_2O. Remove 100 mL of medium and add fetal bovine serum (FBS; *see* **Note 5**) (final 10%), penicillin (100 unit/mL), and streptomycin (100 μg/mL).

2. DMEM-CBCS Medium for 3T3-L1 cells: prepare DMEM medium as for C3H10T1/2 cells, substituting newborn calf serum (NBCS) for FBS.
3. Cell detachment medium: 0.05% Trypsin-EDTA (Cellgro). Store in aliquot at −20°C. Thawed aliquots are kept at 4°C.
4. Differentiation cocktail stock solutions: insulin (Sigma), 1 mg/mL in 0.01 N HCl; Dexamethasone (Dex, Sigma), 1 mM in 100% ethanol; Methyl Isobutyl Xanthine (MIX, Sigma), 50 mM in 0.25 M KOH; troglitazone (Sigma), 5 mM in DMSO.
5. Adipogenic differentiation medium for 10T1/2 cells: DMEM/F12-FBS supplemented with 10 μg/mL insulin, 1 μM Dex, and 0.5 mM MIX.
6. Differentiation medium for 3T3-L1 cells: DMEM-FBS supplemented with 1 μg/mL insulin, 0.25 μM Dex, and 0.5 mM MIX.
7. Lipogenic medium: DMEM/F12-FBS or DMEM-FBS supplemented with only 10 μg/mL or 1 μg/mL insulin for C3H10T1/2 and 3T3-L1 cells, respectively.

2.3 Protein Preparation and Western Blots

1. RIPA lysis buffer: 50 mM Tris-HCl, pH 7.4, 150 mM NaCl, 1 mM EDTA, 1 mM EGTA, 1 mM Na_3VO_4, 1% NP-40, 0.25% deoxycholate, 0.1% SDS, 1 mM PMSF, 1 μg/mL leupeptin, and 1 μg/mL aprotinin.
2. 4x sample buffer: 250 mM Tris-Cl, pH 8.0; 40% glycerol; 8% sodium dodecyl sulfate (SDS); 0.04% bromophenol blue; and 2% DTT.
3. Stock resolving gel buffer and stock stacking gel buffer are from BioRad.
4. 1x Running buffer: 25 mM Tris base, 190 mM glycine and 0.1% SDS (add last). Do not adjust pH.
5. 1x Transfer buffer: 25 mM Tris base, 190 mM glycine. Do not adjust pH. Add methanol to final 20%.
6. Blocking buffer: TBS containing 5% nonfat milk and 0.05% Tween 20.
7. Mouse monoclonal antibody PPARγ, C/EBPα from Santa Cruz; β-actin from Sigma; Erk, pErk, CyclinD, Paxillin, p-Paxillin, FAK, pFAK from Cell Signaling.
8. Anti-rabbit or anti-mouse Horse Radish Peroxidase conjugated secondary antibodies from Promega.
9. Nitocellulose membranes and ECL detection kit from Amersham Biosciences.

2.4 Oil-Red-O and Nile Red

1. Fixation buffer: 10% formalin (Fisher).
2. Oil-Red-O stock solution: dissolve 500 mg of Oil-Red-O (Sigma) in 100 mL of isopropanol.
3. Nile Red stock solution: dissolve 100 mg of Nile Red in 100 mL of acetone.

2.5 Immunocytochemistry

1. FITC- and Rhodamine-conjugated secondary antibodies and DAPI (Molecular Probes Inc.)

2.6 Fluorescence-Activated Cell Sorting (FACS) Analysis

1. 5-Bromo-2'-deoxyuridine (BrdU; Sigma).
2. Anti-BrdU antibody.
3. FITC-conjugated goat anti-mouse antibody (Becton Dickinson).

3 Methods

3.1 Amplification and Storage of Cells

Passaging of these cell lines introduces variations in differentiation characteristics. The tetraploid C3H10T1/2 cells are prone to chromosomal changes in addition to selection of more proliferative cells after repeated culturing. We recommend freezing multiple tubes of cells at a constant passage number to minimize the variability among different batches.

1. Amplify C3H10T1/2 cells in DMEM-F12/FBS medium at 37°C with 5% CO_2. Culture 3T3-L1 cells with DMEM-NBCS medium at 37°C with 10% CO_2.
2. For subcultures, plate C3H10T1/2 or 3T3-L1 cells in a 75-mL flask.
3. Upon reaching 80% confluence, aspirate medium, then add 2 mL of 0.05% trypsin solution and incubate at room temperature for 2–3 min. Check under an inverted microscope to confirm that cells are dissociated.
4. Add 5 mL of medium containing serum to stop trypsinization, and suspend the cells by vigorous pipetting. Transfer the cells to a sterile tube, and centrifuge at 150 g for 5 min at room temperature.
5. To prepare frozen stocks, resuspend the pellet in culture media containing 5% DMSO at 2×10^6 cells per milliliter. Separate cell suspension into tubes each containing no less than 1×10^6 cells. Freeze tubes in a cryo-protection container overnight at −80°C. Subsequently, the frozen cells are transferred into liquid nitrogen for prolonged storage.
6. To expand the culture, a tube of frozen cells is withdrawn from liquid nitrogen and thawed in 37°C water bath. The cells are cultured at 37°C in 15 mL of DMEM/F12-FBS (C3H10T1/2, 5% CO_2) or DMEM-NBCS (3T3L1, 10% CO_2) medium.
7. Change the medium on the following day and then every other day.
8. For differentiation of 10T1/2 cells, additional passaging of the frozen cells should be limited to no more than 4 passages.

3.2   Standard Protocol for Differentiation of C3H10T1/2 Cells

1. To plate out the cells for differentiation follow **Subheading 3.1.6.**
2. Upon reaching 80% confluence, follow **Subheading 3.1.3 and 3.1.4.**
3. Resuspend the cells in DMEM-F12/FBS medium, count the cells, and adjust to 5×10^4 cells/mL.
4. Change the medium every other day until the cells reach confluence (typically about 48 h). This is defined as 0 h for differentiation (*see* **Fig. 13.5**).
5. After reaching confluence, aspirate the old medium and add adipogenic differentiation medium to the cells containing the IDM cocktail.
6. Examine the cells under a light microscope after 24 h. Contraction of the cells should be readily apparent. The contracted cells that are no longer confluent double in number in the next 24 h to restore confluence.
7. Two days after induction, replace the medium with lipogenic medium. Change medium every other day. Lipid droplets in adipocytes are readily visible in the light microscope after 24 h of lipogenic medium addition, and reach optimum levels after 6–8 d (*see* **Note 7**).
8. The protocol for staining lipid droplets with Oil Red O or Nile Red (for fluorescent microscopy) is described in **Subheading 3.6.**
9. Protein and mRNA sampling is performed on cultures in designated wells after 12 h, 24 h, 48 h, and 6 d (*see* **Subheading 3.5**). The following changes occur sequentially; C/EBP β and δ increase after 6–12 h, cyclin D decreases after 24 h, PPARγ1 increases after 12–36 h, and C/EBPα increases after 36–48 h (*see* **Fig. 13.1** *(22,31)*).

3.3   URS Protocol for Differentiation of C3H10T1/2 Cells

Mitotic stimulation of 10T1/2 cells is unnecessary for differentiation of 10T1/2 cells *(22)*. To minimize the contribution of genes involved in cell proliferation and other expression changes induced by growth factors in serum, the standard protocol can be modified by eliminating the use of fresh medium when the initiating cocktail is added. We refer to this as the unrenewed serum protocol (URS) or more specifically URS/IDM. To further diminish the mitotic stimulation, insulin can also be omitted (URS/DM).

1. The standard protocol (*see* **Subheading 3.2**) is modified by first collecting the old medium and then filtering it to remove the cellular debris.
2. Add of insulin, dexamethasone, and Mix from the stock solutions directly to the filtered medium at equal final concentrations as in the standard adipogenic medium (*see* **Note 8**).
3. The MAP kinase stimulation provided by serum is effectively replaced by EGF (stock solution: $10 \mu M$ in 0.1% BSA; final concentration $10 nM$) and FGF (stock solution: $0.06 \mu M$ in 0.1% BSA; final concentration $60 pM$).

All other steps are the same as for the standard procedure (*see* **Subheading 3.2** *(31)*).

3.4   Differentiation of 3T3-L1 Cells

1. Thaw 3T3-L1 cells from liquid nitrogen and culture in DMEM/NBCS with antibiotics at 37°C in a cell incubator containing 10% CO_2. Grow cells to confluence.
2. Two days post confluence, initiate the adipocyte differentiation by adding 3T3-L1 differentiation medium to the cells. This is very similar to the 10T1/2 medium that also functions well. Two days after induction, replace the hormonal cocktail with the 3T3-L1 cell lipogenic medium containing 1.0 mg/L of insulin alone. Subsequently, replace medium with fresh DMEM/NBCS culture medium in the following 2 d and then every other day thereafter (see **Note 9**).

3.5   Protein Analysis by Western Immunoblotting

1. Aspirate medium and wash cells with cold PBS.
2. Add 100 µL of cell lysis buffer to each well in a 6-well plate, incubate on ice for 15 min and then scrape the cells from the well.
3. Transfer the cell lysate to a fresh 1.5-mL tube, vortex for 1 min, and centrifuge at 12,000 g for 5 min at 4°C.
4. Place supernatant in a fresh 1.5-mL tube and measure the protein concentration with a BCA kit.
5. Mix equal amounts of whole cell lysates with SDS loading buffer and resolve by SDS/PAGE on a 4% stacking and 10% separation polyacrylamide gel. Run the gels at constant voltage 200 v in Tris-gliycine system for 45 min. To obtain optimal separation of PPARγ and C/EBP forms including post translational modifications, load 50–60 µg total cell protein onto a 10% acrylamidc/urca gcl (100:1 acrylamidc/bisacrylamide and 4 M urea in the separating gel). For proteins that are expressed as single forms, the urea can be omitted.
6. Transfer protein from gels to nitrocellulose membranes for 90 min at 100 v in cold room with stirring. Block membranes with blocking buffer and treat overnight in a cold room with antibodies that recognize the target proteins indicated in each figure. Wash the membranes and subsequently treat with appropriate secondary antibodies conjugated to horseradish peroxidase. Visualize the peroxidase activity stain by using an ECL system (Amersham Biosciences) according to the manufacturer's instructions (see **Note 6**).

3.6   Visualization of Lipid Droplets in Differentiated Cells

The lipid droplets in adipocytes can be identified with Oil-Red-O, or Nile Red.

3.6.1   Oil-Red-O Staining

1. Aspirate medium and wash cells once in PBS.
2. Fix cells in 10% formalin at room temperature for 15 min.
3. Wash twice with ddH$_2$O for 2 × 5 min.
4. Stain with Oil-Red-O working solution for 15 min at room temperature.
5. Wash with ddH$_2$O for 3 × 5 min.
6. Rinse cells with 50% isopropanol once at room temperature.
7. Rinse cells with ddH$_2$O.
8. The red stained lipid droplets can be visualized by light microscopy. The adipocytes are also readily visualized in normal microscopy through their rounded morphology and markedly enhanced light reflectance of larger lipid droplets.

3.6.2   Nile Red Staining

1. Aspirate medium and wash cells twice in PBS.
2. Fix cells in 1.5% glutaraldehyde in PBS for 5 min.
3. Rinse with PBS.
4. Add Nile Red solution (100 mg/mL in acetone) to final concentration 100 ng/ml for 15 min at room temperature.
5. Rinse with PBS to remove excess dye.
6. Fluorescence microscopy is carried out by Zeiss AXIOSCOP 20 microscopy with excitation wavelength, 515–560 nm, emission wavelength > 590 nm.

This method broadly follows a previously described protocol *(32)*.

3.7   Analysis of Differentiation by FACS

3.7.1   Cell Cycle

The distribution of cells through the cell cycle can be determined by FACS based on the cellular DNA content determined by propidium iodide staining as previously described. Cells can be examined at various time points after stimulation with IDM or DM, with standard conditions or under URS conditions, and with or without addition of EGF.

1. Confluent cells from 5-cm dishes are trypsinized as previously described (*see* **Subheading 3.1**), collected by centrifugation, and counted with the use of a hemocytometer.
2. Fix cells in 95% ethanol at room temperature and stain with propidium iodide medium overnight (propidium iodide 33 µg/L, 1 mg/mL RNase A, and 0.2% NP40).
3. Analyze cells by flow cytometry using laser excitation at 488 nm. We have used FACS Calibur Dual Laser Benchtop Flow Cytometer (Beckton Dickinson). Determine the DNA content per cell by FACS using a CELL FIT software package.

Typical changes produced after 16 hrs are: FBS untreated Go/G1 78%, S 18%, G2/M 4 %; FBS /IDM Go/G1 78%, S 3%, G2/M 19 %. URS conditions or omission of insulin has little affect. The relationship to cell cycle is further studied by addition of DNA synthesis inhibitors along with IDM. These inhibitors block entry into S phase ($10 \mu M$ aphidicolin, $3 mM$ hydroxy urea, $100 \mu M$ L-mimosine or $25 \mu M$ roscovitin) or the transition from G2 to M. ($2.5 \mu g/mL$ nocodazole, $1 \mu M$ paclitaxel, $100 \mu M$ colchicine, or $1 \mu M$ colcimide). For these experiments standard IDM serum conditions are used. G1/S inhibitors completely block PPARγ1 synthesis whereas G2/M inhibitors had no effect *(22)*.

3.7.2 Adipocyte Analysis

A more quantitative analysis of lipid droplets can be provided by FACS analysis using the fluorescent Nile Red lipid soluble dye *(33,34)*.

1. Wash trypsinized cells with PBS, transfer to a polystyrene round bottom tube and fix with 0.88 % paraformaaldehyde (*see* **Note 10**).
2. Stain the fixed cells by adding Nile Red, diluted 1:100 in PBS from 1 mg/ml stock solution to provide a final concentration of 25 ng/mL.
3. Analyze cells by FACS. 10,000 events are collected using the Nile Red emission (488 nm), collected on a 585/482 band pass filter. Forward scatter and side scatter representation of events is used to exclude cell debris.
4. Identify adipocytes by plotting fluorescence emission versus side scatter.
5. Perform data analysis using Cell Quest software (Becton Dickinson). For each sample, 20,000 events are typically collected.

3.8   Immunocytochemistry

Several aspects of the functions of 3T3L1 and 10T1/2 cells can be readily monitored by selective dye or immunofluorescence staining when differentiation is carried out directly on coverslips. This allows the monitoring of multiple functions in individual cells, including adhesion changes, DNA synthesis, and PPARγ induction *(22,35)*.

3.8.1   General Treatment of Cells for Immunofluorescence

1. Culture cells on coverslips placed within 6-well plates. A similar procedure is used for each endpoint, except for BrdU labeling.
2. At the appropriate time point, fix cells by 4% paraformaldehyde for 15 min at room temperature.
3. Apply $0.15 M$ glycine to the fixed cells for 10 min at room temperature.
4. Disrupt membrane to allow access of the primary antibody with 0.2% Triton X-100 (in PBS) for 10 min at room temperature.

5. Incubate with primary antibodies using 10 times the concentration of antibody used for western blots (*see* **Note 11**).
6. Block cells by incubation with 2% BSA and 3% goat serum in PBS for 1 h at room temperature.
7. Visualize the images. The responses are optimally viewed in the same cells by multiple staining with distinguishable wavelengths for fluorescence activation and detection. This allows resolution of the signals with appropriate filters that can be applied to for each selected endpoint. Imagines were visualized with Zeiss AXIOSCOP 20 microscopes.

The culture conditions are essentially the same as for Western Blot analyses and therefore correlation of the two methods is advisable (*see* **Fig. 13.4**).

3.8.2 PPARγ Expression in Individual Cells

1. Incubate permeabilized cells (as described in **Subheading 3.8.1**) with primary PPARγ antibody (1:500 dilution in 2% BSA).
2. Incubate with the fluorescent labeled secondary antibody (FITC-conjugated antibody) at room temperature for 1 h.
3. Add DAPI (Molecular Probes) at 0.1 µg/mL to stain the nuclei. This provides the means to determine what proportion of cells express PPARγ. This is typically a subset that also contains lipid droplets.
4. Wash cells five times after each step with PBS.
5. Mount the slides with antifade solution (Molecular probes).

3.8.3 Cell Morphology

The loss of cell adhesion that occurs after treatment with the IDM cocktail is visualized by immuno-staining of focal adhesion proteins and actin stress fibers (i.e., paxillin and F-actin). The activated forms of these proteins can also be located with specific anti-phospho peptide antibodies.

1. After blocking, add the paxillin antibody (1:100 dilution) and Alexa Fluor 488 phallotoxin (5 units/mL) to coverslips and incubate for 30 min at room temperature.
2. Rinse, then add Rhodamine Red secondary antibody (1:400 dilution) and DAPI (0.1 µg/mL) and incubate for 30 min at room temperature.
3. Dip in dH$_2$O a few times and mount the coverslips with antifade solution.
4. Visualize the location and structure of paxillin focal adhesion sites as well as F-actin fibers in the cells using a fluorescent microscope equipped with filters for fluorescence from Rhodamine (red), FITC (green), DAPI (blue).

Adipocyte Staining can be carried out with Nile Red staining on these same samples.

3.8.4 DNA Synthesis (BrdU Incorporation)

1. For measurement of DNA synthesis, add BrdU (30 µg/mL) for a 3-h pulse period that represents the beginning of the test period (typically about 12 h after IDM treatment).
2. Replace the labeling medium with preconditioned medium generated from a parallel incubation of similarly treated cells.
3. Fix cells on coverslips with 70% ethanol for 30 min.
4. Denature DNA with 1.5 M HCl and neutralize with borate buffer.
5. Permeabilize cell with 0.05% Triton X in PBS
6. Incubate blocked cells with anti-BrdU primary antibody (1:200 dilution).
7. Stain cells with DAPI (*see* **Subheading 3.8.2.3**).

4   Notes

1. 10T1/2 cells and 3T3-L1 cells provide fundamental mechanistic differences that can be very informative. This particularly applies to the delayed increase of PPARγ isoforms in 3T3-L1 cells. This delay may largely derive from the additional need for clonal expansion to provide access to the PPARγ promoter region (*see* **Fig. 13.5**). Other differences in this period include an increase in C/EBPα prior to PPARγ in 3T3-L1 cells (*see* **Fig. 13.2**). C/EBPα and PPARγ2 (increases similar to PPARγ1 in 3T3-L1 cells) may each be more directly stimulated by C/EBPβ than in 10T1/2 cells because of the delay in PPARγ stimulation. Control of the two lines during the morphology change and during the lipogenic stimulation of committed adipocytes appears similar in the two cell lines. A second difference is the larger dependence on an exogenous ligand (BRL/troglitazone) in 10T1/2 cells. The long delay before activation of PPARγ synthesis in 3T3-L1 may provide more time to generate the endogenous ligand than in 10T1/2 cells. This process appears to involve C/EBPβ and xanthine oxidase.
2. 10T1/2 cells appear to function remarkably similarly to primary mouse embyo fibroblasts. This opens up many opportunities for application of these methods in the use of MEFs derived from transgenic mice. Caution needs to be applied because the MEFs play a key role in development that may lead to adaptive changes in the cell regulation. Comparison with more direct siRNA deletions of the same protein will be instructive. However, the 2- to 3-d suppression periods introduce multiple gene changes, which can seriously impact on these complex differentiation processes.
3. In 10T1/2 cells, many changes that occur in the initial 24 hrs (before PPARγ and C/EBPα) of synthesis correlate closely with eventual adipocyte formation. This relationship emphasizes the importance of the early commitment steps. Such correlations may be less clear in 3T3-L1 cells where the relationship of C/EBPβ, C/EBPα and PPARγ isoforms may be more complex.

4. The requirement for clonal expansion in 3T3-L1 cells, but not in 10T1/2 cells leads to important differences in responsiveness to drugs, xenobiotics and siRNA. Agents that inhibit cell proliferation are likely to inhibit differentiation in 3T3-L1 cells but not in 10T1/2 cells. In addition, treatments of 10T1/2 cells that inhibit PPARγ synthesis may impact 3T3-L1 cells at the level of clonal expansion. In this event, the 10T1/2 cells provide a clearer view of the selectivity of these effects. For example, we have shown that TCDD suppression of PPARγ synthesis in 10T1/2 cells is reversed by inhibitors of MEK/Erk kinase pathway *(31)*. This resolution could not be achieved in 3T3-L1 cells because the MEK/Erk inhibitors typically block clonal expansion. Similarly wnt inhibitory effects on 3T3-L1 cells overlap with effects on clonal expansion. We find that PPARγ expression and activity are both suppressed in 10T1/2 cells under URS conditions and thus can be separated from effects on clonal expansion.

5. Sources of FBS can be critical and should be screened in small samples before purchase of larger lots. We have used Atlanta Biologicals (Norcross GA) as our supplier. Suppliers will often provide test samples of FBS. PPARγ and Oil Red O responses (or other responses of interest) should be tested before purchase of large batches of FBS.

6. PPARγ1 typically appears as three clearly resolved and relatively equal bands, while the larger PPARγ2 protein runs much higher on the gel. C/EBP (α and β) appear as the full-length forms and as much more mobile forms in which the co-activation domains have been cleaved. These forms can act as dominant negative regulators of C/EBP activity.

7. Experiments have been conducted with and without addition of thiazolidenedione (TDZ; for example BRL 0.5 μ*M*), an activator of PPARγ. Addition of TDZ makes little impact on the progress up to PPARγ expression but substantially enhances formation of C/EBPα and later lipogenic steps.

8. This modification, particularly with omission of insulin removes all cell division (*see* **Fig. 13.3**). The serum renewal acutely stimulates the MAP kinase pathway as evident by a dramatic rise in Erk activation through phosphorylation. This is measured by western immunoblots on cell lysates using anti-pErk antibodies relative to total protein measured with Erk antibodies. The p-Erk declines to near basal levels between 20 min and 6 h. Under URS conditions, the initial pErk level is barely detectable.

9. The adipocyte differentiation procedure is largely the same as that described for 10T1/2 cells except that the progress after the contraction proliferation phase is delayed (*see* **Fig. 13.2**). Adipocyte formation occurs in nearly all cells. The addition of TZD to 3T3-L1 cells provides at most a small increase in stimulation and is typically omitted. A sub-line of 3T3-L1 cells (we designate these as Type B; *see* **Subheading 1.3**) has been described, which differentiates appreciably more rapidly and does not require cell division to enter the differentiation pathway *(23)*. In most aspects, these cells function like C3H10T1/2 cells. Like 10T1/2 cells, the induction of PPARγ at 48 hrs or

adipocyte markers is substantially retained by the Dexamethasone/MIX combination without insulin *(23,35)*. The differentiation of 3T3-L1 type B cells under the standard IDM procedure (*see* **Subheading 3.4**) is not sensitive to MEK/Erk inhibitor PD98059 *(23)*, which is similar to the observation for 10T1/2 cells *(31)*.

10. It is advisable not to stimulate until there are substantially large adipocytes since these can fragment during fixation thus leading to an underestimate of adipocytes. This can be important when treatments affect droplet sizes.

11. Different epitopes are exposed in the intact proteins to be probed in fixed cells and therefore different antibodies may be optimal for this method. The suppliers should be checked for the appropriateness of a particular antibody.

References

1. Hanlon PR, Cimafranca MA, Liu X, Cho YC, Jefcoate CR (2005) Microarray analysis of early adipogenesis in C3H10T1/2 cells: cooperative inhibitory effects of growth factors and 2,3,7,8-tetrachlorodibenzo-p-dioxin. Toxicol Appl Pharmacol 207:39–58

2. Tang QQ, Otto TC, Lane MD (2003) Mitotic clonal expansion: a synchronous process required for adipogenesis. Proc Natl Acad Sci U S A 100:44–49

3. Tchtonia T, Giorgadze N, Kirkland JL (2006) Fat depot-specific characteristics are retained in strains derived from single human pre-adipocytes. Diabetes 65:2571–2578

4. Voros G, Maquoi E, Demeulemeester D, Clerx N, Collen D, Lijnen RH (2005) Modulation of Angiogenesis during Adipose Tissue Development in Murine Models of Obesity. Endocrinology 146:4545–4554

5. Nakajima I, Yamaguchi T, Ozutsumi K, Aso H (1998) Adipose tissue extracellular matrix: newly organized by adipocytes during differentiation. Differentiation 63: 193–200

6. Liu J, DeYoung SM, Zhang M, Cheng A, Saltiel AR (2005) Changes in integrin expression during adipocyte differentiation. Cell Metab 2:165–177

7. Mueller E, Drori S, Aiyer A, Yie J, Sarraf P, Chen H, Hauser S, Rosen ED, Ge K, Roeder RG, Spiegelman BM (2002) Genetic analysis of adipogenesis through PPARγ isoforms. J Biol Chem 277:41925–41930

8. Zhu Y, Qi C, Korenberg JR, et al. (1995) Structural organization of mouse PPAR gene: alternative promoter use and different splicing yield two mPPAR;γ isoforms. Proc Natl Acad Sci USA 92:7921–7925

9. Taylor SM, Jones PA (1979) Multiple new phonotypes induced in 10T1/2 and 3T3 Cells Treated with Azacytidine. Cell 17:771–779

10. Sordella R, Jiang W, Chen GC, Curto M, Settleman J (2003) Modulation of Rho GTPase signaling regulates a switch between adipogenesis and myogenesis. Cell 113:147–158

11. Rosen ED Spiegelman BM (2000) Molecular regulation of adipogenesis. Annu Rev Cell Dev Biol 16:145–171

12. Cho Y.C., Zheng WC, Yamamoto M, Jefcoate CR (2005) Differentiation of pluripotent C3H10T1/2 cells rapidly elevates CYP1B1 through a novel process that overcomes a loss of Ah Receptor. Arch Biochem Biophys 439:139–153

13. Wang C, Pattabirraman N, Pestell RG (2003) Cuclin D1 Repression of PPARγ expression and Transacivation. Mol Cell Biol 23:6159–6173

14. Liu J, Farmer SR (2004) Regulating the balance between peroxisome proliferator-activated receptor gamma and beta-catenin signaling during adipogenesis. A glycogen synthase kinase 3beta phosphorylation-defective mutant of beta-catenin inhibits expression of a subset of adipogenic genes. J Biol Chem 279:45020–45027

15. Longo K A., Wright WS, Kang S, Gerin I, Chiang SH, Lucas PC, Opp MR, MacDougald OA, (2004) Wnt10b inhibits development of white and brown adipose tissues. J Biol Chem 279:35503–35509

16. Liu FQ, Singh AM, Mofunanya A, Love D, Terada N, Moon RT (2007) Takemaru KI. Chibby promotes adipocyte differentiation through inhibition of {beta}-catenin signaling. Mol Cell Biol 27:4347–4354

17. Gregoire FM, Smas CM, Sul HS (1998) Understanding adipocyte differentiation. Physiol Rev 78:783–809

18. Hamm JK, Park BH, Farmer SR (2001) A role for C/EBPβ in regulating PPARγ activity during adipogenesis in 3T3L1 cells. J Biol Chem 276:18464–18471

19. Tomlinson JJ, Boureau A, Wu D, Atlas E, Hache RJG (2006) Modulation of early human pre-adipocyte differentiation by glucocorticoids. Endocrinology 147:5284–5293

20. Chung KJ, Tzameli I, Pissios P, Rovira I, Gavrilova IO, Ohtsubo T, Chen Z, Finkel T, Flier JS, and Firedman JM (2007) Xanthine oxidase is a regulator of adipogenesis and PPARγ activity. Cell Metab 5:115–128

21. Reichert M, Eick D (1999) Analysis of cell cycle arrest in adipocyte differentiation. Oncogene. 18:459–456

22. Cho YC, Jefcoate CR (2004) PPARgamma1 synthesis and adipogenesis in C3H10T1/2 cells depends on S-phase progression, but does not require mitotic clonal expansion. J Cell Biochem 91:336–353

23. Qui Z, Wei Y, Chen N, Jiang M, Wu J, Liao K (2001) DNA synthesis and mitotic clonal expansion is not a required step for 3t3-l1 pre-adipocyte differentiation into adipocytes. J Biol Chem 276:11988–11995

24. Prusty D, Park BH, Davis KE, Farmer SR (2002) Activation of MEK/ERK signaling promotes adipogenesis by enhancing peroxisome proliferator-activated receptor gamma (PPARgamma) and C/EBPalpha gene expression during the differentiation of 3T3-L1 preadipocytes. J Biol Chem 277:46226–46232

25. Soukas A, Socci ND, Saatkamp BD, Novelli S, Friedman JM (2001) Distinct transcriptional profiles of adipogenesis in vivo and in vitro. J Biol Chem 276:34167–34174

26. Ross SE, Erickson RL, Gerin I, et al (2002) Microarray analyses during adipogenesis: understanding the effects of Wnt signaling on adipogenesis and the roles of liver X receptor alpha in adipocyte metabolism. Mol Cell Biol 22:5989–5999

27. Spiegelman BM, Farmer SR (1982) Decreases in tubulin and actin gene expression prior to morphological differentiation of 3T3 adipocytes. Cell 29:53–60

28. Zhao L, Gregoire F, Sul HS (2000) Transient induction of ENC-1, a Kelch-related actin-binding protein, is required for adipocyte differentiation. J Biol Chem 275: 16845–16850

29. Lilla J, Stickens D, Werb Z (2002) Metalloproteases and adipogenesis, a weighty subject . Am J Pathol 100:1551–1554

30. Turner CE (2000) Paxillin and focal adhesion signaling. Nat Cell Biol 2:E231–E236

31. Hanlon PR, Ganem LG., Cho YC, Yamamoto M, Jefcoate CR (2003) AhR- and ERK-dependent pathways function synergistically to mediate 2,3,7,8-tetrachlorodibenzo-p-dioxin suppression of peroxisome proliferator-activated receptor-gamma1 expression and subsequent adipocyte differentiation. Toxicol Appl Pharmacol 189:11–27

32. Greenspan P, Mayer EP, Fowler SD (1985) Nile Red: a selective fluorescent stain for intracellular lipid droplets. J Cell Biol 100:965–973

33. Sottile V, Seuwen K (2000) Bone morphogenic Protein-2 stimulates adipogenic differentiation of mesenchymal precursor cells in synergy with BRL-49563 (rosiglitazone) FEBS Lett 475:201–204

34. Cimafranca MA, Hanlon PR, and Jefcoate CR (2004) TCDD administration after the pro-adipogenic differentiation stimulus inhibits PPARγ through a MEK-dependent process but less effectively suppresses adipogenesis. Toxicol Appl Pharmacol 196:156–168

35. Liu X, Jefcoate CR (2006) 2,3,7,8 tetrachlorodibenzo-p-dioxin and epidermal growth factor cooperatively suppress PPARγ1 stimulation and restore focal adhesion complexes during adipogenesis: selective contributions of Src, Rho and Erk distinguish these overlapping processes in C3H10T1/2 cells. Mol Pharmacol 70:1902–1915

Chapter 14
Explant Cultures of White Adipose Tissue

Sébastien Thalmann, Cristiana E. Juge-Aubry, and Christoph A. Meier

Summary Obesity is characterized by increased adiposity of visceral and subcutaneous depots as well as other organs, including the vasculature. These fat depots secrete various hormone-like proteins implicated in metabolic homeostasis (e.g., adiponectin, resistin), the central control of appetite (e.g., leptin) and the increased production of cytokines. These molecules act either in a paracrine or endocrine manner, contributing to the metabolic and cardiovascular complications of obesity. Explant cultures of white adipose tissue are an important step in analyzing the secretory mechanisms of adipose tissue by preserving the physiological in vivo cross-talk between the various types of cells.

Key words Obesity; white adipose tissue; tissue culture; cytokines; chemokines; inflammation.

1 Introduction

In the developed world, obesity and its associated metabolic and cardiovascular complications are increasing to endemic levels and contributing to premature mortality and morbidity (1). Obesity has been associated not only with hypertension and coronary heart disease but it affects all organs, including the respiratory system (e.g., obstructive sleep apnea, hypoventilation), gastrointestinal tract (e.g., gallbladder stones, obesity-associated gastroesophageal reflux disease, colon cancer), metabolic homeostasis (development of type 2 diabetes, dyslipidemia), kidney (hyperfiltration, microalbuminuria), as well as the osteoarticular system (osteoarthritis) (2–6).

Obesity is characterized by the increased accumulation of triglycerides in adipose tissue in subcutaneous and intra-abdominal omental depots but also in fat surrounding organs such as kidney, epicardium, skeletal muscle, and blood vessels (7) as the result of hypertrophy as well as hyperplasia of white adipose tissue (WAT).

From: *Methods in Molecular Biology, Vol. 456:*
Adipose Tissue Protocols, Second Edition. Edited by: Kaiping Yang
© Humana Press, a part of Springer Science + Business Media, Totowa, NJ

In evolutionary terms, WAT has evolved to serve as an energy store in case of food deprivation. Hence, adipose tissue has long been thought to be passive and inert; however, this view has changed considerably during the past decade, because plenty of evidence has emerged that WAT is a highly active organ capable of secreting many dozens, if not hundreds, of different peptides, hormones, and inflammatory mediators *(8)*. It has been clearly established that these proteins affect other tissues, as well as metabolic and inflammatory processes, through local and systemic effects. Leptin, adiponectin, visfatin, resistin, retinol-binding protein-4, and other adipose-derived proteins are implicated in the regulation of food intake and insulin sensitivity *(9–11)*. In addition, WAT secretes cytokines, such as tumor necrosis factor-α, interleukins-6 and –8, as well as anti-inflammatory peptides, such as IL-1Ra and IL-10 *(8)*. WAT also secretes chemoattractive cytokines, called chemokines, such as monocyte chemoattractant protein-1, interleukin-8, interferon-γ-inducible protein, and "regulated on activated normal T-cell expressed and secreted" (i.e., RANTES), all of which also contribute to the attraction of inflammatory cells by adipose tissue and possibly to atherogenesis *(12,13)*. For example, we have shown that perivascular adipose tissue, which can be found around blood vessels, secretes MCP-1 and IL-8 and could therefore contribute to obesity associated atherosclerosis and hypertension *(14)*. Human WAT is also a source of many angiogenic factors, such as fibroblast growth factor, vascular endothelial growth factors, and angiogenin (E. Henrichot and C. A. Meier, unpublished data). These factors contribute to neovascularisation of adipose tissue and appear to provide the nutritive support for the expansion of WAT.

Human WAT is a heterogeneous tissue containing preadipocytes, mature adipocytes, monocytes, macrophages, lymphocytes, endothelial cells, and fibroblasts. Although this organ is heterogeneous, cultures of whole adipose tissue (explant cultures) represent an important step in understanding the physiology and pathophysiology of adipose tissue. Such organ cultures preserve the paracrine signals of this tissue by maintaining the existing cross-talk among the different cell types, allowing the investigation of secretory products and their regulation in a physiological microenvironment. The importance of cross-talk among cells is illustrated by various experimental models of cell co-culture *(15–18)*. However, to investigate the detailed regulatory and secretory mechanisms, the different components of adipose tissue have to be separated, purified, and either cultivated independently or co-cultivated in reconstituted systems. These purification steps (e.g., by treatment with collagenase) may destroy the physiological cross-talk and/or lead to the artificial activation of cells. The purification and culture of the different adipose cell types or components of WAT are described in other chapters of this book.

Besides the purification of cells, immunohistological detection of proteins within intact adipose tissue can assist in the determination of the contribution of various cell types to the production of a specific protein of interest *(14)*. In this chapter, we detail the necessary information to prepare and cultivate human WAT explants.

2   Materials

1. Adipose tissue samples are obtained from patients undergoing surgery (approx. 2–500 g; *see* **Notes 1** and **2**).
2. Phosphate buffer saline (PBS, P+S): 136 mM NaCl, 2.7 mM KCl, 10 mM Na$_2$HPO$_4$/KH$_2$PO$_4$, and 1% penicillin and streptomycin (Invitrogen, cat. no. 15140122).
3. Culture medium: M199 (Invitrogen, cat. no. 22340020), 5% FBS, 1% penicillin and streptomycin (see **step 2**).
4. Dissection instruments: scissors (Metzenbaum of 5 5/4″, Wagner of 4 3/4″) and fine dissecting forceps of 4″.
5. 25-cm × 25-cm squares of 350 μm nylon mesh folded into fourths to allow a conical shape (sterilized).
6. Conical 50-mL sterile plastic tubes (Falcon).
7. 10-cm diameter cell culture Petri dishes and 6-well plates (Falcon or Nunc).
8. Sterile plastic pipets (5 mL and 25 mL).
9. Sterile 2- to 10-mL syringes and 40 × 1.2 18G½ or 50 × 0.8 22G2 hypodermic needles.

3   Methods

The whole procedure is undertaken in a laminar flow hood.

1. Immediately after adipose tissue is removed by surgery, it should be placed in a sterile surgery cloth and transport on wet ice (*see* **Note 3**).
2. Place tissue in a 10-cm diameter cell culture Petri dish and dissect into 1- to 2-cm pieces. Place fragments into conical 50-mL plastic tubes containing prewarmed (37°C) PBS supplemented with P+S (with a volume ratio of tissue to PBS of 1:2 to allow sufficient hydration).
3. Further dissect adipose tissue fragments to remove blood vessels and connective tissue, and place into fresh PBS (P+S) buffer.
4. Mince the fragments with scissors into 5- to 10-mg pieces in approx. 2–3 mL of PBS (P+S) in a 50-mL plastic tube (*see* **Note 4**).
5. Filter minced fragments through a 350-μm nylon mesh and rinse with PBS (P+S) (about 5-time the volume of tissue) to remove blood and free fat released by the rupture of some adipocytes. Follow with a final rinse with culture medium (about twice the volume of tissue).
6. Absorb excess medium on sterile towels and weigh the adipose tissue fragments. Distribute fragments into either 10-cm Petri dishes, 6-well plates, or other type of well or dish depending on the available mass of tissue and the further experimental protocol. Distribute explants at a quantity of 0.3 g of explant per milliliter of culture medium.
7. Incubate the explants overnight in a cell culture incubator (5% CO$_2$ atmosphere).

8. Remove medium the next morning by aspiration with a syringe and 40 × 1.2 18G½ or 50 × 0.8 22G2 gauge-needle, and add fresh culture medium to reach the concentration of 0.3 g tissue/mL. The time of the addition of fresh medium is referred to as time 0 for time-course experiments. Various experimental procedures may start at this stage, e.g. with the addition of appropriate stimuli or control media for up to 72 h (typically 24–48 h).

9. After the appropriate time of incubation, recover supernatant by aspiration with a 2-mL syringe and a 40 × 1.2 18G½ or 50 × 0.8 22G2-gauge needle, and transfer to a 1.5-mL conical plastic tube (e.g., Eppendorf). Centrifuge for 5 min at 1500 g, collect supernatant underneath the cell debris floating on the top of the tube, and store at −80°C for analysis. Alternatively, freeze the explants liquid nitrogen and stored at −80°C for further processing (e.g., RNA or protein extraction).

4 Notes

1. The investigator should be aware of the local laws and rules governing the utilization of human tissue and whether the approval of the local ethical committee and informed consent of the patient are necessary.

2. The amount of fat prepared from abdominoplasty depends on the surface of tissue removed from the patient as well as the amount of adipose tissue underneath the skin and on the quality of the tissue (degree of vascularization, quality of connective tissue). Taking account of these parameters, one should expect to obtain about 10% to 30% of the total weight of abdominoplasty at the end of the preparation of explant.

4. The transport of adipose tissue from the operating room to the laboratory must be accomplished during the 30–60 min after surgery. It is very important to avoid adipose tissue contacting glass surfaces, because this disrupts mature adipocytes.

References

1. Adams KF, Schatzkin A, Harris TB, Kipnis V, Mouw T, Ballard-Barbash R, Hollenbeck A, Leitzmann MF (2006). Overweight, obesity, and mortality in a large prospective cohort of persons 50 to 71 years old. N Engl J Med 355:763–778
2. Sharma AM (2003) Obesity and cardiovascular risk. Growth Horm IGF Res 13(Suppl A): S10–S17
3. Kopelman PG (2000) Obesity as a medical problem. Nature 404:635–643
4. National Heart Lung and Blood institute and National Institutes of Health (1998). Clinical guidelines on the identification, evaluation and treatment of overweight and obesity in adults. The Evidence Report
5. Bray GA (2004) Medical consequences of obesity. J Clin Endocrinol Metab; 89:2583–2589

6. Jacobson BC, Somers SC, Fuchs CS, Kelly CP, Camargo CA Jr. (2006) Body-mass index and symptoms of gastroesophageal reflux in women. N Engl J Med 354:2340–2348

7. Montani JP, Carroll JF, Dwyer TM, Antic V, Yang Z, Dulloo AG (2004) Ectopic fat storage in heart, blood vessels and kidneys in the pathogenesis of cardiovascular diseases. Int J Obes Relat Metab Disord 28(Suppl 4):S58–S65

8. Juge-Aubry CE, Henrichot E, Meier CA (2005) Adipose tissue: a regulator of inflammation. Best Pract Res Clin Endocrinol Metab 19:547–566

9. Graham TE, Yang Q, Bluher M, Hammarstedt A, Ciaraldi TP, Henry RR, Wason CJ, Oberbach A, Jansson PA, Smith U, Kahn BB (2006) Retinol-binding protein 4 and insulin resistance in lean, obese, and diabetic subjects. N Engl J Med 354:2552–2563

10. Banerjee RR, Lazar MA (2003) Resistin: molecular history and prognosis. J Mol Med 81:218–226

11. Fukuhara A, Matsuda M, Nishizawa M, Segawa K, Tanaka M, Kishimoto K, Matsuki Y, Murakami M, Ichisaka T, Murakami H, Watanabe E, Takagi T, Akiyoshi M, Ohtsubo T, Kihara S, Yamashita S, Makishima M, Funahashi T, Yamanaka S, Hiramatsu R, Matsuzawa Y, Shimomura I (2005) Visfatin: a protein secreted by visceral fat that mimics the effects of insulin. Science 307:426–430

12. Sartipy P, Loskutoff DJ (2003) Monocyte chemoattractant protein 1 in obesity and insulin resistance. Proc Natl Acad Sci U S A 100:7265–7270

13. Gerhardt CC, Romero IA, Cancello R, Camoin L, Strosberg AD (2001) Chemokines control fat accumulation and leptin secretion by cultured human adipocytes. Mol Cell Endocrinol 175:81–92

14. Henrichot E, Juge-Aubry CE, Pernin A, Pache JC, Velebit V, Dayer JM, Meda P, Chizzolini C, Meier CA (2005) Production of chemokines by perivascular adipose tissue: a role in the pathogenesis of atherosclerosis? Arterioscler Thromb Vasc Biol 25:2594–2599

15. Dietze D, Koenen M, Rohrig K, Horikoshi H, Hauner H, Eckel J (2002) Impairment of insulin signaling in human skeletal muscle cells by co-culture with human adipocytes. Diabetes 51:2369–2376

16. Suganami T, Nishida J, Ogawa Y (2005) A paracrine loop between adipocytes and macrophages aggravates inflammatory changes: role of free fatty acids and tumor necrosis factor alpha. Arterioscler Thromb Vasc Biol 25:2062–2068

17. Sell H, Dietze-Schroeder D, Kaiser U, Eckel J (2006) Monocyte chemotactic protein-1 is a potential player in the negative cross-talk between adipose tissue and skeletal muscle. Endocrinology 147:2458–2467

18. Wang Z, Lv J, Zhang R, Zhu Y, Zhu D, Sun Y, Zhu J, Han X (2006) Co-culture with fat cells induces cellular insulin resistance in primary hepatocytes. Biochem Biophys Res Commun 345:976–983

Chapter 15
Isolation and Culture of Preadipocytes from Rodent White Adipose Tissue

Dorothy B. Hausman, Hea Jin Park, and Gary J. Hausman

Summary Much of the research devoted to understanding adipose tissue development is currently performed in vitro. Several cell culture models, including preadipocyte cell lines and primary culture of adipose-derived stromal vascular precursor cells, are commonly used to study molecular and cellular events and regulatory influences on preadipocyte proliferation and differentiation. Primary preadipocyte culture systems have several distinct advantages over preadipose cell lines. Because they have not been passaged continuously in culture, primary cultures of adipose derived stromal-vascular (SV) cells more closely reflect the in vivo characteristics of the tissue from which they are derived. In addition, primary cells can be obtained from various adipose tissue depots and from animals at different stages of development, from early postnatal life through advanced age. Cells can also be obtained from genetic rodent models of obesity or from rats and/or mice subjected to nutritional or hormonal manipulation. In each case, specific adipose tissue depots are dissected and the SV cells obtained after collagenase digestion. To examine the effect of tissue source or in vivo or in vitro treatment on preadipocyte proliferation, SV cells are labeled by thymidine incorporation during the exponential growth phase and maintained in culture until sufficiently lipid-filled to allow separation by density. Regulatory influences on various stages of preadipocyte differentiation can be examined in rat SV cultures in a controlled environment featuring chemically defined serum-free medium; whereas, more temperamental mouse SV cultures require the presence of serum for optimal differentiation. Alternatively, preadipocytes differentiated in vitro may be used for examining adipocyte metabolic or secretory responses.

Key words Preadipocyte; proliferation; differentiation; ^{3}H-thymidine incorporation; primary culture; paracrine/autocrine.

1 Introduction

Obesity, a disorder that results from excess adipose tissue mass, is a risk factor for many clinical conditions, including diabetes, hypertension, coronary athero-sclerotic heart disease, and some forms of cancer (1). Expansion of adipose

From: *Methods in Molecular Biology, Vol. 456:*
Adipose Tissue Protocols, Second Edition. Edited by: Kaiping Yang
© Humana Press, a part of Springer Science + Business Media, Totowa, NJ

tissue mass during obesity development occurs by both increases in cell size (adipocyte hypertrophy) and cell number (adipocyte hyperplasia) *(1)*, processes involving the recruitment, proliferation, and differentiation of adipose precursor cells.

Regulation of these processes is often studied in vitro in a variety of model systems, including preadipose cell lines and primary cultures derived from adipose tissue stromal vascular precursor cells from various species *(2–7)*. The cellular and molecular mechanisms and regulatory influences on adipocyte differentiation have been primarily investigated using preadipocyte cell lines, which are easy to use requiring only appropriate signals for terminal differentiation into adipocytes. However, because these cells are already committed to the preadipocyte lineage, they are not appropriate for studying preadipocyte recruitment or proliferation, processes that can be examined using primary cultures derived from adipose tissue stromal vascular precursor cells *(1,8)*.

Primary preadipocyte culture systems have several other distinct advantages over preadipose cell lines. Because they have not been passed continuously in culture, primary cultures of adipose derived stromal-vascular cells more closely reflect the in vivo characteristics of the tissue from which they are derived *(2)*. Primary cultures of adipose tissue stromal vascular precursor cells derived from genetically obese rodent models *(9–12)*, transgenic mice *(13)*, or from rats subjected to nutritional *(14)* or hormonal manipulations *(15–17)* represent unique model systems for delineating intrinsic influences on preadipocyte replication, recruitment, and differentiation. Likewise, cultured primary preadipocytes obtained from various adipose tissue depots *(1,15,16,18,19)* and from animals of different ages *(20,21)* provide important insight into mechanisms governing tissue-specific and age-associated alterations in adipose tissue growth.

White adipose tissue is comprised of mature lipid-containing adipocytes of varying size along with blood vessels, lymph nodes, nerves, and stromal-vascular (SV) cells *(1)*. Adipocyte precursor cells (or preadipocytes) reside within the heterogeneous SV cell population, which also contains fibroblasts, erythrocytes, macrophages, endothelial cells and other cell types. Thus, the initial steps of preadipocyte primary culture involve animal preparation and tissue removal, dissociation of the tissue to component cells by enzymatic digestion and plating of the isolated SV fraction. Attachment of the cells is facilitated by the inclusion of fetal bovine serum in the plating medium *(3)* and growth of differentiated adipocytes is optimized through the use of appropriate plating times, media, and media supplements. Relative rates of proliferation can be determined by monitoring ^{3}H-thymidine incorporation into preadipocyte and stromal-vascular cell fractions. We have used this technique to determine the proliferative capacity of various hormones *(22–24)*, serum factors *(24)*, and adipose tissue conditioned media *(23,25,26)*. Alternatively, the effect of endocrine, paracrine/autocrine, or other regulatory factors may be determined in rat preadipocytes during selected stages of differentiation in chemically defined serum-free media. Fully differentiated rat or mouse primary preadipocytes may be used for examining adipocyte metabolic or secretory responses.

2 Materials

2.1 Animal Preparation

1. Rats or mice (see **Note 1**).
2. CO_2 chamber or ketamine:xylazine (90 mg/mL, Ketaset®, Fort Dodge Laboratories: 10 mg/mL, Xyla-Ject®, Phoenix Pharmaceutical).
3. Nair®, Veet® or other depilatory product.
4. Betadine solution, 10% (Purdue Frederick Company).
5. Ethanol:methanol (50:50).
6. Acrylamide boards (~10" × 12") and rubber bands (or similar anchoring device).

2.2 Tissue Collection and Digestion

1. Dulbecco's Modified Eagle's Medium Nutrient Mixture F-12 Ham (Sigma, cat. no. D8900 or D6421; see **Note 2**) containing penicillin/streptomycin (Sigma, cat. no. P4333; a 100X stock solution, add 10 ml/L culture medium). The antibiotic containing medium will subsequently be referred to as DMEM/F12.
2. HEPES solution (1.25X): dissolve 29.8 g of HEPES (N-2-Hydroxyethylpiperazine-N'-2-ethanesulfonic acid), 8.78 g of NaCl, 4.66 g of KCl, 1.12 g of D-glucose, 18.8 g of bovine serum albumin (BSA) and 0.132 g of $CaCl_2$, anhydrous in 1 liter of distilled water. pH to 7.3 with NaOH, and store in 8-mL aliquots at −20°C. Final concentrations of the reagents are as follows: $0.1 M$ HEPES, $0.12 M$ NaCl, $50 mM$ KCl, $5 mM$ D-glucose, 1.5% BSA, $1 mM$ $CaCl_2$.
3. Collagenase, Type I stock solution: prepare at 5000 units activity/mL distilled water, and stored in 1 mL aliquots at −20°C (see **Note 3**).
4. One sterile Petri dish, one autoclaved 50-mL beaker and one 50-mL plastic Erlenmeyer flask per digest.
5. One 0.22-μm filter (Millex-GV PVDF filter unit, Millipore) and two 10- or 20-mL syringes per digest.
6. Sterile surgical instruments (i.e. Metzenbaum scissors curved on flat 7" and flat sharp/sharp 7", forceps with and without prongs).
7. Glass bead sterilizer (Steri 250; Inotech Biosystems International; see **Note 4**).

2.3 Cell Separation and Counting

1. DMEM/F12 (see **Subheading 2.2., step 1**).
2. Red blood cell (RBC) lysis buffer: $155 mM$ NH_4Cl, $10 mM$ $KHCO_3$, 0.1 mM EDTA; sterile filtered through 0.22-μm filter, and stored in 10-mL aliquots at −20°C.

3. Plating medium: DMEM/F12 + 10% fetal bovine serum (FBS; *see* **Note 5**).
4. Rappaport's stain: dissolve 0.1 g of crystal violet and 1.92 g of citric acid in 100 mL of distilled water (*see* **Note 6**).
5. One set of sterile filter holders (25-mm Swinnex®; Millipore) containing 240- and 20-μm mesh filters (Sefar Filtration) (*see* **Note 7**).
6. Hemacytometer (Neubauer, Bright-line).

2.4   Proliferation of Rat Primary Preadipocytes

1. Plating medium (*see* **Subheading 2.3., step 3**).
2. Serum-free DMEM/F12 medium.
3. Treatment medium: DMEM/F12 containing 0.5 to 5.0% porcine serum, test compounds and 0.3 μCi/ml ^{3}H-thymidine (specific activity 6.7 Ci/mmol; *see* **Note 8**); freshly prepared.
4. Thymidine rinse medium: serum-free DMEM/F12 containing 50 mM thymidine; freshly prepared.
5. Lipid filling (LF) medium: DMEM/F12 containing 10% porcine serum, 10 units/mL heparin (Sigma, cat. no. H3149; *see* **Note 9**), and 10^{-9} M insulin (Sigma, cat. no. I6634; prepared as 10^{-6} M stock and stored frozen in aliquots at −20°C, add 1 mL stock/L lipid filling medium).
6. 12.5-cm^2 (25 mL) tissue culture flasks.

2.4.1   Proliferation Harvest

1. Hank's Balanced Salts Solution (HBSS; Sigma, cat. no. H1387): prepare at 1.2X concentration by mixing one bottle (9.8 g) of HBSS with 800 mL of distilled H$_2$O, add 0.35 g of sodium bicarbonate and pH cold to 7.3.
2. Hanks 1.2X + 0.5% BSA: prepare by slowly dissolving 0.005 g/mL of BSA into Hanks 1.2X (*see* **step 1**). Avoid rapid stirring and excess bubbling. pH cold to 7.3.
3. Enzyme solution: prepare by dissolving 0.5 mg/mL of trypsin (Sigma, cat. no. T8003) and 125 units/mL type I collagenase (use 5000 units/mL stock solution, as described in **Subheading 2.2., step 3**).
4. Hank's Balanced Salts with Phenol Red (10X solution): add one vial (9.8 g) of Hanks with phenol red (Sigma, cat. no. H6136) in 100 mL of distilled water, add 0.35 g of sodium bicarbonate and pH cold to 7.3 (*see* **Note 10**).
5. Percoll solution: add 9 mL of Percoll (Sigma, cat. no. P1644) and 1 mL of 10X Hanks with Phenol Red to 90 mL of Hanks 1.2X + 0.5% BSA.
6. 1% Triton solution: add 1 mL of Triton-X 100 (Sigma) per 100 mL of distilled water.
7. Scintisafe scintillation cocktail (Fisher Scientific).
8. 12 × 75-mm tubes, plastic or siliconized glass (*see* **Note 11**).
9. Scintillation counting vials.
10. Liquid scintillation counter.

2.5 Differentiation of Rat Primary Preadipocytes

1. DMEM/F12 (*see* **Subheading 2.2., step 1**).
2. T_3 (3,3′,5-Triiodo-L-thyronine sodium salt; Sigma, cat. no. T5516) stock solution (20 µg/mL or 30 µM): add 1 mL of sterile 1 N NaOH, swirl gently to dissolve, then add 49 mL of sterile culture medium. Aliquot to sterile micro-storage vials (30 µL) and store at −20°C.
3. ITS stock solution: prepare by reconstituting ITS supplement (25 mg insulin, 25 mg transferrin and 25 µg of sodium selenite; Sigma, cat. no. I1884) with 5 mL of sterile acidified H_2O. Aliquot to sterile micro-storage vials (30–300 µL) and store at −20°C.
4. Serum-free differentiation medium: add 1 µL of T_3 stock soluition and 15 µL of ITS stock solution per 15 mL of DMEM/F12. Final concentration of supplements in DMEM/F12 is: T_3 1.33 ng/mL (2 nM); insulin, 5 µg/mL (872 nM); transferrin, 5 µg/mL (65 µM), and sodium selenite 5 ng/mL (29 nM; *see* **Note 12**).
5. AdipoRed™ Lipid Assay Reagent (Cambrex).
6. Phosphate Buffered Saline (PBS; Sigma, cat. no. D8537).
7. Baker's Formalin:
   10 mL of 37% formaldehyde
   10 mL of 10% calcium chloride aqueous solution
   80 mL of distilled water
   Mix, cool to 4°C for at least 1 h before use. Store at 4°C.
8. Oil Red O stock solution: add 0.7 g of Oil Red O (Sigma, O0625) to 200 mL of isopropyl alcohol.
9. HBSS (*see* **Subheading 2.4.1, step 1**).
10. Harris Hematoxylin (Sigma, cat. no. HHS-16).
11. Glycerol gelatin (Sigma, cat. no. GG-1).
12. 35-mm dishes; 12-, 24-, or 96-well tissue culture plates.
13. Plate Shaker.
14. Fluorometer.

2.6 Differentiation of Mouse Primary Preadipocytes

1. DMEM/F12 + 5% FBS
2. Induction medium: 17 nM insulin (Sigma I6634), 0.1 µM dexamethasone (Sigma D9902), 250 µM 3-Isobutyl-1-methylxanthine (IBMX (Sigma I5879)), and 60 µM indomethacin (Sigma I7378) in DMEM/F12 + 5% FBS. The induction medium is prepared from stock solutions of the individual media supplements (1.7 µM insulin, 10 µM dexamethasone, 25 mM IBMX, and 6 mM indomethacin) which are diluted 100 times. Preparation of the stock solution requires dilution or serial dilution as follow:

   a. Insulin stock:
      9.75 mg insulin + 1 mL of 0.02 M HCl (1.7 mM). Store at −20°C.

Day of media prep: 100 µL of 1.7 mM stock + 900 µL of medium (170 µM)
10 µL of 170 µM stock + 990 µL of medium (1.7 µM)
   b. Dexamethasone stock:
      3.925 mg of dexamethasone + 1 mL of ethanol (10 mM); store at −20°C.
      Day of media prep: 100 µL of 10 mM stock + 900 µL of PBS (1 mM)
      10 µL of 1 mM stock + 990 µL of medium (10 µM)
   c. IBMX stock: 5.56 mg of IBMX + 1 mL of 0.5 N KOH (25 mM) Make fresh
      every time.
   d. Indomethacin: 2.147 mg of indomethacin + 1 mL of DMSO (6 mM)

Caution: Indomethacin is VERY TOXIC and should be handled under the fume
hood using suitable protective clothing and eye and face protection.

3. Insulin containing medium: DMEM/F12 + 10% FBS containing 17 nM insulin.
4. Maintenance medium: DMEM/F12 + 10% FBS.
5. AdipoRed™ Lipid Assay Reagent (Cambrex).
6. Fluorometer.

3   Methods

Inguinal fat pads of young rats or mice typically serve as the cell source for preadipocyte primary culture, though older animals and/or other fat depots may also be
used. The tissue is removed from anesthetized animals in non-survival surgery,
under aseptic conditions. The tissue is minced, digested with collagenase to isolate
the SV cells, composed of preadipocytes and other cell types, and the SV cells are
plated according to assay requirements as detailed in subsequent sections. The plating
medium typically contains up to 10% FBS, which aids in attachment to the culture-ware.
After attachment, the serum content of the medium is either reduced (2) or eliminated
(4,5) to aid in differentiation. Experimental treatments may be applied during
proliferation or during various stages of differentiation.

Because preadipocyte primary cultures represent a heterogeneous population of
cells, it is critical to isolate and study specific subpopulations of cells in these cultures. To that end, we have utilized the ^{3}H-thymidine incorporation assay to determine
the effect of treatment on the proliferation of preadipocytes, one subpopulation of
cells. Though more laborious than the MTT reduction (27) or BrdU incorporation
(28) assays frequently used in 96-well plate format to quantitate cellular proliferation
in various culture systems, the described procedure has the distinct advantage of dis-
tinguishing proliferation of preadipocytes from that of the other SV cells.

In this procedure, cells are labeled by thymidine incorporation during the exponential
growth phase early in culture, and the cells are maintained in culture until the preadipocytes have differentiated and filled sufficiently with lipid to allow separation by density
(2). The measurement of radioactive thymidine in the two cell fractions provides an indication of the rates of DNA synthesis (proliferation) by the preadipocytes and mixed stromal fraction at the time of labeling. Rates of thymidine incorporation on a particular day,

or days, may be compared among treatments. Alternatively, thymidine incorporation from several days can be used to determine growth or doubling rates *(2)*.

Protocols for differentiating preadipocyte precursors to lipid filled adipocytes vary considerably dependent on the species under investigation *(6,7)*. Growth of cells in a controlled environment in chemically defined serum-free medium provides optimal conditions for identifying factors regulating adipose differentiation. Rat preadipocytes can be successfully differentiated in serum-free medium containing specific supplements *(4,5)*. We have used the serum-free differentiation protocol described below (*see* **Subheading 3.5.**) to determine the effect of hormonal and paracrine/autocrine factors, including insulin growth factor-I *(8)*, insulin *(8)*, tumor necrosis factor-α *(29,30)*, and leptin *(24)* on the recruitment, differentiation, and apoptosis of rat preadipocytes. In contrast, mouse preadipocytes are difficult to maintain and differentiate in primary culture under serum-free conditions. They require a more complex protocol (*see* **Subheading 3.6.**) to induce cells to differentiate to lipid containing adipocytes, similar to the murine 3T3-L1 cell line and human primary preadipocytes.

3.1   Animal Preparation

These steps are performed on the bench top, preferably adjacent to a sink. Wear gloves and disposable plastic apron to protect your lab coat from Betadine stains.

1. Anesthetize animals by exposure to CO_2 or by intraperitoneal. injection with ketamine:xylazine (~0.1 mL/100 g of body weight).
2. Liberally apply Nair® or other depilatory solution over the entire lower abdominal area of anesthetized animal, taking care to fully saturate the fur. After 5–7 min, rinse the depilatory solution under warm tap water. If necessary, reapply depilatory.
3. Place two sets of rubber bands approximately 1" apart on acrylamide boards. Liberally apply Betadine disinfectant to acrylamide boards and under the attached rubber bands. Fasten the animal on the board by placing the limbs and tail under the rubber bands. Apply Betadine over depilated area of the animal.
4. After several minutes, rinse off Betadine with the 50:50 ethanol:methanol solution.

3.2   Tissue Collection and Digestion

These steps are performed aseptically in the biosafety cabinet wearing appropriate laboratory clothing and sterile gloves.

1. Fill a sterile Petri dish with prewarmed DMEM/F12 medium (~20 mL).
2. Using sterile instruments, perform a mid-line ventral incision through the animal's skin, taking care not to cut into the peritoneal cavity.

3. Gently separate the skin and attached inguinal fat pad from the body using blunt tip scissors. Locate the end of one white inguinal fat pad and gently grasp it using the pronged forceps. Carefully snip the fat pad from the skin, along its entire length. Tissue can be removed either in toto or in several smaller pieces.

4. Pool tissue from two fat pads per animal in a sterile Petri dish containing DMEM/F12 medium (*see* **Note 13**).

5. When dissection is complete, transfer the tissue to a sterile 30-mL beaker, using sterile forceps.

6. Using a sharp sterile pointed scissors, mince the tissue to a very fine consistency. To avoid contamination by air, be sure to hold the beaker containing the tissue in the center of the biosafety cabinet during mincing. This is a key, but time-consuming step. To optimize subsequent tissue digestion, the finished minced product should be homogeneous and have the approximate consistency of applesauce.

7. Prepare HEPES:collagenase solution by adding two 1 mL aliquots of collagenase / 8 mL aliquot of HEPES (*see* **Note 14**). Attach a 0.22-μm filter to a 10-mL syringe, add the HEPES:collagenase solution to the syringe, and sterile filter the solution into the beaker containing the minced tissue (*see* **Note 15**).

8. Gently swirl the beaker to mix the minced tissue with the HEPES:collagenase solution and carefully transfer to a sterile 50-mL plastic Erlenmeyer digestion flask. Cap the flask and give one vigorous shake to get all the tissue into the collagenase solution.

9. Incubate for 45 min to 1 h in a shaking water bath or rotating shaker set at 37°C, 115 RPM (*see* **Note 16**).

3.3 Cell Separation and Counting and Plating

Steps 8-12 may be conducted on the benchtop, and the remaining steps should be performed inside the biosafety cabinet.

1. When digestion is complete, add 10 mL of DMEM/F12 to the digestion flask and attach a 10- or 20-mL syringe to the filter unit containing the 240-μm mesh filter. Pour the digested fat solution through the filter unit into a sterile 50-mL plastic centrifuge tube. Allow the solution to flow through by gravity, or gently push with the syringe plunger. Discard undigested tissue retained on the filter.

2. Add additional DMEM/F12 to the digested fat solution, to approximately the 30-mL mark on the centrifuge tube and centrifuge at 50 g for 5 min to sediment clumps *(31)*.

3. Remove the infranatant from beneath the floating cell layer, transfer to a sterile 50-mL plastic centrifuge tube, cap and centrifuge at 200 g for 10 min to pellet the SV cells.

4. Remove and discard all except 5 mL of the supernatant and add 10 mL of sterile RBC lysis solution. Cap the tube and gently vortex to resuspend the cell pellet (*see* **Note 17**).

5. After a 5-min incubation at room temperature, add an equal volume of plating medium (DMEM/F12 + 10% FBS). To remove endothelial cell clumps, filter the cell suspension through a sterile 20-μm mesh filter into a sterile 50-mL plastic centrifuge tube.

6. Centrifuge at 200 g for 5 min, remove all but ~2.5 mL of the supernatant, add ~7.5 mL of plating medium and resuspend the pellet by gentle vortexing.

7. Thoroughly mix the cell suspension and remove a small (< 0.1 mL) aliquot with a sterile 1 mL pipet.

8. Transfer 20 μL of the cell suspension aliquot to a small tube and mix with 40 μL of Rappaport's stain (dilution factor will be 3:1 part cells, 2 parts stain). Thoroughly mix cells and the stain.

9. Put the cover slip on the hemocytometer and carefully load the cell:stain mixture to both sides.

10. To assure penetration of stain into cells, allow loaded hemocytometer to rest in a humidified atmosphere for at least 5 min.

11. Count the number of cells in four large squares (outside corners, each comprised of 4 × 4 smaller squares). Cells to be counted should be rounded with intact outer membranes and distinct nuclei, having a "fried egg" appearance.

12. Calculate the cell number as follows:

   a. [cell count per square/number of squares counted] × dilution factor × conversion factor* = cells / ml

   b. *The conversion factor, which accounts for the volume of cell suspension counted within the counting area, is typically 10,000

3.4   Monitoring Preadipocyte Proliferation with ^{3}H-thymidine Incorporation

These steps are performed aseptically in the biosafety cabinet wearing appropriate laboratory clothing and sterile gloves. See **Fig. 15.1** for schematic overview of the procedure.

1. Plate isolated SV cells in 12.5-cm^2 flasks at a density of 4.8×10^3 cells/cm^2. Cells are added to flasks in a 2-mL plating medium containing 3×10^4 cells /mL. Before plating, add the required amount of cells (plus a little extra) to a sterile flask containing a stir bar. Add an appropriate amount of plating medium and mix well both before and throughout the plating process. For greater consistency, aliquot cells to one flask at a time using a two mL disposable pipette. Gently rotate the flask to assure that medium is spread over the entire bottom surface of the flask.

2. Place flasks in a humidified cell culture incubator set at 37°C and 5% CO_2.

3. Approximately 24 h after plating, remove the plating medium and nonadherent cells and replace with serum-free DMEM/F12 medium.

4. After 24 h, remove the serum-free medium and replace with a 2 mL/flask of treatment media (DMEM/F12, serum, test article, and 0.3 μCi/mL ^{3}H-thymidine).

Protocol for Proliferation Assay

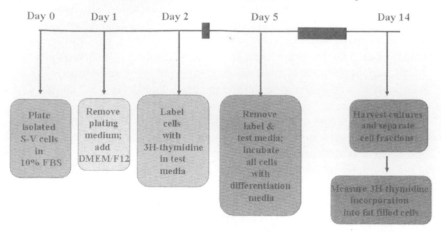

Fig. 15.1 Schematic overview of the ³H-thymidine proliferation assay

Reserve a small aliquot of each treatment for determining the specific activity of the added ³H-thymidine (*see* **Note 18**).

5. Expose the cells to ³H-thymidine from days 2 to 5 of culture, or as appropriate within the exponential growth phase of culture.

6. After the treatment period, remove the test media and apply thymidine rinse medium to all flasks for at least 30 min. Remove the thymidine rinse medium and add 2 mL / flask of lipid filling (LF) medium. **Important**: The ³H-thymidine treatment medium and thymidine rinse medium will contain significant amounts of radioactivity, please dispose of according to local institutional regulations.

7. Change (remove and replace) LF medium every 2–3 days until cells become sufficiently lipid-filled (day 13–15 of culture).

8. Harvest cells by separation of lipid-filled and SV cells through Percoll density gradients (*see* **Subheading 3.4.1.**).

3.4.1 Harvest of ³H-Labeled Preadipocytes

These steps may be performed on the benchtop. Keep all solutions and collection tubes on ice.

1. Bring 10–12 culture flasks at a time to room temperature (~12 min).

2. Remove and discard caps from flasks and aspirate off the medium.

3. Rinse the flasks twice with 2 mL of 1.2X Hank's. Completely aspirate off the last rinse.

4. Add 0.6 mL of the trypsin/collagenase enzyme solution and rotate the flask to assure that the solution completely covers the cell layer on the bottom of the flask.

5. After 12 min of digestion at room temperature, place the flasks on ice.
6. Lift the cells off the flasks by gently vortexing the bottom of the flasks. Examine under the microscope to assure complete removal of the cells.
7. Remove the detached cells with a 1 mL pipet and transfer to a labeled 12- × 75- mm siliconized tube. Rinse the flask by adding 0.8 mL of Hank's 1.2X+0.5% BSA and vortex as above. Remove the rinse to the same tube (above). Discard the flask and pipet.
8. Rinse a 6-inch Pastuer pipet with the Percoll solution and place it in the 12- × 75- mm siliconized tube containing the cells. Add 1 mL of the Percoll solution into the tube through the pipet. Allow the Percoll to drain completely into the tube, then place the pipet into a corresponding labeled scintillation vial.
9. Immediately, centrifuge the tubes containing cells and Percoll solution at 330g and 4°C for 10 min.
10. As soon as centrifugation is complete, use the Percoll-coated pipet to carefully remove the top layer of the gradient containing the floating lipid filled cells from the tube (*see* **Note 19**). Transfer to scintillation vials designated to contain the 'adipose' fraction.
11. Add 1 mL of the Percoll solution to the tube and vortex to resuspend the SV pellet.
12. Centrifuge tubes at 485g and 4°C for 10 min.
13. When centrifugation is complete, pour the supernatant into the corresponding 'adipose' vial and add 9 mL of Scintisafe scintillation cocktail to each vial.
14. To collect the pelleted SV cells from the tube, add 0.75 mL of 1% Triton solution to the tube, vortex to resuspend the pellet, and pour the cell suspension into scintiallation vials designated to contain the 'SV' fraction. Add 1.5 mL of H$_2$O to rinse the tube, vortex; add to the 'SV' vial.
15. Add 9 mLof Scintisafe scintillation cocktail to each 'SV' vial. Wipe all vials and allow them sit in the dark for at least 1 h before counting in a liquid scintillation counter programmed to read ^{3}H at ten minutes per sample.
16. Calculate the rate of ^{3}H-thymidine incorporation in each cell fraction using the counts per minute (CPM) determined above (*see* **Methods 3.4.1., step 15**) and the specific activity of the treatment media (*see* **Methods 3.4., step 4.;** *see* **Note 18**).

3.5   Differentiation of Rat Primary Preadipocytes in Serum-Free Medium

1. Plate isolated SV cells at a density of 1.5 × 10^4 cells/cm^2 in 12-well plates or 35 mm dishes (oil red O staining), 96 well plates (AdipoRed^{TM}) or other culture plates or dishes as required. Cells are added to 12-well plates (1 well = 4.0 cm^2) in 1.0 mL of plating medium containing 6 × 10^4 cells/mL. Before plating, add the required amount of cells (plus a little extra) to a sterile beaker or 10-cm culture dish containing a stir bar. Add an appropriate amount of plating medium and mix well both before and throughout the plating process. For greater consistency,

aliquot cells to one well at a time using a sterile 1 mL pipet. Gently rotate the dish to assure that medium is spread over the entire bottom surface of the wells.

2. One day after plating, remove all plating medium and replace with serum-free differentiation media containing positive or negative controls (*see* **Note 20**) or experimental treatments.

3. Replace media every 2 to 3 d by removing 70% of the spent media and replacing with fresh media. **Caution:** Do not remove all of the media from lipid containing cells, as they are very delicate and will burst if exposed to air. After eight to ten days, the cells will be lipid filled and ready for end-point assay (*see* **Fig. 15.2**).

4. To determine the effect of treatment on lipid deposition in a simple high-throughput manner, cells can be grown in 96-well plates and intracellular triglyceride accumulation quantified using the AdipoRed™ Assay Reagent (*see* **Subheading 2.5., step 5**):

   a. Before the assay, remove treatment media (*see* **Note 21**), rinse one time with 200 µL per well of PBS, and add an additional 200 µL of PBS to all wells, including blanks (no cells).

   b. Place assay plate on a plate shaker, add 5 µL of AdipoRed™ Assay Reagent to each well using a multichannel pipetter and mix briefly after each addition.

   c. Hold plate in the dark at room temperature for 15–20 min before reading on a standard microplate fluorometer at 572 nm.

5. To provide a visual record of the effect of treatment on lipid deposition or to determine effects of treatment on adipose cell size, cells can be grown in 12-well or 35 mm dishes and subsequently stained for lipid using the oil red O staining procedure:

   a. On the day of staining, make a 60% Oil Red O solution with water (i.e., 30 mL of Oil Red O + 20 mL of dH$_2$O); let stand for 15 min; filter through Whatman filter paper. Filter Harris Hematoxylin through Whatman filter paper.

   b. Remove treatment media and rinse dishes 2 to 3 times with HBSS, blot excess. **Caution:** Take care that dishes do not dry-out at any time during the staining procedure; this will adversely affect staining.

   c. Fix dishes with Baker's Formalin (2 mL/35 mm dish) for 30 min at 4°C.

   d. Remove the formalin to a waste container; blot residual liquid onto absorbent paper.

   e. Apply the 60% Oil Red O solution to dishes (2 mL/35 mm dish); leave on for 10 min.

   f. Remove the stain to a waste container and rinse dishes in running water until the water runs clear (~3–5 min; *see* **Note 22**).

   g. Blot the excess water and counterstain the cultured cells, if desired, with filtered hematoxylin (2 mL/35 mm dish) for 2 min.

   h. Remove the hematoxylin (can be reused up to three times) and rinse dishes in running water until the water runs clear (~5 min).

   i. Drain and blot the excess water.

   j. Mount cover-slips with two to three drops of warm glycerol gelatin. Dishes are ready for preparing photomicrographs or for image analysis.

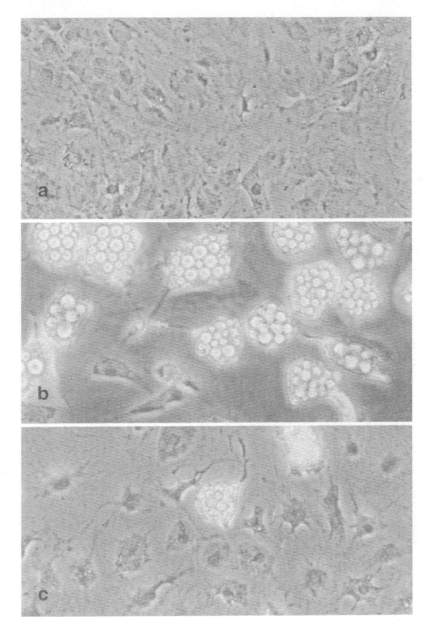

Fig. 15.2 Micrographs of cultured rat primary preadipocytes in the undifferentiated state (**A**; maintained in DMEM/F12 + 5% serum) or differentiated for 10 d in serum-free DMEM/F12 medium containing 850 nM ITS and T$_3$ (**B**) or in serum-free DMEM/F12 medium containing 850 nM ITS and T$_3$ + 0.2 nM tumor necrosis factor-α (**C**). Note the inhibitory effect of tumor necrosis factor-α on preadipocyte differentiation

3.6 Differentiation of Mouse Primary Preadipocytes

1. Plate isolated SV cells at a density of 2.5×10^4 cells/cm^2 in 96-well plates (AdipoRed™), 12-well plates or 35-mm dishes (oil red O staining) or other culture plates or dishes as needed. Cells are added to 96-well plates (1 well = 0.32 cm^2) in 0.2 mL of plating medium containing 4×10^4 cells/mL. Before plating, add the required amount of cells (plus a little extra) to a sterile beaker or 10-cm culture dish containing a stir bar. Add an appropriate amount of plating medium and mix well both before and throughout the plating process. Using a multichannel pipetter, aliquot cells to the culture plates by gently adding the cell suspension to the side of the wells. Gently rotate the dish to assure that medium is spread over the entire bottom surface of the wells.
2. One day after plating, remove all plating medium and replace with DMEM/F12 containing 5% FBS.
3. Replace the medium every two days until the cells reach confluency (typically 5–6 d).
4. When the cells reach confluency, differentiation is induced by the addition of induction medium (see **Materials 2.6.**; see **Note 23**).
5. After 48 h, the induction medium is replaced with insulin containing medium.
6. After 48 h, the insulin containing medium is removed and the cultures fed every other day with maintenance medium. Cells are fed by removing 70% of the spent medium and replacing with fresh medium. **Caution:** Do not remove all of the media from lipid containing cells. They are very delicate and will burst if exposed to air. After 6–8 d, the cells will be lipid filled and ready for use.
7. To quantitate or visualize the effect of treatment on lipid accumulation, perform the AdipoRed™ and/or Oil Red O procedure as described previously (see **Subheading 3.5.**).

4 Notes

1. To reduce inter-culture variability, use animals of a similar age and weight for an experimental series. For rat preadipocyte cultures, we typically order 35- to 49-g (3- to 4-wk-old) Sprague–Dawley rats and allow them to acclimate to the facility for approximately one week before use. Preadipocytes from younger or older animals may be cultured dependent on experimental objectives; however, preadipocytes derived from older animals will have a reduced differentiation capacity (21) and are more susceptible to lipotoxicity (32) compared with cells derived from younger animals.
2. We have purchased DMEM/F12 either as a powdered formulation (Sigma, Cat# D8900) or as a pre-made liquid solution (Sigma, cat. no. D6421). The powdered DMEM/F12 is reconstituted following manufacturer's instructions

using water that has a resistivity of 18.2 MΩ-cm. The liquid formulation requires supplementation with 0.365 g/L of L-glutamine (i.e., 1:80 dilution of 200 mM L-glutamine stock solution; Sigma, cat. no. G7513). Store medium in the dark at 2–8°C. Use within one month of reconstitution or L-glutamine addition.

3. There is considerable lot-to-lot variation in the activity and performance of crude enzyme preparations such as the type I collagenase used in this protocol. It is advisable to pre-test the particular lot of enzyme you plan to use. Many vendors will provide samples of different lots of collagenase for in-lab evaluation. Once a satisfactory lot is identified it should be purchased in bulk and used for all cultures within an experimental series.

4. Alternatively, a disinfectant solution (ethanol, iodine, or other) can be used to hold instruments during use.

5. There is considerable variation in cell culture performance associated with different sources and lots of FBS. We typically test several lots to determine that which optimizes plating efficiency or gives results most consistent with our previous experiments prior to purchasing in bulk and storing at −20°C.

6. For cell counting, we use Rappaport's stain, which delineates the cell membrane and nuclei. Alternatively, or additionally, an exclusion stain such as a 0.4% solution of trypan blue (Sigma, cat. no. T8154) may be used to verify the number of viable cells.

7. The filter units are prepared by cutting 1" circles of 240- and 20-μm nylon mesh, placing the mesh within the individual filter units and assembling the two units within a package for autoclaving, with the unit containing the 240-μm mesh positioned for first access.

8. ^{3}H-thymidine decomposes slowly during storage (2–4% decomposition per month at 2°C) and should be discarded or analyzed for radiochemical purity if used more than 2 mo after receipt.

9. Heparin is sold in unit quantities of 25,000 to 1,000,000 units. Activity, which is specified to be ≥ 140 USP units/mg, varies from lot to lot. The amount of heparin required for each batch of LF medium is determined as follows:

volume of medium (in mL) × 10 (units/mL) / heparin activity
(units/mg) listed on the bottle = heparin amount (mg)

10. HBSS + Phenol Red, 10 X may be stored at 2–8°C for several months. Discard if a precipitate forms.

11. To prevent premature breakage of lipid-filled adipocytes, glass tubes used for the proliferation harvest must be pre-treated with a protective coating of siliconizing solution (i.e. Sigmacote®; Sigma, cat. no. SL-2, or 10% dimethyldichlorosilane [Sigma, cat. no. 440272] in toluene). Thoroughly coat the interior surface of the tube with the siliconizing solution, drain the excess (can be reused), and allow the tubes to air-dry overnight. Wash and rinse the

tubes repeatedly with tap water, then deionized water and dry thoroughly before use.

12. The supplements added to the serum-free differentiation medium are based on the simple serum-free differentiation medium for rat adipose precursor cells originally defined by Deslex et al. *(4)*. We use the ITS supplement mixture for convenience. Alternatively, insulin and transferrin can be added as individual components and/or the concentration of individual components may be varied to obtain a lower baseline differentiation rate.

13. The procedure, as outlined, is for use with rats weighing 80–100 g at tissue collection, which yield approximately one gram of inguinal fat per rat. If using larger or smaller animals, the number pooled should be adjusted accordingly.

14. As noted above, the procedure assumes approximately two grams of minced fat per digest. Adjust the amount of HEPES:collagenase solution accordingly if using larger or smaller pools of rat or mouse tissue.

15. Do not hold the syringe with attached 0.22-μm filter over the beaker of minced tissue during addition of HEPES:collagenase solution to the syringe barrel. Spillage of non-sterile solution into the beaker during the transfer process is a potential source of contamination.

16. To assure more complete digestion of fat tissue and/or increase the yield of preadipocytes a longer digestion time, up to 120 min, may be required. Alternatively, type I elastase (0.3 mg/mL; Sigma, cat. no. E1250) and/or type I hylauronidase (0.5 mg/mL; Sigma, cat. no. H3506) may be added to aid tissue dissociation *(2)*. We have generally found the additional enzymes to be unnecessary when using tissue from young animals containing minimal connective tissue.

17. To avoid discarding loosely pelleted SV cells, do not remove all of the supernant during this and in subsequent steps.

18. A specific activity calculation (fmol/dpm) is required to determine the amount of thymidine (fmol) represented by each unit of radioactivity (dpm). Thus, duplicate 50- to 100-μL aliquots of radiolabeled test media are added to Scintisafe scintillation cocktail and counted on a liquid scintillation counter programmed to detect ^3H. Radioactivity of the counted sample (in counts per minute, CPM) is converted to disintegrations per minute (DPM) based on the counting efficiency of the instrument, and total DPM per liter is subsequently determined. This value is divided by 1.5, the μmolar concentration of thymidine in DMEM/F12, to obtain DPM/μmole. The number is inverted (1/DPM per μmole) to obtain μmole/DPM and subsequently multiplied by 10^9 to obtain fmol thymidine represented by each DPM of radioactivity. This value is used in conjunction with radioactivity counts determined at proliferation harvest to determine ^3H thymidine incorporation in each sample.

19. For best results, remove Percoll gradients containing lipid-filled cells immediately after centrifugation. Remove the top layer of the gradient containing the lipid-filled cells with the pre-coated Pastuer pipet and attached rubber bulb. Be

careful not to disturb the gradient during transfer. The residual portion of the gradient (containing Hank's plus phenol red and Percoll) of approximately 1 ml will be light pink. Viewing against a white background will assist in separation.

20. Tumor necrosis factor-α (0.1 – 0.2 nM *(29,33)*) or transforming growth factor-β (0.1 nM *(33)*) typically serve as negative controls for preadipocyte differentiation; whereas carbaprostacyclin (cPGI$_2$, 200 nM *(33)*) may serve as a positive control.

21. Although not specified by manufacturer's instructions, we typically include one 200 µL PBS rinse following removal of treatment media and prior to AdipoRed™ assay, to achieve more consistent results. To avoid breaking the lipid filled cells, use the plate cover to minimize exposure to air and remove and replace media from only two or three rows of cells at a time during the rinse process.

22. To optimize staining, it is important that the dishes be retained in running water for the specified periods of time (or longer).

23. To determine effects on preadipocyte differentiation, include varying doses of test compounds in treatment media during the early differentiation phase (day 0–2), late differentiation (day 4–6 or 4–8) or throughout the differentiation process.

References

1. Hausman DB, DiGirolamo M, Bartness TJ, Hausman GJ, Martin RJ (2001) The biology of white adipocyte proliferation. Obes Rev 2:239–254
2. Novakofski JE (1987) Primary culture of adipose tissue. In: Hausman GJ, Martin R (eds): Biology of the adipocyte: research approaches. Van Nostrand Reinhold, New York, New York
3. Ramsay TG, Rao SV, Wolverton CK (1992) In vitro systems for the analysis of the development of adipose tissue in domestic animals. J Nutr 122:806–817
4. Deslex S, Negrel R, Ailhaud G (1987) Development of a chemically defined serum-free medium for differentiation of rat adipose precursor cells. Exp Cell Res 168:15–30
5. Serrero G, Mills D (1987) Differentiation of newborn rat adipocyte precursors in defined serum-free medium. In Vitro Cellular Dev Biol 23:63–66
6. Gregoire F, Smas C, Sul HS (1998) Understanding adipocyte differentiation. Physiol Rev 78:783–809
7. Gregoire F (2001) Adipocyte differentiation: from fibroblast to endocrine cell. Exp Biol Med 226:997–1002
8. Kras KM, Hausman DB, Hausman GJ, Martin RJ (1999) Adipocyte development is dependent upon stem cell recruitment and proliferation of preadipocytes. Obes Res 7:491–497
9. Landron D, Dugail I, Ardouin B, Quignard-Boulange A, Postel-Vinay MC (1987) Growth hormone binding to cultured preadipocytes from obese fa/fa rats increases during cell differentiation. Horm Metab Res 19:403–406
10. Lau DC, Shillabeer G, Wong KL, Tough SC, Russell JC (1990) Influence of paracrine factors on preadipocyte replication and differentiation. Int J Obes 14:193–201
11. Shillabeer G, Forden IM, Russell JC, Lau DC (1990) Paradoxically slow preadipocyte replication and differentiation in corpulent rats. Am J Physiol 258:E368–E376

12. Gregoire FM, Johnson PR, Greenwood MR (1995) Comparison of the adipoconversion of preadipocytes derived from lean and obese Zucker rats in serum-free cultures. Int J Obes Relat Metab Disord 19:664–670

13. Simon MF, Daviaud D, Pradere JP, Gres S, Guigne C, Watitsch M, Chun J, Valet P, Saulnier-Blache JS (2005) Lysophosphatidic acid inhibits adipocyte differentiation via lysophosphatidic acid 1 receptor-dependent down-regulation of peroxisome proliferator-activated receptor gamma2. J Biol Chem 280:14656–14662

14. Shillabeer G, Lau DC (1994) Regulation of new fat cell formation in rats: the role of dietary fats. J Lipid Res 35:592–600

15. Lacasa D, Garcia E, Henriot D, Agli B, Giudicelli Y (1997) Site-related specificities of the control by androgenic status of adipogenesis and mitogen-activated protein kinase cascade/c-fos signaling pathways in rat preadipocytes. Endocrinology 138:3181–3186

16. Garcia E, Lacasa M, Agli B, Giudicelli Y, Lacasa D (1999) Modulation of rat preadipocyte adipose conversion by androgenic status: involvement of C/EBPs transcription factors. J Endocrinol 161:89–97

17. Lacasa D, Garcia Dos Santos E, Giudicelli Y (2001) Site-specific control of rat preadipocte adipose conversion by ovarian status: possible involvement of CCAAT/enhancer-binding protein transcription factors. Endocrine 15:103–110

18. Kirkland IM, Hollenberg CH, Gillon WS (1996) Effects of fat depot site on differentiation-dependent gene expression in rat preadipocytes. Int J Obes Relat Metab Disord 20: S102–S107

19. Gesta S, Bluher M, Yamamoto Y, Norris AW, Berndt J, Kralish S, Boucher J, Lewis C, Kahn CR (2006) Evidence for a role of developmental genes in the origin of obesity and body fat distribution. Proc Natl Acad Sci U S A 103:6676–6681

20. Kirkland IM, Hollenberg CH, Gillon WS (1993) Ageing, differentiation, and gene expression in rat epididymal preadipocytes. Biochem Cell Biol 71:556–561

21. Kirkland IM, Tchkonia T, Pirtskhalava T, Han J, Karagiannides I (2002) Adipogenesis and aging: does aging make fat go MAD? Exp Geronotol 37:757–767

22. Jones DD, Ramsay TG, Hausman GJ, Martin RJ (1992) Norepinephrine inhibits rat pre-adipocyte proliferation. Int J Obes Relat Metab Disord 16:340–354

23. Marques BM, Hausman DB, Latimer AM, Kras KM, Grossman BM, Martin RJ (2000) Insulin-like growth factor mediates high-fat diet-induced adipogenesis in Osborne-Mendel rats. Am J Physiol Integr Comp Physiol 278:R654–R662

24. Wagoner B, Hausman DB, Harris RBS (2006) Direct and indirect effects of leptin on preadipocyte proliferation and differentiation. Am J Physiol Regul Integr Comp Physiol 290: R1557–R1564

25. Marques BG, Hausman DB, Martin RJ (1998) Association of fat cell size and paracrine growth factors in development of hyperplastic obesity. Am J Physiol 275:R1898–1908

26. Hausman DB, Bartness TJ, DiGirolamo M, Fine JJ, Plunkett S, Martin RJ (2005) Proliferative activity of adipose tissue conditioned media correlates with fat cell size in animal models of obesity. Adipocytes 1:25–34

27. CellTiter 96® AQ$_{ueous}$ One Solution Cell Proliferation Assay (2005) Promega Corporation. Madison, WI

28. Cell Proliferation ELISA, BrdU (2006) Roche Diagnostics Corporation. Indianapolis, IN

29. Kras KM, Hausman DB, Martin RJ (2000) Tumor necrosis factor-a stimulates cell proliferation in adipose tissue-derived stromal-vascular cell culture: promotion of adipose tissue expansion by paracrine growth factors. Obes Res 8:186–193

30. Qian H, Hausman DB, Compton MM, Martin RJ, Della-Fera MA, Hartzell DL, Baile CA (2001) TNFalpha induces and insulin inhibits caspase 3-dependent adipocyte apoptosis. Biochem Biophys Res Commun 284:1176–1183

31. Yan H, Aziz E, Shillabeer G, Wong A, Shanghavi D, Kermouni A, Abdel-Hafez M, Lau DC (2002) Nitric oxide promotes differentiation of rat white preadipocytes in culture. J Lipid Res 43:2123–2129

32. Guo W, Pirtskhalava T, Tchkonia T, Xie W, Thomou T, Han J, Wang T, Wong S, Cartwright A, Hegardt FG, Corkey BE, Kirkland JL (2007) Aging results in paradoxical susceptibility of fat cell progenitors to lipotoxicity. Am J Physiol Endocrinol Metab 292:E1041–E1051
33. Vassaux G, Negrel R, Ailhaud G, Gaillard D (1994) Proliferation and differentiation of rat adipose precursor cells in chemically defined medium: differential action of anti-adipogenic agents. J Cell Physiol 161:249–256

Chapter 16
Flow Cytometry on the Stromal-Vascular Fraction of White Adipose Tissue

Danett K. Brake and C. Wayne Smith

Summary Adipose tissue contains cell types other than adipocytes that may contribute to complications linked to obesity. For example, macrophages have been shown to infiltrate adipose tissue in response to a high-fat diet. Isolation of the stromal-vascular fraction of adipose tissue allows one to use flow cytometry to analyze cell surface markers on leukocytes. Here, we present a technical approach to identify subsets of leukocytes that differentially express cell surface markers.

Key words Flow cytometry; adipose; stromal-vascular; CD11b; MACS.

1 Introduction

The structure and morphology of adipose tissue, with its large lipid vacuoles and honeycomb structure, creates challenges in identifying and studying smaller, often widely dispersed nonadipose cells. These cells, such as infiltrating macrophages, may play key roles in regulating local inflammatory responses when exposed to a high-fat diet [1]. There are strong links between the expression of proinflammatory cytokines such as tumor necrosis factor-α and the progression of insulin resistance associated with obesity [2,3]. Identifying which cell types are responsible for the regulation and secretion of these inflammatory molecules is an area of active investigation. Flow cytometry allows investigation of size, granularity, and extracellular and intracellular expression of proteins on single cells isolated from adipose tissue. Antibodies against cell markers such as CD3, CD11b, F4/80, CD11c, and CD34 can be used to identify T cells, myeloid cells, macrophages, dendritic cells, and stem cell-like cells, respectively. Recent publications have used flow cytometry to identify CD3[+] lymphocytes, CD34[+] cells, and subsets of F4/80[+] macrophages in the stromal-vascular fraction of white adipose tissue [4–7]. Flow cytometry on fresh adipose cells has been used to investigate the potential of transplanted bone marrow derived progenitor cells to migrate to adipose tissue and differentiate into adipose cells [8]. Work has also been performed to characterize freshly isolated cells from human tissue [9].

From: *Methods in Molecular Biology, Vol. 456:*
Adipose Tissue Protocols, Second Edition. Edited by: Kaiping Yang
© Humana Press, a part of Springer Science + Business Media, Totowa, NJ

Analysis of fresh adipose tissue presents difficulty because of fewer stromal-vascular cells per gram of adipose tissue as well as the relative fragility of these cells. For this procedure, cells must be in a single-cell suspension, which requires collagenase digestion of the adipose tissue. Optionally, the cells can be purified to study a specific population through magnetic bead-assisted cell sorting (MACS). The following protocol presents a method of isolating and preparing freshly isolated stromal-vascular tissue for flow cytometric analysis.

2 Materials

2.1 Adipose Tissue Isolation and Stromal-Vascular Fractionation

1. Krebs-Ringer bicarbonate (KRB) buffer supplemented with glucose and 4% bovine serum albumin (BSA) is prepared fresh on the day of the experiment. Stock reagents for KRB may be prepared ahead of time with sterile ddH$_2$O and stored at room temperature. One 100 mL batch can be made up as follows, 13.16 mL of 0.77 M NaCl, 0.52 mL of 0.77 M KCl, 0.132 mL of 0.77 M KH$_2$PO$_4$, 0.132 mL of 0.77 M MgSO$_4$, 2.76 mL of 0.77 M NaHCO$_3$, 4.16 mL of 0.3 M Glucose and 0.38 mL of 0.275 M CaCl$_2$ (see **Note 1**). Glucose is made fresh and added to the KRB solution. The solution is then gassed with 95% O$_2$–5% CO$_2$ for 20–30 min. BSA is added to a final concentration of 4% and mixed well. Solution is kept on ice for isolation and washing procedures.
2. Type I Collagenase (Worthington Chemicals) is added to minced adipose tissue at a concentration of 280 U/mL of KRB (see **Note 2**).
3. A 37°C shaking incubator is used for the collagenase digestion of the adipose tissue.
4. Chiffon or a 25-μm nylon mesh is cut into 2- to 3-inch squares (see **Note 3**).

2.2 Positive Selection of Cells From Stromal-Vascular Fraction Using Magnetic Cell Sorting

1. MicroBeads conjugated to monoclonal rat anti-mouse CD11b (Mac-1) antibody. Clone M1/70, isotype IgG2b. Store at 4°C, protected from light. (Miltenyi Biotec, Auburn, CA).
2. MACS MS column (for positive selection) and miniMACS separator (Miltenyi Biotec; see **Note 4**).
3. MEC buffer: PBS without Ca^{2+} or Mg^{2+}, 0.5% BSA, and 2 mM EDTA. Filtering though a 0.22-μm pore filter is recommended. Store at 4°C.

Table 16.1 Antibodies used to stain stromal-vascular cells from adipose tissue

Antibody	Clone	Fluorophore	Amount used (µg)
CD11b	M1/70	Fluorescein isothiocyanate (FITC)	1
ICAM-1	YN1/1.7.4	Phycoerythrin (PE)	1
CD14	Sa2-8	FITC	2
IgG2b	–	FITC	1
IgG2b	–	PE	2
IgG2a	–	FITC	2

2.3 Analysis of Cell Surface Markers by Flow Cytometry

1. Wash buffer: phosphate-buffered saline (PBS) with 10 mM glucose.
2. 1x FACS Lysing Solution (BD Bioscience, San Jose, CA) to remove red blood cells.
3. Fixative: 4% paraformaldehyde.
4. Antibodies: CD11b clone: M1/70, Fluorophore-conjugated isotype control IgG2a and IgG2b (BD Bioscience), ICAM-1 clone: YN1/1.7.4, CD14 clone: Sa2-8 (eBioscience, San Diego, CA). See also **Table 16.1**.
5. BD-FACScan flow cytometer or similar equipment with a laser that emits at 488 nm.

3 Methods

3.1 AT Isolation and Stromal-Vascular Fractionation

1. Adipose depots (perigonadal, periepidiymal, retroperitoneal, or subcutaneous) are isolated with sterile techniques. Any lymph nodes are carefully removed if present and the tissue placed on ice in KRB solution. (*see* **Note 5**).
2. Tissue is then washed with KRB solution, weighed, and minced with scissors, razor blades, or disposable scalpels (*see* **Note 6**).
3. Minced adipose tissue is added to KRB buffer in a ratio of 1 g/ 3 mL of KRB. Collagenase is then added at a concentration of 280 U/mL of KRB. The solution is mixed gently by light vortexing or hand mixing.
4. The collagenase solution is put into a shaking incubator at ~200–300 rpm at 37°C for 40 min (*see* **Note 7**).
5. Cells are filtered through chiffon or nylon mesh and washed with 2x initial volume of KRB solution.
6. Adipose cells are separated from stromal-vascular (S-V) cells by centrifugation at 500 g for 5 min.

7. Adipose cells will float while S-V fraction will pellet. S-V cells are removed from the bottom of a conical tube and resuspended and washed twice with 15 mL of KRB solution and centrifuged again at 500 g for 5 min.
8. S-V cells may be resuspended in 1 mL of PBS + glucose and counted with a Coulter counter or a cell counting chamber.

3.2 Positive Selection of Cells From S-V Fraction Using Magnetic Cell Sorting

1. Centrifuge cell suspension at 300 g for 10 min and gently aspirate supernatant as pellet will be easy to disturb (see **Note 8**).
2. Protocol closely follows manufacturer's recommendations. For 10^7 or fewer cells use the following concentrations. For more cells, scale proportionally. Resuspend cells in 90 μL of of MEC buffer (see **Note 9**).
3. Add 10 μL of CD11b MicroBeads.
4. Mix well by flicking tube or very gentle vortexing and incubate on ice for 15 min.
5. Wash cells by adding 1 mL of MEC buffer and centrifuge at 300 g for 10 min. Aspirate off supernatant.
6. Resuspend in 500 μL of MEC buffer for use in MS separation columns.
7. Place column in magnetic field of the MACS separator and equilibrate column with 500 μL of MEC buffer.
8. Apply cell suspension to column.
9. Collect unlabeled cells that pass through and wash with 500 μL of MEC buffer 3x, allowing the column to empty each wash, but do not allow to dry. This fraction is the CD11b negative fraction that can be assayed by flow cytometry if desired.
10. Remove column from the magnetic field of the separator and place in a 1-mL microfuge tube.
11. Add 1 mL of MEC buffer to the column and expel the fraction into the microfuge tube with the provided plunger. This is the CD11b-positive fraction that can be assayed by flow cytometry. An example of the selection by CD11b can be seen in **Fig. 16.1**.
12. Centrifuge cell fractions at 300 g for 5 min, aspirate off supernatant, and resuspend in 300 μL of PBS (With Ca^{2+} and Mg^{2+}) +glucose (see **Note 10**).

3.3 Analysis of Cell Surface Markers by Flow Cytometry

1. Add 5×10^5 cells per test to 400 μL of PBS.
2. Add antibodies according to **Table 16.1** to separate cell samples. If two-color staining is desired, incubate primary antibodies at the same time (see **Note 11**).
3. Vortex to mix and place on ice for 30 min protected from light (see **Note 12**).
4. Add 1 mL of PBS and centrifuge at 300 g for 2 min. Decant gently and repeat wash twice.

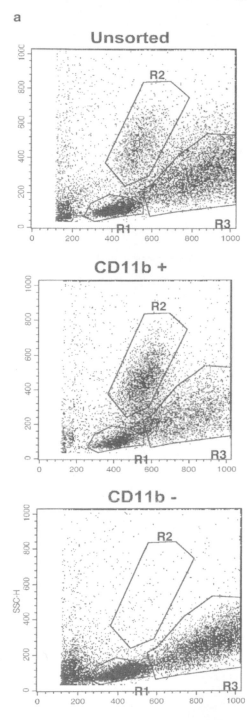

Fig. 16.1 Sorting of adipose stromal-vascular cells into myloid (CD11b⁺) fraction and non-myloid (CD11b⁻) fraction. (**A**) Forward versus side scatter shows three populations of cells labeled R1-R3. Sorting by CD11b expression shows that most of the cells in Region 2 (R2) express CD11b and only appear on the CD11b positive plot

b

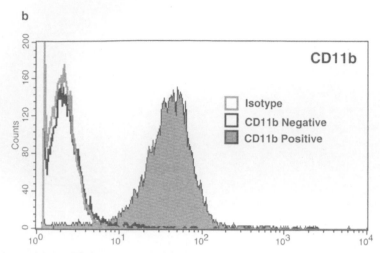

Fig. 16.1 (continued) **(B)** Histogram of CD11b expression of the two fractions show a >98% efficiency of sorting

5. Resuspend in 700 μL of FACS Lysing Solution and incubate at room temperature for 10 min in the dark.
6. Add 1 mL of PBS and centrifuge at 300 g for 2 min. Decent gently and repeat wash twice.
7. Resuspend in 400 μL of 4% paraformaldehyde (*see* **Note 13**).
8. Keep samples on ice while analyzing samples on flow cytometer. Vortex occasionally to keep cells from settling. It is possible to gate, or select, certain regions and look at cell surface expression of only those populations. See **Fig. 16.2** for sample results of two-color staining and differential expression of CD14 by two-gated regions.

4   Notes

1. CaCl$_2$ can cause the solution to precipitate and therefore should be added slowly and lastly, while swirling the flask before gassing the solution. The solution may appear slightly clouded. Be sure to add BSA after gassing, because it will cause the solution to bubble over if added before.
2. Lot-to-lot variance in collagenase can widely affect results. It is recommended that one tries several lots or different manufacturers and use only one lot of collagenase for all experiments.
3. Chiffon is a low-cost alternative to nylon mesh. It can be bought at any fabric store in large quantities. If culturing of cells is desired, it may be sterilized by washing with 70% ethanol and exposing to UV light inside a laminar flow hood overnight. For immediate antibody staining and flow cytometric analysis, sterilization is not required. Chiffon is not recommended if experiment is endotoxin sensitive.

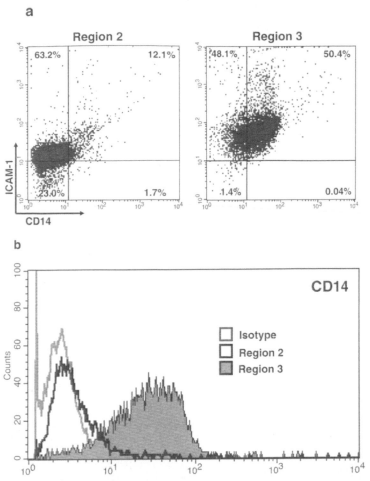

Fig. 16.2 CD11b⁺ cells from Region 2 and Region 3 express differing levels of CD14 expression. (**A**) Stromal-vascular cells were sorted positively for CD11b expression and then assayed for ICAM-1 and CD14 expression. (**B**) Histogram for CD14 expression between Region 2 (grey line) and Region 3 (solid area) shows higher levels of CD14 expression in Region 3

4. Miltenyi Biotec has a wide variety of MACS columns and separators that will work for this procedure. The MS columns described are designed for positive selection for cell quantities less than 10^7. That is, the primary cells of interest will remain in the column attached to the antibody of interest, while all unlabeled cells will fall through. Once removed from the magnetic field, the magnetically conjugated antibodies and the cells will easily and with high efficiency release from the column.

5. The minimal amount of adipose tissue recommended for this procedure is 1 g. Depending on the fat pad isolated, it is recommended to pool tissue from 3 or more mice. Mice 8–12 weeks of age on a normal chow diet have between 0.1 and 0.4 g of perigonadal fat.

6. Cell yield is highly dependent on how finely minced the tissue is when added to the collagenase solution. The greater the surface area, the more efficiently the collagenase will work, resulting in higher yield.

7. Incubation time of collagenase may vary. Incubation times longer than 1 h are not recommended because of increased cell death and apoptosis. Alternatively, the collagenase solution may be fractionated after 30 min, with undigested tissue being mixed with fresh collagenase and KRB and incubated for another 30 min. The single cell supernatant can be stored on ice during that time.

8. After the stromal-vascular fraction has been isolated, all subsequent centrifugation steps are done at 4°C.

9. MEC buffer contains an ion chelator, EDTA, and no Ca^{2+} or Mg^{2+}. Although this greatly increases the efficiency of antibody binding resulting in highly pure populations, this buffer makes the cells fragile and prone to apoptosis. Move without delay through these steps and keep the solutions and cells on ice at all times. High centrifugation speeds at this stage will greatly reduce yield.

10. Be sure to switch back to PBS containing Ca^{2+} and Mg^{2+} for all subsequent steps. MEC buffer will greatly reduce yield if used for incubation and wash steps. If forward versus side scatter plots show a high percentage of debris in the lower left corner, this could be due to being in MEC buffer too long or due to too high centrifugation speeds.

11. When staining with multiple antibodies in the same sample, ensure that the fluorophores do not emit at the same wavelengths. Fluorophores phycoerythrin (PE) and fluorescein isothiocyanate (FITC) are commonly paired together because of their emission spectra. However, light emissions from PE will be detected in the FITC channel, therefore compensation will be required. PE and FITC conjugated isotypes can be used to set the appropriate compensation settings. For this reason, it is recommended that the brighter or more highly expressed protein be FITC labeled.

12. The quantity of antibody and duration of incubation should always be optimized for each antibody and the amount of cells being stained. Usually, less than the manufacturer's recommended amount may be used with satisfactory results. Always run the appropriate isotype control antibody for each set of experiments as this is essential in establishing the background fluorescence for nonspecific binding. Isotypes should be run for each condition, as a treatment may alter nonspecific binding, resulting in potential false-positives.

13. The final resuspension volume can vary depending on the limits of the flow cytometer. If the flow rate, or counts per second, is low, resuspend in a smaller volume.

Acknowledgments This work was supported by grants from the United States Department of Agriculture 6250-51000-046 (CWS). Partially supported through grant numbers DGE-0086397 and DGE-0440525 of the National Science Foundation (DKB).

References

1. Weisberg SP, McCann D, Desai M, Rosenbaum M, Leibel RL, Ferrante AW Jr (2003) Obesity is associated with macrophage accumulation in adipose tissue. J Clin Invest 112:1796
2. Hotamisligil GS, Shargill NS, Spiegelman BM (1993) Adipose expression of tumor necrosis factor-alpha: direct role in obesity-linked insulin resistance. Science 259:87
3. Xu H, Barnes GT, Yang Q, Tan G, Yang D, Chou CJ, Sole J, Nichols A, Ross JS, Tartaglia LA, Chen H (2003) Chronic inflammation in fat plays a crucial role in the development of obesity-related insulin resistance. J Clin Invest 112:1821
4. Robker RL, Collins RG, Beaudet AL, Mersmann HJ, Smith CW (2004) Leukocyte migration in adipose tissue of mice null for ICAM-1 and Mac-1 adhesion receptors. Obes Res 12:936
5. Brake DK, Smith EO, Mersmann H, Smith CW, Robker RL (2006) ICAM-1 expression in adipose tissue: effects of diet-induced obesity in mice. Am J Physiol Cell Physiol 291:C1232
6. Caspar-Bauguil S, Cousin B, Galinier A, Segafredo C, Nibbelink M, Andre M, Casteilla L, Penicaud L (2005) Adipose tissues as an ancestral immune organ: site-specific change in obesity. FEBS Lett 579:3487
7. Wu H, Ghosh S, Perrard XD, Feng L, Garcia GE, Perrard JL, Sweeney JF, Peterson LE, Chan L, Smith CW, Ballantyne CM (2007) T-cell accumulation and regulated on activation, normal T cell expressed and secreted upregulation in adipose tissue in obesity. Circulation 115:1029
8. Crossno JT Jr, Majka AM, Grazia T, Gill RG, Klemm DJ (2006) Rosiglitazone promotes development of a novel adipocyte population from bone marrow-derived circulating progenitor cells. J Clin Invest 116:3220
9. Yoshimura K, Shigeura T, Matsumoto D, Sato T, Takaki Y, Aiba-Kojima E, Sato K, Inoue K, Nagase T, Koshima I, Gonda K (2006) Characterization of freshly isolated and cultured cells derived from the fatty and fluid portions of liposuction aspirates. J Cell Physiol 208:64

Chapter 17
Application of Electrophoretic Mobility Shift Assay and Chromatin Immunoprecipitation in the Study of Transcription in Adipose Cells

Melina M. Musri, Ramon Gomis, and Marcelina Párrizas

Summary Chromatin, long thought to be no more than a scaffold supporting DNA compaction inside the cell nucleus, has emerged in the last few years as a major regulatory element involved in the control of gene expression both acutely during interphase and programmatically throughout complex processes of development and differentiation. Adipogenesis is the result of an intertwined network of transcription factors and coregulators with chromatin-modifying activities and offers an excellent model for the study of transcriptional regulation. In this regard, electrophoretic mobility shift assay and immunoprecipitation of chromatin are complementary methods that can be used to study the binding of nuclear proteins to DNA and to characterize how these proteins interact with and modify chromatin to regulate gene expression and, more globally, cell differentiation. This chapter provides some strategies to perform these two assays using 3T3-L1 cells and rodent primary preadipocytes and adipocytes.

Key words Transcription; chromatin; histones; adipocytes; 3T3-L1; ChIP; EMSA.

1 Introduction

The continuing study of adipose tissue for the last two decades has completely transformed the perception that most investigators had of the organ. Nowadays, it is well known that adipose tissue plays a crucial role in the regulation of energy balance and acts as an active endocrine organ that mediates a number of physiological and pathological processes *(1)*. The renewed interest in this tissue has warranted that the study of the molecular basis of the development of the adipose cell has become the main area of research for many investigators. The mechanisms driving adipocyte differentiation can be easily dissected using well-characterized cell lines such as 3T3-L1 preadipocytes, as well as primary preadipocytes or adipocytes isolated from different sources. Adipogenesis involves not only a superbly coordinated network of transcription factors, but also a host of cofactors endowed with chromatin modifying activities *(2)*, including chromatin-remodeling complexes *(3)* as well

From: *Methods in Molecular Biology, Vol. 456:*
Adipose Tissue Protocols, Second Edition. Edited by: Kaiping Yang
© Humana Press, a part of Springer Science+Business Media, Totowa, NJ

as enzymes with histone modifying activities (4–6), which are crucial for the establishment of the gene expression pattern characteristic of the mature adipocyte. These coregulators lack DNA binding specificity and are ubiquitously expressed, hence their selectivity during adipogenesis is afforded by their interacting with sequence-specific transcription factors which recruit them to the appropriate gene promoters at the proper time points (2,7).

The study of transcriptional processes in higher eukaryotes has long been curtailed by a shortage of direct technical approaches. Until recent times, the study of transcription was mostly restricted to in vitro analysis using electrophoretic mobility shift assays (EMSA) and reporter gene assays. These methods have become increasingly inappropriate in light of the advancing knowledge of the role of chromatin as a major regulatory element in transcription. For this reason, the development of new techniques has been an important step forward in allowing the unraveling of the molecular basis of transcriptional regulation in a number of cell types, including adipocytes. The advent of the immunoprecipitation of chromatin (ChIP), in particular, afforded an opportunity to eavesdrop on what is happening inside the living cell at close quarters. In the last few years, ChIP has been established as a powerful method to examine the access of nuclear proteins to their target promoters in the natural chromatin environment, as well as to study the covalent modifications of the histones that constitute the nucleosomes spanning genomic regions of interest.

The ChIP procedure (see **Fig. 17.1A**) is based on the ability of formaldehyde to reversibly crosslink amino and imino groups of both amino acids and DNA that are found within a maximal distance of 2 Å from each other (8). This short range of action warrants that the crosslinks generated in this way represent direct interactions taking place in the cell at a determined time-point (9). By using a specific antibody directed against a particular transcription factor or a post-translationally modified histone, it is possible to precipitate the protein of interest, pulling along with it the DNA sequences to which this protein is bound. To achieve this, the DNA should have been previously fragmented randomly into small pieces ensuring that the co-precipitated DNA actually represents the sequences found in the close vicinity of the selected protein. Once the formaldehyde-generated crosslinks are reverted, the co-precipitated DNA can be analyzed by semi-quantitative PCR using primers designed to detect the presence or absence of a region of interest in the precipitate. The major limitations of the ChIP assay are the need for a not-always-available specific antibody and the inability to determine exactly where the binding is occurring, because it is difficult to shear the fixed chromatin into fragments of less than a few hundred base pairs. Moreover, this assay does not allow discernment of whether the binding of the transcription factor of interest to chromatin is direct or mediated by its interaction with other DNA binding proteins. In this respect, EMSA remains the standard assay for determining which specific nucleotides within a region of DNA constitute a binding site for a particular factor. Although in later times it has been all but displaced by ChIP assays, EMSA is a relatively easy method that can be used to establish if the binding of a protein to DNA is direct and determine precisely where it is occurring.

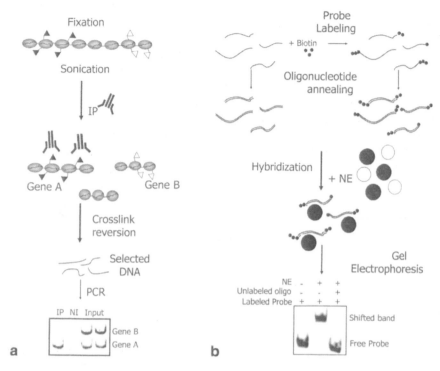

Fig. 17.1 Schematic representation of ChIP and EMSA. (**A**) ChIP. Cells are fixed with formaldehyde and chromatin is sonicated to generate 0.5- to 2.0-kb DNA fragments. Immunoprecipitation using an antibody directed against either transcription factor or modified histone is performed to obtain a DNA fraction enriched in regions that were interacting with this protein. Once the crosslinking is reverted, the selected DNA is analyzed by PCR. (**B**) EMSA. Single-stranded oligonucleotides comprising the binding site of interest are labeled with biotin to generate the probe. The oligonucleotides are then annealed and the hybridization is performed by mixing the probe and nuclear extracts. Protein–DNA binding is detected by a shift in the migration of the labeled probe subjected to gel electrophoresis. The specificity of the binding is ascertained by adding an unlabeled oligonucleotide competitor

A schematic representation of EMSA is shown in **Fig. 17.1B**. The EMSA procedure is based on the observation that although an unbound DNA fragment migrates rapidly through a nondenaturing polyacrylamide gel, protein binding to the fragment significantly slows its migration *(10,11)*. Because of the appearance of shifted or retarded bands as a consequence of protein binding, the assay is also often referred to as gel shift or gel retardation assay. Using an excess of an unlabeled oligonuleotide that competes with the labeled probe, the disappearance of the shifted band can be observed, allowing a quantitative calculation of the thermodynamic and kinetic parameters of the binding *(10,12)*. In contrast to the ChIP assay, the use of EMSA does not provide information about the actual occurrence of the observed binding in a specific cell line. Moreover, sites that require multiprotein complex formation to stabilize protein–DNA interactions, as well as the binding of non sequence-specific coregulators, are difficult to study in vitro. Thus, by providing different kinds of information, ChIP and EMSA are actually complementary methods.

The purpose of this chapter is to provide guidelines to carry out ChIP and EMSA experiments using freshly isolated rodent preadipocytes or adipocytes as well as cell culture lines such as 3T3-L1 at different times during the differentiation process. The general protocol is detailed with indications of some modifications that should be applied depending on the source of the cells used.

2 Materials

2.1 Chromatin Immunoprecipitation (ChIP)

2.1.1 Formaldehyde Crosslinking and DNA Fragmentation

1. Formaldehyde (36.5%), stabilized with 10% methanol (Merck).
2. Formaldehyde dilution buffer: 50 mM HEPES, pH 8.0;100 mM NaCl, 1 mM EDTA, 0.5 mM EGTA. Can be stored at 4°C for long periods of time.
3. Glycine 1.25 M prepared in distilled water. Can be stored at 4°C for several weeks.
4. Adipocyte lysis buffer: 5 mM PIPES, pH 7.9;80 mM KCl, 1% Igepal. Store at 4°C.
5. SDS lysis buffer: 50 mM Tris-HCl, pH 8.0; 10 mM EDTA; 1% SDS. Can be stored at 4°C for long periods of time, but it needs to be equilibrated at room temperature before use.

2.1.2 Immunoprecipitation

1. ChIP dilution buffer: 16.7 mM Tris-HCl, pH 8.0; 0.01% SDS; 1.1% Triton X-100; 1.2 mM EDTA; 167 mM NaCl. Store at 4°C. Add 1 mM PMSF and protease inhibitors before use.
2. Salmon sperm DNA/Protein A-agarose (50% slurry): Swell 0.15 g of Protein A CL-4B (GE Healthcare) with 2 mL of dH$_2$O for 15 min at room temperature on a rotating wheel. Pellet beads by centrifugation at 1,000 g for 1 min and discard supernatant. Repeat washing two more times with dH$_2$O and once with TE. Resuspend the final pellet in 1 mL of TE and supplement with 200 μg of sonicated salmon sperm DNA and 500 μg of bovine serum albumin. It can also be purchased already prepared from Upstate. For mouse monoclonal antibodies, protein A can not be used as it displays low affinity binding to mouse IgG1s. In these cases, Protein G (Upstate) or M-280 sheep anti-mouse IgG DynaBeads (Dynal Biotech) may be used instead.
3. Anti-di-methylated histone H3, Lys4 (Upstate, cat. no. 07–030).
4. Low-salt immune complex wash buffer: 20 mM Tris-HCl, pH 8.0; 0.1% SDS; 1% Triton X-100; 2 mM EDTA; 150 mM NaCl. Store at 4°C.

5. High-salt immune complex wash buffer: 20 mM; Tris-HCl, pH 8.0; 0.1% SDS; 1% Triton X-100; 2 mM EDTA; 500 mM NaCl. Store at 4°C.
6. LiCl immune complex wash buffer: 10 mM Tris-HCl, pH 8.0; 0.25 M LiCl; 1% Igepal; 1% sodium deoxycholate; 1 mM EDTA. Store at 4°C.
7. TE buffer: 10 mM; Tris-HCl, pH 8.0 ; 1 mM EDTA.
8. Elution buffer: 1% SDS, 0.1 M NaHCO$_3$. Prepare fresh before use.

2.1.3 Reversal of Crosslinking and Sample Analysis

1. 5 M NaCl.
2. 1 M Tris-HCl, pH 6.5.
3. 0.5 M EDTA, pH 8.0.
4. Proteinase K diluted in dH$_2$O at 10 mg/mL. Store in small aliquots at −20°C.
5. Phenol:Chloroform:Isoamyl alcohol (25:24:1).
6. Ethanol 100% (Merck).
7. Sodium acetate 3 M, pH 5.5 adjusted with acetic acid.
8. Glycogen (20 mg/mL; Roche).
9. PCR primers and reagents.
10. Equipment: We use a Branson sonifier model 150 for chromatin fragmentation. Any sonifier may be used, preferably one that is equipped with a microprobe, because the volumes of the samples are often small. A rotating wheel is used for the immunoprecipitation step. For the analysis of DNA fragmentation we perform 1% agarose gel electrophoresis. ChIP is analyzed by multiplex semi-quantitative PCR run in 12% acrylamide gels (Mini-Protean 3, Bio-Rad).

2.2 Electrophoretic Mobility Shift Assay (EMSA)

2.2.1 Nuclear Extract Preparation

1. Schreiber buffer A: 10 mM HEPES, pH 8.0; 10 mM KCl; 0.1 mM EDTA; 0.1 mM. EGTA. Store at 4°C. At the moment of use, add 1 mM PMSF, 1 mM DTT, and protease inhibitors.
2. Schreiber buffer C: 20 mM HEPES, pH 8.0; 0.4 M NaCl; 1 mM EDTA; 1 mM EGTA, 20% glycerol (v/v). Store at 4°C. At the moment of use, add 1 mM PMSF, 1 mM DTT, and protease inhibitors.
3. Igepal 1%.

2.2.2 Probe Labeling

1. Oligonucleotides containing the binding site of interest (*see* **Note 1**). In the case of PPARγ, we use a blunt-ended double-stranded synthetic oligonucleotide con-

taining the PPAR binding element (PPRE) from the mouse aP2 promoter (underlined):

Mouse aP2 PPRE, forward oligo: 5′-AGAAGAT<u>GGGGCA</u>A<u>AAGTCA</u>AAAC-CAC-3′

Mouse aP2 PPRE, reverse oligo: 5′-GTGGTTTTGACTTT<u>TGCCCC</u>ATCTTCT-3′

2. Biotin-N4-dCTP.
3. Terminal deoxynucleotidyl transferase (TdT) enzyme and 5x reaction buffer. Biotin-N4-dCTP and TdT can be conveniently purchased together as a 3′ End Biotin Labeling Kit available from Pierce (Rockford, IL; cat. no. 89818). In addition, the kit includes a labeled control oligonucleotide. which can be used to assess the efficiency of the labeling.
4. Chloroform:IAA (24:1).

2.2.3 Hybridization and Electrophoresis

1. Buffer: Either 100 mM HEPES, pH 7.9, or 100 mM Tris-HCl, pH 7.5-pH 8.0, may be used depending on the protein. The function of the buffer is to maintain the optimal pH to favor the DNA-protein interaction of interest. In the case of PPARγ, we use a final concentration of 10 mM Tris-HCl, pH 7.5.
2. Monovalent cations: 1 M NaCl or 1 M KCl. The concentration of monovalent cations establishes the astringency of the conditions. For PPARγ, we use a final concentration of 100 mM KCl.
3. Poly(DI.DC) 1 mg/mL prepared in TE. Poly(DI.DC) is a nonspecific competitor DNA which is used to minimize the binding of nonspecific proteins to the labeled probe by adsorbing proteins that will bind to any general DNA sequence. Store in aliquots at −20°C.
4. BSA 10 mg/ml in dH$_2$O. Store in small aliquots at −20°C.
5. DTT 100 mM. Store in aliquots at −20°C.
6. TBEx5: 0.45 M Tris base, 0.45 M boric acid, 10 mM EDTA.
7. Acrylamide:Bis-acrylamide 37.5:1 (30%).
8. Ammonium peroxodisulfate 10%.
9. TEMED.
10. Loading bufferx10: 250 mM Tris-HCl, pH 7.5; 40% glycerol; 0.2% bromophenol blue; 0.2% xylene cyanol. Store at −20°C.
11. Nylon membranes for nucleic acid hybridization, such as Biodine (Pierce)
12. Chemiluminiscent Nucleic Acid Detection Module (Pierce; cat. no. 89880). Store at 4°C.
13. Equipment: We use the Mini-Protean 3 vertical electrophoresis and transfer system equipped with a PowerPac Basic power supply (all from Bio-Rad) for gel electrophoresis and transfer. DNA is crosslinked to the membrane using a UVP crosslinker model CL-1000.
14. Chemiluminiscence is detected using a LAS3000 Lumi-Imager (Fuji Photo Film Inc.).

3 Methods

3.1 Chromatin Immunoprecipitation (ChIP)

3.1.1 Formaldehyde Crosslinking and DNA Fragmentation

1. Prepare crosslinking solution freshly by diluting 36.5% formaldehyde in formaldehyde dilution buffer to reach a final concentration of 11%.
2. Add crosslinking solution to the cells maintained in cell culture medium to reach a final concentration of 1% formaldehyde (*see* **Note 2**).
3. Incubate the cells for 10 min at room temperature (*see* **Note 3**).
4. Stop fixation by adding 1.25 M glycine to attain a final concentration of 125 mM. Glycine quenches the fixation reaction by providing excess amino groups. This step is optional and can be skipped if the samples are quickly processed.
5. When using adherent 3T3-L1 cells, discard the cell culture medium containing formaldehyde and wash cells twice with ice-cold phosphate-buffered saline (PBS). Scrape the cells with a cell scraper into 1 mL of PBS and pellet them in a microfuge at 10,000 g for 1 min. Discard the supernatant and resuspend the cell pellet in an appropriate volume of SDS Lysis buffer (*see* **Note 4**). Proceed to **step 8**.
6. When using freshly isolated primary pre-adipocytes or adipocytes, the cells will be in suspension. For preadipocytes, centrifuge the samples after fixation in 50-mL tubes at 4,000 g for 5 min to pellet the cells. Remove the supernatant and wash the pellet twice with 1 ml ice-cold PBS (the cells can at this point be transferred to a microfuge tube). Resuspend the final pellet of pre-adipocytes in 7–10 volumes of SDS Lysis buffer (*see* **Note 5**). Proceed to **step 8.**
7. In the case of primary adipocytes, centrifuge the cells after fixation in a 50-mL tube at 4000 g for 5 min. After centrifugation, the cells form a dense superior phase. We discard the infranatant by aspiring it with a 19-gauge syringe needle. After two washes with 10 mL of ice-cold PBS, primary adipocytes are resuspended in approximately 10 volumes of adipocyte lysis buffer and incubated on a rotating wheel at 4°C for 1 h. This incubation results in rupture of the cell membrane and release of the nuclei. The samples are then centrifuged at 20,000 g for 10 min and the pellet of nuclei obtained (usually a volume of 30–50 μL) is resuspended in 5–7 volumes of SDS lysis buffer. Proceed to **step 8**.
8. Incubate the samples at room temperature for 10 min on a rotating wheel. The high concentration of SDS results in breakage of the nuclear membrane, exposing the fixed chromatin and thus facilitating sonication.
9. Sonicate the samples to generate 0.5- to 2-kb fragments. Maintain the samples on ice during sonication to minimize foaming and avoid overheating that may denature the chromatin. Insert the sonicator tip to a distance of approximately 10 mm from the surface of the liquid if possible to reduce aerosoling and foaming, and avoid touching the walls of the tube with the probe which will cause the energy to dissipate unproductively. Sonication is performed at medium

power for eight 15-s intervals at continuous setting. Wait several minutes in between pulses for foaming to subside (*see* **Note 6**).

10. Centrifuge samples in a microfuge at maximum speed ($20,000\,g$) for 1 h at 4°C to eliminate cellular debris and high-molecular weight DNA-protein aggregates.

3.1.2 Immunoprecipitation

1. Recover supernatants after centrifugation and transfer to new tubes. Dilute samples 10-fold using ChIP dilution buffer to decrease the concentration of SDS in the samples.

2. Preclear the samples prior to immunoprecipitation by incubating every 1 mL with 60 µL of salmon sperm DNA/Protein A or Protein G for 30 min to 1 h at 4°C with gentle rotation (*see* **Note 7**).

3. Centrifuge samples and recover supernatants in new microfuge tubes. Use 0.5 mL for each individual immunoprecipitation.

4. Add 2.5 µL of antibody in the case of anti-dimethylated histone H3, Lys4. For other antibodies check the conditions recommended by the manufacturer or test several concentrations, usually 1–5 µg should be enough. Add non immune IgGs or an unrelated serum to one sample, which will be the negative control to check for specificity of the assay (*see* **Note 8**).

5. Incubate overnight at 4°C on a rotating wheel.

6. Centrifuge samples at $11,000\,g$ in microfuge for 2 min and transfer supernatants to new tubes. This step is intended to precipitate any aggregates that may have formed and that would otherwise be nonspecifically precipitated in the following steps.

7. Add 30 µL of salmon sperm DNA/Protein A or Protein G to each 0.5-mL sample and continue incubation for 1–4 h at 4°C to allow for the antibody-protein A/ Protein G complexes to form (*see* **Note 9**).

8. Centrifuge the samples for 2 min in microfuge at $11,000\,g$ and discard supernatants. The supernatant of the control immunoprecipitation (the one performed with non-immune serum) can be kept to be used later on as input DNA control for PCR analysis (*see* **Note 10**).

9. Wash pellets with 1 mL of low salt immune complex wash buffer for 3–5 min at room temperature on a rotating wheel. Centrifuge for 1 min in at $11,000\,g$ and discard supernatants.

10. Wash pellets with 1 mL of high salt immune complex wash buffer for 3–5 minutes at room temperature on a rotating wheel. Centrifuge for 1 min at $11,000\,g$ and discard supernatants. These washing steps are intended to eliminate nonspecific protein–antibody interactions.

11. Wash pellets with 1 mL of LiCl immune complex wash buffer for 3–5 min at room temperature on a rotating wheel. Centrifuge for 1 min at $11,000\,g$ and discard supernatants (*see* **Note 11**).

12. Wash pellets three times with 1 mL of TE Buffer to eliminate excess LiCl in the samples. Centrifuge for 1 min at 11,000 g and discard supernatants.
13. Elute samples by adding 200 µl of elution buffer prepared fresh, and incubating for 15 min at room temperature on a rotating wheel. Spin samples for 1 min at maximum speed and recover the supernatant in fresh microfuge tubes.
14. Repeat elution and combine eluates (total 400 µL per sample).

3.1.3 Reversal of Crosslinking and Sample Analysis

1. Add 16 µL of 5 M NaCl to the samples and input DNA (final concentration of 200 mM NaCl) and revert crosslinks by incubating at 65°C at least 4 h to overnight. Formaldehyde crosslinks are easily reverted by incubation at high temperatures in the presence of salts and detergents.
2. To digest proteins, add 8 µL of 0.5 M EDTA, 16 µL of 1 M Tris-HCl, pH 6.5; and 1.6 µL of 10 mg/mL Proteinase K and incubate the samples for 1–2 h at 45°C or overnight at 37°C.
3. Recover DNA by phenol:chloroform extraction and ethanol precipitation. Add 1 volume of phenol:chloroform:IAA (25:24:1) to the samples and vortex. Centrifuge for 5 min in microfuge at 12,000 g and recover the aqueous (superior) phase in new microfuge tubes.
4. Precipitate DNA by adding 1/10th volume 3 M sodium acetate, pH 5.5;20 µg glycogen as carrier and 2 volumes 100% ice-cold ethanol. Incubate 1 h-overnight at −20°C.
5. Centrifuge samples in microfuge at 12,000 g for 10 min at 4°C. Discard supernatants and wash pellets with 1 ml of 70% ethanol. Repeat centrifugation and discard supernatants. Allow the pellets to air-dry for several minutes.
6. Resuspend DNA in TE using the same volume for all samples (*see* **Note 12**).

The samples are ready to be analyzed by semiquantitative PCR. We use multiplex primer pairs including a positive control (a promoter that is not modified by the experimental conditions tested) and a negative control for specificity (a promoter that is not expected to be precipitated) *(5)*. Perform a series of dilutions of the input DNA to work within the same range as experimental samples (*see* **Note 13**). **Figure 17.2** shows a ChIP analyzed by multiplex PCR with β-actin and pancreatic transcription factor pdx1 as positive and negative controls, respectively, in 3T3-L1 cells and primary preadipocytes and adipocytes.

3.2 Electrophoretic Mobility Shift Assay (EMSA)

3.2.1 Nuclear Extract Preparation

The method described by Schreiber et al. *(13)* is an easily performed, two-step procedure that results in the purification of a crude nuclear extract fraction by first

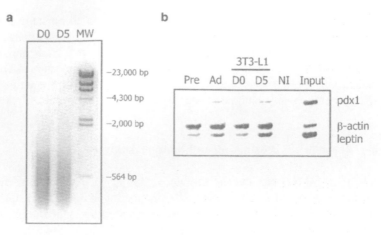

Fig. 17.2 ChIP assay using 3T3-L1 cells and primary mouse pre-adipocytes and adipocytes.
(**A**) DNA fragmentation. The size of the sonicated DNA of 3T3-L1 pre-adipocytes (D0) and adipocytes (D5) is checked by 1% agarose gel electrophoresis. (**B**) Multiplex PCR analysis. D0 and D5
3T3-L1 cells, as well as mouse primary preadipocytes and adipocytes, are subjected to ChIP as
described in the text using an antibody directed against dimethyl(Lys4) histone H3, a well-known
mark of transcriptional activity. The selected DNA is analyzed by multiplex PCR using three
primer pairs, namely: the pancreatic transcription factor pdx1, which is not expressed in 3T3-L1
cells and thus acts as a negative control to ascertain the specificity of the amplification; the β-actin
promoter, which is used as a positive control; and the gene of interest, which in this case is leptin.
The results show selective amplification of the β-actin and leptin promoters, indicating that both
genes are dimethylated in preadipocytes and adipocytes. NI indicates a control precipitation
performed with nonimmune serum. The input lane shows the pattern of amplification of the three
amplicons using unselected DNA

swelling the cells in an hypotonic buffer preserving nuclei intact and then extracting
the nuclear pellet with a high-salt buffer.

1. Precool buffers and equipment at 4°C.
2. When using 3T3-L1 cells in a p100 dish, discard cell culture medium and wash
   cells twice with ice-cold PBS. Scrape the cells into 1 mL of PBS and transfer to
   a microfuge tube. Centrifuge the cells for 30 s at 8000 g at 4°C. Resuspend the
   pellet in 400 μL of Buffer A. Proceed to **step 4**.
3. When using primary mouse pre-adipocytes or adipocytes, centrifuge the samples
   at 4,000 g in a 50 ml tube. Wash the cells twice with 10 mL of ice-cold PBS.
   Resuspend the final pellet in 1 ml of PBS and transfer to a microfuge tube.
   Centrifuge the cells for 30 ss at 8000 g at 4°C. Resuspend the pellet in 400 μL of
   Buffer A (*see* **Note 14**). Proceed to **step 4**.
4. Incubate samples on ice for 15 min to allow cell swelling.
5. Add 25 μL of 1% Igepal to cells. Vortex for 15 s. Centrifuge in microfuge at
   maximum speed for 30 s and discard supernatant (Otherwise, supernatant can be
   kept as the cytoplasmic fraction).
6. Add 50 μL of Buffer C to pellets (*see* **Note 15**).

7. Incubate for 15 min at 4°C on a rotating wheel.
8. Spin samples at maximum speed for 5 min in a microfuge at 4°C. Recover the supernatant, which represents the nuclear fraction, and measure protein concentration with a suitable method, such as Bradford. Samples may be stored at −20°C until the moment of use.

3.2.2 Probe Labeling

We perform 3′ end labeling of synthetic oligonucleotides with terminal deoxynucleotidyl transferase and biotin-N4-CTP (*see* **Note 1**). Single-stranded oligonucleotides are the best substrate for TdT, although blunt ended or 3′ overhang double-stranded oligonucleotides may also be used. Thus, when using synthetic oligonucleotides, it is convenient to label them separately and then anneal the labeled oligos.

1. Dissolve the oligonucleotides in TE buffer to reach a concentration of 10 μM. From these stocks prepare a working concentration of 1 μM.
2. Prepare the labeling reactions by adding the components in the following order:

   25 μL of dH$_2$O
   10 μL of 5xTdT reaction buffer
   5 μL of single-stranded oligonucleotide (1 μM)
   5 μL of Biotin-N4-CTP (5 μM)
   5 μL of TdT (2 U/μL)

3. Incubate reactions for 30 min at 37°C.
4. Stop reactions by adding 2.5 μL of 0.2 M EDTA.
5. Add 50 μL of chloroform:IAA to each reaction mixture to extract TdT. Vortex briefly and centrifuge 5 min in microfuge at maximum speed to separate the phases.
6. Recover and save the top aqueous phase (50 μL), which contains the labeled oligonucleotide at 100 nM. Optionally, labeling efficiency may be estimated by dot blot (*see* **Note 16**).
7. Anneal the labeled oligonucleotides (final concentration: 20 nM) by mixing together:

   50 μL of Forward oligonucleotide-Biotin (100 nM)
   50 μL of Reverse oligonucleotide-Biotin (100 nM)
   150 μL of TE buffer

Incubate 3 min at 97°C and 2 h at 37°C.

8. Anneal the competitor unlabeled oligonucleotide using the stock concentrations (10 μM) to obtain a final concentration of 2 μM (*see* **Note 17**). Mix the following:

   50 μL of Forward oligonucleotide (10 μM)
   50 μL of Reverse oligonucleotide (10 μM)
   150 μL of TE buffer

   Incubate for 3 min at 97°C and 2 h at 37°C.

3.2.3 Hybridization and Electrophoresis

1. Prepare a native electrophoresis gel in 0.5 × TBE.

   7.3 mL of dH$_2$O
   1.62 mL of acrylamide:bisacrylamide 30% (30:0.8)
   1 mL of 5xTBE
   60 μL of ammonium peroxodisulfate 10%
   20 μL of TEMED

(These are adequate quantities for two MiniProtean 3 gels [0.75 mm]. If other equipment is used, adjust volumes in consequence)

2. Once the gel is cast, pre-electrophorese it for 30–60 min at 100 V (*see* **Note 18**).
3. Prepare the following hybridization mixes (*see* **Note 19**):

   8 μL of dH$_2$O
   2 μL of Tris-HCl 100 m*M*, pH 7.5
   2 μL of DTT 10 m*M*
   1 μL of KCl 1 *M*
   1 μL of Poly(DI.DC) 10 mg/mL
   2 μL of unlabelled oligonucleotide 2 μ*M* (4 pmol)/ dH2O
   2 μL of NE (5–10 μg protein)/ dH$_2$O (*see* **Note 20**)
   2 μL of labelled oligonucleotide 20 n*M* (40 fmols)

4. Incubate samples for 15 min at room temperature.
5. Add 2 μL of 10xloading buffer to each sample and load the gel (*see* **Note 21**).
6. Electrophorese samples at 100 V until the bromophenol blue is three quarters down the length of the gel. This will take less than an hour in a Mini Protean 3 system.
7. Soak nylon membrane in 0.5xTBE for at least 10 min (*see* **Note 22**).
8. Sandwich the gel and nylon membrane in an electrophoretic transfer unit according to the instructions of the manufacturer and transfer in 0.5X TBE for 1–2 h at 380 mA (approx 100 V).
9. Crosslink the DNA to the membrane at 120 mJ/cm^2 using a UV-light crosslinker equipped with 254-nm bulbs. At this point the membrane can be stored dry at room temperature until the time of analysis.
10. Block the membrane for 15 min in 20 mL of blocking buffer (Pierce).
11. Prepare conjugate solution by adding 66.7 μL of stabilized streptavidin-HRP conjugate to 20 mL of blocking buffer. Streptavidin binds to biotin and the horseradish peroxidase conjugated to it allows its detection by conventional chemiluminiscence methods. Decant the blocking solution from the membrane and replace with conjugate solution. Incubate for a further 15 min at RT.
12. Change the membrane to a new container and wash it with 1x wash buffer for 5 min. Decant the liquid and repeat a total of four times.
13. Change the membrane to a new container. Add 30 mL of substrate equilibration buffer and incubate for 5 min at room temperature.

Fig. 17.3 EMSA performed with 3T3-L1 cells. Nuclear extracts from D0 and D5 3T3-L1 cells were prepared as described and incubated with a biotin-labeled double-stranded oligonucleotide comprising the PPRE of the mouse aP2 gene. To assess the specificity of the interaction, the shifted band was displaced with a 100x excess of the same oligonucleotide unlabeled (PPRE) or an unrelated oligonucleotide (RIPE3b, comprising the enhancer element from the mouse insulin promoter). The complexes were separated in a 5% native acrylamide gel in 0.5× TBE.

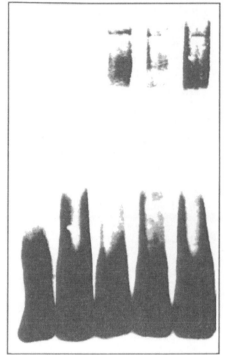

14. Prepare substrate working solution by mixing 50:50 luminol/enhancer solution with stable peroxide solution.
15. Transfer the membrane to a new container. Add the substrate working solution to the membrane and incubate for 5 min at room temperature.
16. Remove membrane from the substrate working solution and blot excess liquid with a filter paper. Do not allow the membrane to become dry.
17. Expose the membrane for 1–5 min to an X-ray film or use a CCD camera. Exposure time may be adjusted to obtain the desired signal. **Figure 17.3** shows a PPARγ EMSA performed with NE prepared from D0 and D5 3T3-L1 cells.

4   Notes

1. The probes to be used in EMSA assays may be generated in a number of ways, including digestion from DNA cloned in plasmids with appropriate enzymes to yield the region of interest, PCR amplification of an adequate

template, or by synthesizing and annealing complementary oligonucleotides. The labeling of these probes can also be performed by a selection of
techniques. Thus, radioactive labeling may be achieved by incubation of
double-stranded probes containing at least one 5′ overhang in the presence
of Klenow enzyme and deoxynucleotides including, usually, α^{32}P-dCTP
(the 5′ overhang, which will act as template, should then have at least an
unpaired G residue). In the case of blunt-ended probes, or probes with a 3′
overhang, they may be labeled with polynucleotide kinase and γ^{32}P-ATP.
Alternatively, non-isotopic methods may be used for the labeling and detection of the probe.

2. Formaldehyde 36.5% may also be added directly to the cells to reach a final
concentration of 1%, but it has been shown that pH influences the final
result of the crosslinking reaction (*14*). If fixation is allowed to proceed at
a physiological pH (7.4) protein-DNA crosslinks are favored in detriment
of protein-protein crosslinks. Avoid using any cell media containing Tris
during fixation, because it provides a source of amine groups that will
quench the reaction.

3. Fixation is one of the critical steps of the process. When analyzing nucleosomal
proteins such as histones, relatively weak fixation conditions can be used, but
longer fixation times may be required for transcription factors. Thus, a time-course
experiment to assess the optimal extent of fixation is a sensible first step when
testing, for the first time, an antibody in ChIP assays. Excess crosslinking can
turn the samples refractory to sonication and results in reduced antigen availability,
thus complicating the immunoprecipitation step.

4. Adjust the volume of SDS lysis buffer used according to cell number. When
using 3T3-L1 cells, calculate about half confluent p100 dish for each immunoprecipitation. Take into account that a complete experiment includes at least
two immunoprecipitations, e.g., the antibody of interest and a control immunoprecipitation performed with nonimmune serum or an unrelated antibody.
Ideally use the correspondence 50–60 μL for each immunoprecipitation (thus
100–150 μL for each p100 dish). If the volume is too small to be sonicated,
increase as necessary.

5. We use the epididymal fat pads of 2–3 mice. We generally obtain 2–3 × 10^6
pre-adipocytes, which represent a pelleted volume of 30–40 μL. We resuspend
these cells in 200–300 μL of SDS buffer.

6. Adequate DNA fragmentation is essential for a successful ChIP assay.
Fragment size can be checked by reverting the crosslink in an aliquot taken
from each sample and running a 1% agarose gel electrophoresis. This is an
important control step when first performing ChIP experiments with a
given cell line, but once the conditions have been optimized it is no longer
necessary to check the quality of sonication every time (an aliquot of the
samples can still be reserved prior to immunoprecipitation to be used as the
input control for PCR analysis later on). The samples can be stored at 4°C
while DNA size is checked, if necessary. Fixed chromatin samples can be

stored at 4°C for at least one week, or they can be kept at −80°C for longer-term storage.

7. When using anti-mouse Dynabeads, we usually skip pre-clearing.

8. Use preimmune serum if available, non-immune IgGs, or an unrelated serum, preferably against a non-nuclear protein. This negative control is intended to detect the DNA fragments that are nonspecifically precipitated.

9. For monoclonal antibodies, 30 μL of Protein G or 20 μL of washed anti-mouse Dynabeads have to be used because protein A binds only weakly to mouse IgG1s. When using Dynabeads, the centrifugation steps are substituted by placing the tubes for 2 min in a magnet to concentrate the beads.

10. Alternatively, an aliquot may be taken prior to immunoprecipitation (*see* **Note 6**), but this saves material if limiting. Note that this sample can not be used to check DNA size, as it contains salmon sperm DNA which will mask the sample in agarose gel electrophoresis.

11. If lower-astringency conditions are needed, this and the previous step can be excluded.

12. The volume of resuspension depends on the initial cell number. Resuspend in as small a volume as possible (e.g., 20–30 μL) and test PCR. Generally input DNA needs to be diluted at least 50–100 times prior to PCR to be in the same working range as the samples.

13. Real-time PCR may also be used to analyze ChIP.

14. We usually extract the epididymal fat pads of two to three mice for nuclear extract preparation. For primary pre-adipocytes we use $2–4 \times 10^6$ cells for each nuclear extract preparation. In the case of primary adipocytes, we use a pelleted cell volume of 70–100 μL for each preparation.

15. It is convenient to dry the walls of the tube completely with a cotton swab without disturbing the pellet not to dilute the hypertonic buffer C with remains of hypotonic buffer A.

16. To assess the efficiency of the biotinylation reaction, the detection of the labeled probe is compared to that of a series of dilutions of a control labeled oligonucleotide (for instance, that provided by the 3' End Labeling Kit) by means of a dot blot assay.

17. The unlabeled double-stranded oligonucleotide thus generated represents a 100x concentration compared with the labeled probe.

18. This step is intended to remove the ammonium peroxodisulfate in the gel and minimize temperature changes taking place during the actual electrophoresis of the samples, which may affect the stability of the bound complexes.

19. Depending on the transcription factor whose binding to DNA is to be tested, other components can be added to the hybridization mix. Some proteins may need the presence of divalent cations such as Mg^{2+} or Zn^{2+} to stabilize their binding to DNA. The astringency of the conditions can also be increased by adding Igepal to the mix (usually to a final concentration of 0.05%). The time and temperature of the binding reaction also influence the end result.

20. NE samples are diluted in hypertonic buffer C and thus contain $0.4M$ NaCl, resulting in an increase by $40\,mM$ of the total content of salts in the reaction mix when $2\,\mu L$ of NE is used in a final reaction volume of $20\,\mu L$. If more than $2\,\mu L$ of NE is needed in order to provide $5-10\,\mu g$ protein, samples may be dialyzed against a low-salt $0.2M$ NaCl-containing buffer (otherwise, additional KCl may be left out of the hybridization mix).

21. The ability of the protein–DNA complexes to withstand, as a whole, the short time required to migrate into the gel, is a critical step on which the major part of the success of the procedure rests. The interactions are stabilized by the low ionic strength of the electrophoresis buffer used (0.5X TBE) and the DNA-protein complexes are quickly resolved upon entry on the gel.

22. In the case of radioactive labeling, the gel is placed on top of an $3\,MM$ Whatmann filter paper, covered in saran wrap and dried in a vacuum gel drier at 80°C for $30\,min$-$1\,h$. The dried gel can then be exposed overnight to an X-ray film or analyzed using a Phosphorimager.

Acknowledgments M.P. is a recipient of a Ramón y Cajal contract from the Ministerio de Educación y Ciencia (Spain). This work was supported by grants SAF2003–06018 and BFU2006–14251/BMC from the M.E.C. (Spain) awarded to R.G. and M.P., respectively.

References

1 Kershaw EE, Flier JS (2004) Adipose tissue as an endocrine organ. J Clin Endocrinol Metab 89:2548–2556
2. Farmer SR (2006) Transcriptional control of adipocyte formation. Cell Metab 4:263–273
3. Salma N, Xiao H, Mueller E, Imbalzano AN (2004) Temporal recruitment of transcription factors and SWI/SNF chromatin-remodeling enzymes during adipogenic induction of the peroxisome proliferator-activated receptor gamma nuclear hormone receptor. Mol Cell Biol 24:4651–4663
4. Fajas L, Egler V, Reiter R, Hansen J, Kristiansen K, Debril MB, Miard S, Auwerx J (2002) The retinoblastoma-histone deacetylase 3 complex inhibits PPARgamma and adipocyte differentiation. Dev Cell 3:903–910
5. Musri MM, Corominola H, Casamitjana R, Gomis R, Parrizas M (2006) Histone H3 lysine 4 dimethylation signals the transcriptional competence of the adiponectin promoter in preadipocytes. J Biol Chem 281:17180–17188
6. Yoo EJ, Chung JJ, Choe SS, Kim KH, Kim JB (2006) Down-regulation of histone deacetylases stimulates adipocyte differentiation. J Biol Chem 281:6608–6615
7. Hassan AH, Neely KE, Vignali M, Reese JC, Workman JL (2001) Promoter targeting of chromatin-modifying complexes. Front Biosci 6:D1054–D1064
8. Orlando V (2000) Mapping chromosomal proteins in vivo by formaldehyde-crosslinked-chromatin immunoprecipitation. Trends Biochem Sci 25:99–104
9. Katan-Khaykovich Y, Struhl K. (2002) Dynamics of global histone acetylation and deacetylation in vivo: rapid restoration of normal histone acetylation status upon removal of activators and repressors. Genes Dev 16:743–752
10. Fried MG (1989) Measurement of protein-DNA interaction parameters by electrophoresis mobility shift assay. Electrophoresis 10:366–376
11. Revzin A (1989) Gel electrophoresis assays for DNA-protein interactions. Biotechniques 7:346–355

12. Fried MG, Crothers DM (1984) Kinetics and mechanism in the reaction of gene regulatory proteins with DNA. J Mol Biol 172:263–282
13. Schreiber E, Matthias P, Muller MM, Schaffner W (1989) Rapid detection of octamer binding proteins with 'mini-extracts', prepared from a small number of cells. Nucleic Acids Res 17:6419
14. Jackson V (1999) Formaldehyde cross-linking for studying nucleosomal dynamics. Methods 17:125–139

Chapter 18
Application of RNA Interference Techniques to Adipose Cell Cultures

Shigeki Shimba

Summary RNA interference (RNAi) is a powerful, quick, and easy technique to reduce the expression of a particular gene. However, investigators need to consider several steps for the experiments, including the design of the siRNA, an efficient delivery method, and a means for monitoring the biological effects of the siRNA introduced. Adipocytes have long been recognized as one of the most difficult cell types in which to perform gene delivery. In this chapter, three distinct transfection methods to obtain high efficiency of gene delivery and gene knockdown are described. The transfection efficiency of siRNA and shRNA is also compared.

Key words Adipocytes; RNAi; gene expression; transfection.

1   Introduction

Suppression of gene expression in cells is one of the best strategies for analyzing the roles of a particular gene on cellular functions. Knockout mice have proven to be a powerful way of studying the specific role of a gene in vivo. However, this knockout strategy takes a long time and requires someone with great technical expertise. RNA interference (RNAi) is the process of using specific sequences of double-stranded RNA (dsRNA) to "knockdown" the expression of complementary genes *(1)*. This technique allows a powerful and easy reduction in the expression of a particular gene, often by 90% or greater, allowing the analysis of the effect which that gene has on cellular functions *(1–4)*. Currently, a variety of commercial resources are available to start up RNAi experiments. Therefore, this technique is now widely used in vitro and even in vivo. For the best performance of this technique, investigators need to consider several steps, including the design of siRNA, an efficient delivery method, and a means to monitor the biological effects of the siRNA introduced. Adipocytes have long been recognized as one of the most difficult cell types in which to perform gene delivery. There are many excellent, recent reviews covering the mechanism and theoretical basis for the identification and

From: *Methods in Molecular Biology, Vol. 456:*
Adipose Tissue Protocols, Second Edition. Edited by: Kaiping Yang
© Humana Press, a part of Springer Science + Business Media, Totowa, NJ

selection of effective and specific siRNAs *(4–8)*. In this chapter, different transfection methods to obtain high efficiency of gene delivery and knockdown in adipocytes are described and compared.

2 Materials

2.1 Cell Culture

1. 3T3-L1 cells.
2. Growth medium: Dulbecco's modified Eagle medium (DMEM) supplemented with 10% calf serum.
3. Differentiation medium: 3:1 mixture of DMEM and Ham's F12 containing 10% fetal bovine serum, 1.6 mM insulin, 0.0005% transferrin, 180 μM adenine, and 20 pM triiodothyronine.
4. Isobutylmethylxanthine (IBMX): 500 mM stock solution in dimethyl sulfoxide, store in aliquots at −20°C.
5. Dexamethasone (DEX): 250 μM stock solution in ethanol, store in aliquots at −20°C.

2.2 Transfection

1. OPTI-MEM I (Invitrogen).
2. siRNA: 20 μM siRNA solution with RNase free water; store in aliquots at −20°C (*see* **Notes 1–3**).
3. Transfection reagents: X-tremeGENE (Roche Diagnostics) and Lipofectamine 2000 (Invitrogen) are stored at 4°C. Deliver X and the supplied buffers are stored at −20°C and room temperature, respectively.
4. Bioruptor Sonicator (Cosmo Bio, Tokyo, Japan).
5. Electroporation apparatus: Nucleofector device and solution V (Amaxa Inc.).
6. 0.25% Trypsin (Becton Dickinson) and 0.5% collagenase (Sigma, cat. no. C6885): prepare in phosphate-buffered saline (PBS) containing 0.02% ethylenediamine tetra-acetic acid.

3 Methods

In this section, three distinct transfection methods are compared. The first method is performed with conventional liposome-type transfection reagents such as X-tremeGENE and Lipofectamine 2000. The second method is similar to the first one, but distinct in the composition of transfection reagent. The last method is electrophoration with Nucleofector (Amaxa). Comparison of transfection efficiency and toxicity is described in the Notes section (*see* **Note 4**).

3.1 Cell Culture

3T3-L1 cells are maintained in DMEM supplemented with 10% calf serum. For induction of adipocyte differentiation, the cells are grown to confluence. The cells are then fed with differentiation medium supplied with $0.25\,\mu M$ DEX and $500\,\mu M$ IBMX. After 3 d, the cells are refed with fresh differentiation medium without DEX and IBMX and maintained over the following days. The transfection experiments described in this section are performed 8 days after the induction of differentiation.

3.2 Transfection with Conventional (Liposome-Based) Transfection Reagents (X-tremeGENE, Lipofectamine 2000)

The amount of reagents described below is for one well in a 12-well plate.

1. Mix the reagent gently before use, and dilute the reagent with OPTI-MEM-I medium (1–10 μL of reagent in 49–40 μL of medium). Mix gently and incubate for 5 min at room temperature.
2. Dilute siRNA (1–3 μL of $20\,\mu M$ stock solution) with OPTI-MEM-I medium.
3. Mix gently and incubate for 20 min at room temperature to allow complex formation to occur. The solution may appear cloudy, but this will not inhibit transfection.
4. Add 400 μL of differentiation medium without antibiotics.
5. Remove the culture vessel from the incubator.
6. Replace the medium with the siRNA/reagent containing the medium prepared previously.
7. After 24–48 h, total cellular RNA or protein is extracted (*see* **Note 5**).

3.3 Transfection with Polypeptides-Based Reagents (Deliver X)

The amount of reagents described below is for one well in a 12-well plate.

1. Thaw siRNAs and DeliverX Transfection Reagent and store on ice.
2. Sonicate the DeliverX Transfection Reagent at maximum speed and continuous power for 4 min to achieve a homogenous solution.

3. Dilute the reagent (4.8 µL) with buffer 2 (45.2 µL).
4. After diluting, vortex briefly and sonicate again at maximum speed and continuous power for 4 min.
5. Dilute siRNA (3.2 µL of 20 µM of stock solution) with buffer 1.
6. Mix the siRNA/buffer 1 solution (50 µL) with the reagent/buffer 2 solution (50 µL).
7. Vortex for 3 s or pipet up and down several times.
8. Incubate tubes at 37°C for 20 min.
9. Add 400 µL of OPTI-MEM-I medium.
10. Remove the culture vessel from the incubator.
11. Replace the medium with the siRNA/Deliver X containing the medium prepared previously.
12. After 24–48 h, total cellular RNA or protein is extracted (*see* **Note 5**).

3.4   *Transfection by Electroporation (Nucleofector)*

1. Wash cells (cultured in 6-well plate) with PBS, and treat with 0.25% trypsin and 0.5 % collagenase for 2 min at 37°C.
2. Remove trypsin and collagenase, and tap culture vessel several times. Collect cells in proper amounts of PBS and centrifuge at 1000 g for 5 min.
3. Add 100 µL of solution V and suspend the cells completely.
4. Add 5–10 µL of 20 µM siRNA solution.
5. Transfer the mixture into the cuvette.
6. Choose program U-28 and press the start button.
7. Resuspend cells into DMEM containing 10% fetal bovine serum and seed into a well in 12-well plate.
8. After 4–6 h, replace medium with the differentiation medium.
9. After 24–48 h, extract total cellular RNA or protein (*see* **Note 5**).

4   Notes

1. In the first step of the RNAi experiments, the investigator has to decide whether to use siRNA or shRNA. Generally, the transfection efficiency of adipocytes with plasmid DNA is less than 10% (*see* **Fig. 18.1A**). In contrast, short dsRNA can be efficiently delivered into adipocytes (*see* **Fig.18.1B**). Also, siRNA is stable, and significant amounts of siRNA exist in adipocytes even 10 days after transfection (*see* **Fig. 18.2**). Therefore, siRNA is the choice for mature adipocytes, unless investigators have a particular reason for using shRNA.
2. There are many companies providing "guaranteed to silence" or "validated" siRNAs. However, we simply purchase Stealth siRNA from Invitrogen because of minimum induction of interferon responses (*see* **Fig. 18.3; Note 3**). You may select a couple of sequences. At least one of them knocks down the target gene. If you would like to customize your siRNA, several web sites support the design

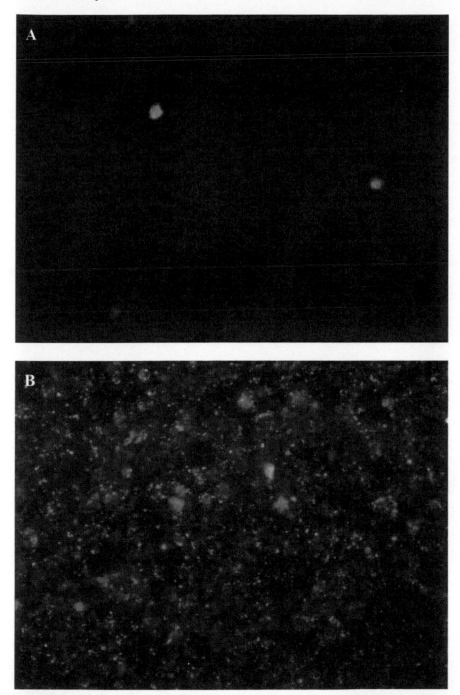

Fig. 18.1 Comparison of transfection efficiency of plasmid DNA and siRNA. 3T3-L1 adipocytes were transfected with either plasmid DNA expressing GFP (Amaxa) **(A)** or Alexa Fluor 555-siRNA (Invitrogen) **(B)** with the use of lipofectamine 2000. After 48 h of transfection, siRNA was detected on fluorescence microscopy

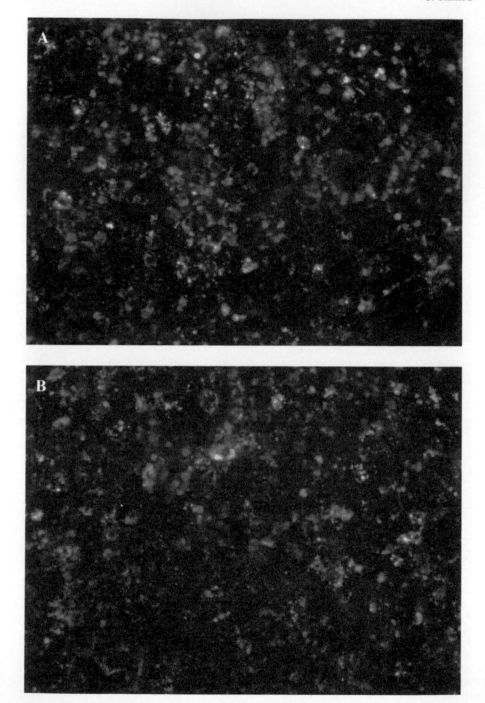

Fig. 18.2 Stability of siRNA in 3T3-L1 adipocytes. 3T3-L1 adipocytes were transfected with Alexa Fluor 555-siRNA (Invitrogen) with the use of X-tremeGENE (Roche Diagnostics). After 2 d (**A**) or 10 d (**B**) of transfection, siRNA was detected on fluorescence microscopy

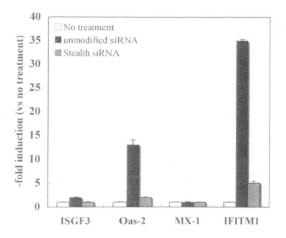

Fig. 18.3 Induction of interferon response gene expression in 3T3-L1 adipocytes. 3T3-L1 adipocytes were transfected with either unmodified siRNA or stealth siRNA (Invitrogen) by using X-tremeGENE (Roche Diagnostics). After 48 h of transfection, total RNA was extracted and the expression of interferon response genes was determined by qRT-PCR. ISGF-3, IFN-stimulated gene factor-3; Oas-2:2′-5′ oligo synthetase-2; Mx1, interferon-induced GTP-binding protein; IFITM1, interferon-induced transmembrane protein 1

Table 18.1 Selected web sites for the design of siRNA

	URL
BLOCK-iT RNAi Designer	https://rnaidesigner.invitrogen.com/rnaiexpress/
siSearch	http://sisearch.cgb.ki.se/
siDESIGN Center	http://www.dharmacon.com/sidesign/default.aspx
RNAi Central	http://katahdin.cshl.org:9331/RNAi_web/
siRNA Design Tool	http://www1.qiagen.com/Products/GeneSilencing/ CustomSiRnaDesigner.aspx
siRNA Sequence Selector	http://bioinfo.clontech.com/rnaidesigner
siRNA Target Finder	http://www.ambion.com/techlib/misc/siRNA_finder.html
siRNA Selection Server	http://jura.wi.mit.edu/bioc/siRNA/

of siRNA (*see* **Table 18.1**). If possible, try at least two distinct sequences for your target gene to exclude the possibility of off-target effects.

3. When dsRNA is delivered into cells, cellular responses called "interferon responses" occur. The induction of interferon responses depends on many factors, such as cell type, the length, and sequence of dsRNA, etc. (*1*). Initially, it was thought that only longer dsRNA molecules (>30 nt) can trigger an interferon response and shorter (21 nt) siRNAs do not stimulate this cellular response. Recent extensive studies have revealed that even 21 nt siRNA can induce interferon responses. As shown in **Fig. 18.3**, several interferon target genes are induced in 3T3-L1 adipocytes transfected with commercial 21 nt length-siRNA. Interferon triggers the degradation of all messenger RNA and, eventually, cell

Table 18.2 Efficiency of gene knockdown by different transfection methods 3T3-L1 adipocytes were transfected with either control siRNA or BMAL1 siRNA. After 48 h of transfection, total RNA was extracted and expression levels of BMAL1 mRNA and 36B4 mRNA were determined by qRT-PCR. Each data point represents the relative BMAL1 mRNA level normalized to the 36B4 mRNA level

	Control siRNA (A)	BMAL1 siRNA (B)	(B) / (A) %
Lipofectamine 2000	2.00	0.53	26
X-tremeGENE	1.15	0.13	11
Deliver-X	2.80	0.42	15
Nucleoefector	0.79	0.22	28

death *(9,10)*. These effects are unrelated to reduction of target gene expression. Therefore, investigators are advised to use modified siRNA such as Stealth siRNA to reduce interferon responses, or at least use control siRNA from the same company where the target siRNA is obtained. In any case, the effects resulting from the RNAi-mediated knockdown of target genes will have to be verified by other experiments.

4. We compared three transfection methods. As conventional methods, we chose two liposome-based reagents, X-tremeGENE and Lipofectamine 2000. As a target gene, BMAL1 was selected. This gene is abundantly expressed in 3T3-L1 adipocytes and plays a role in the regulation of mature adipocyte functions *(11,12)*. The siRNA used in this study almost completely knocked down the BMAL1 expression in proliferating cells (data not shown). X-tremeGENE and Deliver X (polypeptides-based reagent) efficiently reduced the expression of target gene. Comparison of X-tremeGENE and Lipofectamine 2000 indicated that the lipid composition of the reagent might influence transfection efficiency. Electroporation has long been used for transfection of genes into adipocytes. In fact, although efficiency of gene knockdown is slightly lower than other methods, electroporation can deliver a substantial amount of siRNA into adipocytes and knockdown the target gene (*see* **Table 18.2**). However, one of the biggest disadvantages of this method is that less than half of the cells can attach on the culture dish after 1 day of experiments. Therefore, this method is no longer the first choice of RNAi experiments with adipocytes. Both X-tremeGENE and Deliver-X effectively reduced the expression of the target gene in specific siRNA transfected cells (*see* **Table 18.2**). However, the absolute expression level in control cells transfected with X-tremeGENE is less than those in Deliver-X or Lipofectamine 2000 transfected cells (*see* **Table 18.2**). These results suggest that the X-tremeGENE may have efficient gene delivery activity, but with high toxicity. Those who want to study metabolic changes by knockdown of a particular gene should carefully analyze the side effects of transfection reagents, especially liposome-based reagents. The lipids in the reagents may influence the metabolic pathway. In conclusion, polypeptide-based transfection reagents such as Deliver X may be most effective and less toxic in RNAi experiments for adipocytes, although this method requires sonification of the reagent every time. If

investigators require ease of use, then conventional reagent will suffice for the experiments.

5. For evaluation of gene knockdown, detection of protein level and/or activity is preferable. If neither antibodies nor assay methods for activity are established, the mRNA level of the target gene should be determined by qRT-PCR.

Acknowledgments This work is supported in part by a grant from the Ministry of Education, Sciences, Sports, and Culture of Japan; a General Individual Research grant from Nihon University; a grant from Ono Medical Research Foundation; an Interdisciplinary General Joint Research grant from Nihon University; and the Academic Frontier Project for Private Universities, a matching fund subsidy from the Ministry of Education, Culture, Sports, Science, and Technology 2002–2006.

References

1. Meister G, Tuschl T (2004) Mechanisms of gene silencing by double-stranded RNA. Nature 431:343–349
2. Echeverri CJ, Perrimon N (2006) High-throughput RNAi screening in cultured cells: a user's guide. Nat Rev Genet. 7:373–384
3. Elbashir SM, Harborth J, Weber K, Tuschl T (2002) Analysis of gene function in somatic mammalian cells using small interfering RNAs. Methods 26:199–213
4. Pei Y, Tuschl T (2006) On the art of identifying effective and specific siRNAs. Nat Methods 3:670–676
5. Sontheimer EJ (2005) Assembly and function of RNA silencing complexes. Nat Rev Mol Cell Biol 6:127–138
6. Tomari Y, Zamore PD (2005) Perspective: machines for RNAi. Genes Dev 19:517–529
7. Filipowicz W, Jaskiewicz L, Kolb FA, Pillai RS (2005) Post-transcriptional genesilencing by siRNAs and miRNAs. Curr Opin Struct Biol 15:331–341
8. Valencia-Sanchez MA, Liu J, Hannon GJ, Parker R (2006) Control of translation and mRNA degradation by miRNAs and siRNAs. Genes Dev 20:515–524
9. Reynolds A, Anderson E M., Vermeulen A, Fedorov Y, Robinson K, Leake D, Karpilow J, Marshall WS, Khvorova A (2006) Induction of the interferon response by siRNA is cell type and duplex length dependent. RNA 12:988–993
10. Stark GR, Kerr IM, Williams BR, Silverman RH, Schreiber RD (1998) How cells respond to interferons. Annu Rev Biochem 67:227–264
11. Aoyagi T, Shimba S, Tezuka M (2005) Characteristics of circadian gene expressions in mice white adipose tissue and 3T3-L1 adipocytes. J Health Sci 51:21–32
12. Shimba S, Ishii N, Ohta Y, Ohno T, Watabe Y, Hayashi M, Wada T, Aoyagi T, Tezuka M (2005) BMAL1, a component of the molecular clock, regulates adipogenesis. Proc Natl Acad Sci U S A 102:12071–12076

Chapter 19
RNA Isolation and Real-Time Quantitative RT-PCR

Haiyan Guan and Kaiping Yang

Summary Adipose tissue has emerged as a major endocrine organ producing a wide spectrum of hormones and factors that play crucial roles in regulating cell turnover and function, not only locally within the adipose tissue but also in the brain and other key metabolic organ systems. It is known that gene activity is controlled at both transcriptional and post-transcriptional levels. Consequently, one of the most important means by which the activity of a gene is assessed is through the determination of levels of the corresponding messenger ribonucleic acid (mRNA). This process involves the isolation of total cellular RNA and subsequent analysis of the mRNA of interest. Given the unique nature of adipose tissue and adipocytes (i.e., containing high amounts of lipid), special RNA isolation techniques that have been tested in both white adipose tissue and isolated mature adipocytes from rats and mice will be presented. Although several methods are available for mRNA quantitation, we will describe a real-time quantitative reverse transcription polymerase chain reaction protocol because of its superior sensitivity and reliability.

Key words White adipose tissue; adipocyte; total RNA; mRNA; qRT-PCR.

1 Introduction

Accumulating evidence suggests that white adipose tissue (WAT) plays a key role in the pathogenesis of obesity and its associated metabolic disorders via the production of a wide variety of hormones, cytokines, growth factors, and enzymes (*1–4*). They are collectively known as adipokines, the majority of which are produced by adipocytes. These lipid-filled cells, often referred to as fat cells, are the main constituents of WAT. Therefore, the study of gene expression patterns of adipokines in adipocytes and/or WAT under normal/basal and various treatments (e.g., drugs, hormones, and dietary manipulations), as well as pathological conditions (e.g., obesity), has taken the center stage during the past decade.

Traditionally, levels of gene expression (i.e., messenger ribonucleic acid [mRNA]) in tissues/cells of interest have been determined with northern blot analysis, RNase

From: *Methods in Molecular Biology, Vol. 456:*
Adipose Tissue Protocols, Second Edition. Edited by: Kaiping Yang
© Humana Press, a part of Springer Science + Business Media, Totowa, NJ

protection assays, solution hybridization, and/or in situ hybridization techniques. All of these assays/techniques are time-consuming and labor intensive. Furthermore, with the exception of in situ hybridization, they all require relatively large quantities (at least 10 μg) of total or poly A+ RNA. With the advent of polymerase chain reaction (PCR) technology, semi-quantitative reverse transcription PCR (RT-PCR) protocols have been developed to assess mRNA levels in samples as small as individual cells (5–7). More recently, real-time PCR has revolutionized the way by which DNA and RNA levels are quantified, because of the ability to monitor the PCR reaction/product as it occurs in real time, rather than at the end of the PCR (i.e., conventional PCR (8,9)). There are numerous reviews and online resources that describe in detail the theories, the principles and the applications of real-time PCR. This chapter will focus on the practical aspects of real-time PCR as it applies to the detection of mRNAs, known as real-time quantitative RT-PCR (qRT-PCR), which is the most reliable and sensitive method for mRNA quantitation.

As its name implies, qRT-PCR involves two major steps: RT and real-time PCR. Similar to other mRNA detection methods, the success of qRT-PCR relies primarily on the quality (i.e., purity) of total RNA (10), which must be free from genomic DNA contamination. Although there are numerous in-house protocols as well as commercial kits for isolation of total cellular RNA from tissues/cells of interest, the unique nature (i.e., extremely high lipid content) of WAT and adipocytes requires specially developed reagents, protocols and kits. Here, we describe an in-house developed protocol that makes use of conventional RNA extraction reagents and conventional RNA isolation kits, which has given us consistently high yield and high purity total RNA (11). In addition, this protocol also eliminates the need to purchase additional reagents and kits that have been developed specifically for lipid-rich tissues/cells.

For real-time PCR, we describe both the probe- (i.e., TaqMan probe) and dye-based (i.e., SYBR Green I dye) PCR protocols. The primary advantage of the probe method is that the PCR specificity is guaranteed (i.e., there is no need to perform post-PCR analysis to verify that there are no nonspecific PCR products), because specific hybridization between probe and target is required to generate fluorescent signal. The major disadvantage is the requirement for the synthesis of different probes for different target sequences. However, the major advantage of the dye-based method is the reduced cost, especially when running multiple assays for multiple mRNAs, because no probes are required. However, the main disadvantage is that it may generate false-positive signals because the dye binds to any double-stranded DNA (i.e., both the target and nonspecific sequences). Consequently, post-PCR analysis (e.g., melting curve and sequencing of the PCR product) must be performed to validate the PCR specificity. In conclusion, both methods have distinct advantages and disadvantages, and it will be the decision of researchers as to which one is best suited for their particular study.

The next step in designing and conducting qRT-PCR is the choice of quantitation method. The quantitation of mRNA can be achieved by one-step or two-step RT-PCR. The major drawback of one-step RT-PCR is the need to use fresh RNA sample and conduct RT-PCR for every target gene. As a consequence, we only present a two-step RT-PCR protocol in this chapter. When calculating the results of quantitation assays, either absolute or relative quantitation may be used. The absolute quantitation assay is used to determine the absolute amount of the target mRNA by interpolating its quantity from a standard curve. However, the absolute quantities of the standards must first be known by independent means. In contrast, a relative quantitation assay is used to analyze changes in gene expression (i.e., mRNA levels) in a given sample relative to a reference sample (e.g., an untreated control sample). Given that it is often unnecessary to know the absolute amount of a target mRNA, relative quantitation assays are preferred and described here.

There are two calculation methods used for relative quantitation: standard curve and comparative C_T (threshold cycle number). Although these two methods give equivalent results, we have routinely used the standard curve method because it requires the least amount of optimization and validation. We run the target and reference (i.e., house-keeping gene) amplifications in separate wells using their respective standard curves for the following two reasons. First, to use the comparative C_T method, a validation experiment must be performed to show that the efficiencies of the target and reference amplifications are roughly equal, a prerequisite that can be difficult, if not impossible, to meet. Second, to amplify the target and reference genes in the same tube, limiting primer concentrations must be identified and shown not to affect C_T values. However, one caveat of using the standard curve method is the need to construct a standard curve for every assay, and the potential adverse effect of any dilution errors made in creating the standard curve samples.

2 Materials

2.1 Total RNA Isolation and Purification

1. Fresh or frozen white adipose tissue (WAT), and fresh or frozen isolated mature adipocytes.
2. TRIzol® Reagent (Invitrogen, cat. no. 12183–555) and TRIzol® LS Reagent (Invitrogen, cat. no. 10296–010) for WAT and isolated mature adipocytes, respectively.
3. Polytron (for WAT only).
4. GenEluteTM Mammalian Total RNA Miniprep Kit (Sigma, cat. no. RTN-70; *see* **Note 1**).
5. Chloroform.
6. 70% ethanol (in RNase-free water).
7. DNase I Stock solution: Gently dissolve DNase I (supplied with RNase-free DNase Set; Qiagen, cat. no. 79254) in 550 µL of RNase-free water. Can be store at 4°C for up to 6 wk (*see* **Note 2**).
8. DNase I working solution: Add one part of DNase I stock solution to seven parts of Buffer RDD (supplied with the RNase-free DNase Set). Mix by gently inverting the tube. Do not vortex, because DNase I is sensitive to physical denaturation.

2.2 Reverse Transcription

1. High Capacity cDNA Reverse Transcription Kit (Applied Biosystems, cat. no. 4368813).
2. Sterile 0.5-mL thin-wall PCR tubes.
3. Thermal-cycler.

2.3 Real-time PCR

1. Probe-based PCR reaction master mix: TaqMan Universal PCR Master Mix (2X; Applied Biosystems, cat. no. 4304437).
2. Dye-based PCR reaction master mix: Platinum SYBR Green qPCR SuperMix-UDG with ROX (2X; Invitrogen, cat. no. 11744; *see* **Note 3**).
3. TaqMan probes and PCR primers (*see* **Note 4**).
4. Real-time PCR plate and cover: 384-well Clear Optical Reaction Plates (Applied Biosystems, cat. no. 4309849), and Optical Adhesive Covers (Applied Biosystems, cat. no. 4311971; *see* **Note 5**).
5. Real-time PCR machine: ABI PRISM 7900HT Sequence Detection System (Applied Biosystems).

3 Methods

3.1 Total RNA Isolation and Purification From WAT

1. Add up to 300 mg of rat WAT (*see* **Note 6**) to 1 mL of *Trizol*® Reagent (*see* **Note 7**) in a sterile tube. Homogenize with a Polytron at medium speed for 15 seconds (*see* **Note 8**), and incubate for 5 min at room temperature to allow complete dissociation of nucleoprotein complexes.
2. Centrifuge the homogenate at 12,000 g for 10 min at 4°C.
3. Remove and discard the top layer of excess lipid. Then, carefully transfer the light pink color solution containing RNA to a fresh tube, leaving behind the pellet (*see* **Note 9**).
4. Add 0.2 mL of chloroform, and cap the tubes tightly. Shake tubes vigorously by hand for 15 sec. Do not vortex (*see* **Note 10**). Incubate at room temperature for 3 min.
5. Centrifuge at 12,000 g for 15 min at 4°C. After centrifugation, the solution separates into distinct phases: a lower, red phenol-chloroform phase, inter-phase and the colorless upper aqueous phase. RNA remains exclusively in the aqueous phase, whereas DNA and proteins are in the inter-phase and organic phase, respectively. The volume of the aqueous phase is about 60% of the volume of

TRIzol Reagent used for homogenization. Collect the colorless upper aqueous phase that contains RNA into a fresh tube.

6. Add an equal volume of 70% ethanol to the aqueous phase (a final ethanol concentration of 35%). Vortex or pipet thoroughly to mix. All the remaining steps are performed at room temperature.
7. Transfer up to 700 μL of the aforementioned solution containing RNA into a spin column supplied with the RNA extraction kit (Sigma).
8. Centrifuge at 12,000 g for 15 s. Discard the flow-through.
9. Transfer the remaining solution to the same spin column, and centrifuge at 12,000 g for 15 s. Discard the flow-through.
10. Add 250 μL of Wash Solution 1 to the column, and centrifuge at 12,000 g for 15 s. Discard the flow-through.
11. Remove genomic DNA by adding 80 μL of DNase I working solution to each column. Incubate for 15 min.
12. Wash the column once with 250 μL of Wash Solution 1, and transfer the column to new collection tube. Wash twice with 500 μL of Wash Solution 2 (*see* **Note 11**). Washes are carried out by centrifugation at 12,000 g for 15 s, discarding the flow-through.
13. Centrifuge the empty column at 12,000 g for 2 min to remove ethanol (*see* **Note 12**). Transfer the column to new collection tube.
14. Elute RNA by adding 80–100 μL of prewarmed elution solution (50–60°C) to the center of the column membrane. To maximize RNA recovery, gently vortex the tube containing the column, and incubate in a dry heating block at 50–60°C for 3 min (*see* **Note 13**).
15. Centrifuge the column for 2 min at 12,000 g. Discard the column, keep the eluted RNA on ice, and quickly perform the next step before storing RNA at −80°C (*see* **Note 14**).
16. The concentration and integrity of purified total RNA are assessed by UV absorbance spectrophotometry and standard agarose gel electrophoresis, respectively (*see* **Note 15**).

3.2 Total RNA Isolation and Purification from Primary Adipocytes (see Note 16)

1. Add 0.75 mL of Trizol-LS® Reagent to 0.25 ml of the sample containing freshly isolated or frozen mature adipocytes from rats or mice (*see* **Note 17**). Completely lyse the cells by mixing several times with a pipette, and incubate the lysates for 5 min at room temperature to permit the complete dissociation of nucleoprotein complexes.
2. Centrifuge the lysates at 12,000 g for 10 min at 4°C.
3. Remove and discard the top layer of excess lipid. Then, carefully transfer the light pink color solution containing RNA to a fresh tube, leaving behind the pellet (*see* **Note 9**). Follow **Subheading 3.1., steps 4–16**, as described previously.

3.3 Reverse Transcription (RT)

The High Capacity cDNA Reverse Transcription Kit contains all components necessary for the quantitative conversion of up to 2 µg of total RNA in a 20-µL reaction volume to single stranded cDNA.

1. Prepare 2 X RT Master Mix (100 rections) as follows:

   RNase-free water: 420 µL
   10X RT buffer: 200 µL
   10X random primers (*see* **Note 18**):   200 µL
   25X dNTP mix (100 mM)   80 µL
   MultiScribe™ Reverse Transcriptase (*see* **Note 19**):   100 µL

   Mix well, and store in aliquots at −20°C.

2. Prepare up to 2 µg of total RNA in 10 µL of RNase-free water in a sterile microfuge tube, and add 10 µL of 2X RT master mix (a total of 20 µL of reaction volume). Mix gently and spin briefly.
3. Place the tubes in a thermal-cycler and run RT reactions with the following program (based on the manufacturer's instructions): 10 min at 25°C, 2 h at 37°C, and 5 s at 85°C. Store RT at −20°C.

3.4 Real-time PCR

3.4.1 TaqMan Assays

1. Optimal concentrations of primers and probe are determined following the guidelines developed for sequence detection systems by Applied Biosystems (*see* **Note 20**).
2. Set up negative controls (*see* **Note 21**): Two different negative controls should be prepared for each assay; one is no template control (NTC), and the other is no reverse transcriptase control (NRTC).
3. Choose appropriate reference genes: Common reference genes, also known as housekeeping genes, include GAPDH, β-actin, α-tubulin, 28 S rRNA, and 18 S rRNA (*see* **Note 22**).
3. Construct standard curves: Using a known starting concentration of template from one of a variety of sources, a dilution series is performed. The standard curve should consist of at least four dilutions (*see* **Note 23**).
4. Prepare PCR Master Mix (without cDNA templates). The following compositions are based on a 20-µL reaction volume per well in triplicate (a total volume of 60 µL for each sample) for 384-well plates. Assuming that 1.0 µL of RT products (*see* **Note 24**) will be used per well, the total volume of PCR Master Mix for each sample will be 57 µL. The following example assumes that you have 20 unknown samples, one NTC, one NRTC, 4 standard curve samples, and one extra reaction volume to accommodate reagent losses during pipetting (for a

total of 27 samples). Carefully pipette the following solutions into a sterile 2.0-mL microfuge tube in the order as shown below:

Sterile ddH$_2$O	21 µL × 27 = 567 µL
TaqMan PCR Master Mix	30 µL × 27 = 810 µL
TaqMan probe (*see* **Note 25**)	3 µL × 27 = 81 µL
Primer Mix (*see* **Note 26**)	3 µL × 27 = 81 µL

Mix gently by vortexing, and keep on ice.

5. Prepare each PCR in a sterile 1.5-mL microfuge tube by adding 57 µL of the PCR Master Mix to the tube containing 3 µL of RT products, mix well by vortexing and spin briefly.
6. Load 19 µL of each PCR reaction into triplicate wells on a 384-well plate, and do so consistently and carefully for all the samples. Immediately cover the wells completely by placing one piece of the Optical Adhesive Cover onto the plate, following the manufacturer's instructions.
7. Spin the plate at 4000 g for 2 min. Run the plate on an ABI PRISM 7900HT Sequence Detection System immediately (*see* **Note 27**), or store at −20°C for up to 2 wk.
8. At the end of PCR, select Analyze from the Analysis menu, and examine the semi-log view of the amplification plots. Then, adjust the default baseline setting to accommodate the earliest amplification plot, and select a threshold above the noise close to the baseline, but still in the linear region of the semi-log plot. Upon completion of the Analysis (*see* **Note 28**), export data to Excel for further analysis.

3.4.2 SYBR Green I Assay

The procedures for SYBR Green I assays are identical to the TaqMan probe assay as described except that: (i) SYBR Green I, instead of TaqMan, PCR Master Mix is used; (ii) No TaqMan probe is used (use sterile water to make up the volume difference when preparing the SYBR Green I PCR Master Mix); and (iii) a melting (dissociation) curve and sequence (and/or restrictions enzyme digestion) analyses are performed at the end of PCR (*see* Note **29**).

4 Notes

1. Although the Sigma kit is used as an example in this chapter, we have also successfully used other commercially available mini RNA isolation kits (i.e., Invitrogen and Qiagen).
2. DNase I stock solution may be stored in aliquots at −20°C for up to 9 mo. Do not refreeze after thawing.
3. We have found SYBR Green master mix from other companies work equally well, including Qiagen and Applied Biosystems.
4. We routinely use Primer Express 2.0 (Applied Biosystems) to design probes and primers for both probe- and dye-based real-time PCR. When designing primers

for SYBR Green I assays, we keep all the default settings except for the Amplicon Requirements under which we increase the minimum length from 50 to 100, and the maximum length from 150 to 200. Our reasoning is that a larger amplicon size (i.e., PCR product) makes subsequent sequencing and/or restriction enzyme digestion analysis easier.

5. For first time users, an Optical Adhesive Starter Kit should be ordered because it includes one compression pad and one applicator in addition to 20 optical adhesive covers. Although the compression pad is not needed with the ABI 7900HT machine, the applicator is required regardless of the machine type.

6. Successful RNA isolation requires fast processing and careful handling of the tissues or cells before isolation. Endogenous RNases are released from cellular compartments immediately after harvesting tissue. It is essential to inactivate these RNases as soon as possible to prevent RNA degradation. Therefore, either homogenize samples immediately after harvesting, or flash-freeze samples in liquid nitrogen and store at −80°C. To prevent RNA degradation, it is important that the tissue be cut into small pieces to allow rapid and thorough freezing of the entire tissue.

7. RNA isolation from WAT or mature adipocytes using TRIzol or TRIzol-LS® Reagent followed by purification with silica-based columns allows you to obtain relatively large amounts of pure total RNA. Using this protocol, we routinely obtain 60–80 µg of total RNA from 300 mg of rat WAT.

8. Clean off tissue homogenizer with large volumes of water before and after each use to avoid cross-contamination.

9. After centrifugation, there should be three layers:

   a. Top layer: colorless containing excess lipid, which should be discarded.
   c. Middle layer: light pink color solution containing RNA, which should be collected.
   b. Bottom layer: pellet containing extracellular membranes, polysaccharides, and high molecular weight DNA, which should be discarded.

10. Do not vortex because it may increase DNA contamination of your RNA sample.

11. Before use, add ethanol to the concentrated Wash Solution 2 according to the manufacturer's instructions.

12. Traces of ethanol must be removed, since ethanol may inhibit downstream enzymatic reactions.

13. Vortexing will spread the elution solution evenly on the membrane, and incubation at 50–60°C will significantly increase RNA yield. If smaller amounts of fat tissues or adipocytes are used, decrease the RNA elution volume to 40 µL, and/or re-elute RNA using the same elution solution.

14. For short-term storage, RNA can be stored at −20°C; for long-term storage, it should be stored at −80°C. Although RNA can be stored in water or buffer, it is more stable in 70% ethanol at −80°C. We recommend storing RNA samples in aliquots to prevent damage to the RNA from successive freeze-thaw cycles and to reduce the risk of introducing RNases into the tube.

15. The concentration of RNA is first estimated by UV absorption spectrophotometry. A relatively pure RNA sample should yield an O.D. ratio of 1.80–2.0 at

260 nm to 280 nm. Samples diluted in water may give lower A260/280 O.D. ratios. We routinely use 10 mM Tris-HCl (pH 7.5) to dilute our samples for O.D. measurements. Before gene expression analysis, RNA integrity should be checked by standard agarose gel electrophoresis. We routinely run 0.5 – 1 μg of total RNA on a 0.8% agarose gel containing ethidium bromide. Intact total RNA will show 2 distinct and sharp bands representing 28 S and 18 S rRNA. The 28 S rRNA band should be approximately twice as intense as the 18 S rRNA band. This 2:1 ratio (28 S:18 S rRNA) is a good indication that the RNA is intact. Partially degraded RNA will have a smeared appearance, and will lack the sharp rRNA bands, or will have a lower 28 S:18 S rRNA ratio. Highly degraded RNA will appear as a very low molecular weight smear.

16. This protocol is specifically designed for freshly isolated (or frozen) mature adipocytes. For all other adipose cells (e.g., preadipocytes, stromal-vascular cells, and in vitro differentiated primary and 3T3-L1 adipocytes), follow conventional total RNA extraction protocols.

17. If the sample volume is less than 0.25 mL, adjust the volume to 0.25 mL with RNase-free water. If it is greater than 0.25 mL, scale up the reagent proportionally (the volume ratio of TRIzol LS Reagent to sample should always be 3:1).

18. Although oligo-dT primers can also be used for RT, we recommend the use of random primers, because this will allow you the freedom of choosing whatever internal control you may wish to utilize in subsequent real-time PCR assays, including 18 S rRNA and 28 S rRNA as well as other house-keeping genes (e.g., GAPDH, β-actin, and α-tubulin).

19. The reverse transcriptase is substituted with RNase-free water for preparing the negative control master mix. For every batch of RT reations, one negative control (-RT) must be included.

20. Based on our experience, the optimal concentrations of primers for house keeping genes and genes of your interest are between 50 and 150 nM and 200 and 300 nM, respectively. Consequently, we recommend that you focus on evaluating the optimal concentrations of primers between 50 and 300 nM and 150 and 450 nM, respectively for house-keeping genes and genes of your interest. When optimizing primers for SYBR Green I assays, it is also crucial to analyze the melting curve data for each primer concentration pair to ensure a single homogenous product is being generated. If several primer combinations give very similar results, choose the primer combination with the lowest concentration (to avoid primer dimer formation). Probe optimization is similar to primer optimization; select the probe and primer combination that results in the lowest C_T value and the highest delta Rn value.

21. At least two negative controls (NTC and NRTC) are used for quality control purposes. Ideally, signal amplification should not be observed in the NTC wells, and when observed, C_T values should be at least five and preferably more than ten cycles from the C_T values of your least concentrated samples. NRTC serves as an indicator of genomic DNA contamination, and their C_T values should be at least five cycles more than those of your least concentrated samples. Another negative control to consider is NAC (i.e., no amplification control), which includes all the

PCR components except for the DNA polymerase. This is useful if you suspect that you may be seeing an increase in fluorescence in your reaction that is not due to actual amplification (e.g., your probe is degrading).

22. Although using the same amount and quality of input RNA (*see* **Note 15**) in each sample ensures that equivalent amounts of RNA are compared, it cannot compensate for variations in the efficiency of reverse transcription, which is required to produce cDNA for subsequent PCR. Therefore, it is imperative that researchers normalize expression levels of genes of interest to that of a reference gene *(12,13)*. This step can remove inaccuracies due to variations in reverse transcription efficiency because RNA of the reference gene is reverse transcribed along with that of the gene of interest. Housekeeping genes such as GAPDH, α-tubulin, β-actin, cyclophilin, 28 S rRNA, and 18 S rRNA have often been used as reference genes for normalization *(14)*, with the assumption that the expression of these genes is constitutively high and that a given treatment will have little effect on their expression. However, this assumption must be validated empirically, as the expression of housekeeping genes can vary under certain conditions *(15–17)*. In any case, it is crucial to select a reference or even multiple reference genes whose expression has been empirically tested to be constant across all experimental conditions in your study.

23. Samples for constructing standard curves are chosen based on their anticipated levels of mRNA. Ideal samples should contain higher levels of the mRNA. If it is not possible to make a prediction, a small portion of cDNA from a few samples may be pooled, or cDNA from untreated control may be used. Alternatively, a separate RT reaction may be set up specifically for standard curves. The dilution series should encompass a large range of concentrations, ideally covering the expected levels of target in experimental samples. To accomplish this objective, a 3- to 10-fold dilution series over several orders of magnitude should be generated. For example, a typical serial dilution would consist of five points of a 5-fold serial dilution, starting with 100 ng of total RNA per reaction (or the cDNA equivalent amount).

24. The amount of cDNA templates (i.e., RT products) to be used in real-time PCR reactions should be determined according to their expected C_T values, which can be estimated from those obtained during primer optimization. As a general rule, more RT should be used for lower abundance mRNAs. However, the volume of RT must not exceed 10% of the final PCR reaction volume, because greater than 10% RT will inhibit the PCR.

25. After the optimal concentration of your probe has been determined, adjust its concentrations such that 1 μL of probe will be required per 20 μL of PCR.

26. After optimal concentrations of your primers have been determined, make a ready to use Primer Mix containing both sense and antisense primers at a concentration (according to their predetermined optimal concentrations, and diluted with appropriate volumes of sterile water) such that 1 μL will be required per 20 μL of PCR.

27. PCRs are run with a standard program: incubation at 50°C for 2 min, activation at 95°C for 10 min, and 40 cycles of denaturation at 95°C for 15 s and

annealing/extension (detection) at 72°C for 1 min. For SYBR Green I assays, a dissociation/melting curve should be run at the end of your amplification reaction. The purpose of the dissociation curve is to determine if anything other than the gene of interest is amplified in the PCR. Because SYBR green I will bind to any double-stranded DNA, nonspecific amplifications in your unknown wells will artificially increase fluorescence signal and make it impossible to accurately quantitate your sample. In addition, to facilitate post-PCR analyses, such as sequencing and restriction enzyme digestions, add a 5-min extension at 72°C at the end of PCR program.

28. The slope and Rsq values of the standard curve help determine the sensitivity of a given assay. PCR cycles that generate a linear fit with a slope between approximately −3.1 and −3.6 are considered acceptable. The linearity is denoted by the R squared (Rsq) value, which should be very close to 1 (> 0.985). Another quality indicator of your assay is that there should be a difference of approximately 3.3 in C_T values between two standard curve points with a 10-fold dilution.

29. Most PCR products will melt somewhere in the range of 80–90°C, although this melting point can vary with the size and sequence of your specific target. Ideally, the experimental samples should yield a single sharp peak within this temperature range, and the melting temperature should be the same in all the reactions. Furthermore, both NTC and NRTC should not generate significant fluorescent signal. If the dissociation curve reveals a series of peaks, it indicates that there is not enough discrimination between specific and non-specific reaction products, which would render optimization of the qRT-PCR necessary. Finally, the validity of your SYBR Green I assay should be verified by sequence and/or restriction enzyme digestion analysis of the PCR product.

References

1. Trayhurn P, Beattie JH (2001) Physiological role of adipose tissue: white adipose tissue as an endocrine and secretory organ. Proc Nutr Soc 60:329–339
2. Kershaw EE, Flier JS (2004) Adipose tissue as an endocrine organ. J Clin Endocrinol Metab 89:2548–2556
3. Laclaustra M, Corella D, Ordovas JM (2007) Metabolic syndrome pathophysiology: the role of adipose tissue. Nutr Metab Cardiovasc Dis 17:125–139
4. Fruhbeck G (2008) Overview of adipose tissue and its role in obesity and metabolic disorders. In: Yang K (ed) Adipose Tissue Protocols. Humana, Totowa, NJ, pp 1–22.
5. Rappolee DA, Wang A, Mark D, Werb Z (1989) Novel method for studying mRNA phenotypes in single or small numbers of cells. J Cell Biochem 39:1–11
6. Brady G, Iscove NN (1993) Construction of cDNA libraries from single cells. Methods Enzymol 225:611–623
7. Steuerwald N, Cohen J, Herrera RJ, Brenner CA (1999) Analysis of gene expression in single oocytes and embryos by real-time rapid cycle fluorescence monitored RT-PCR. Mol Hum Reprod 5:1034–1039
8. Higuchi R, Fockler C, Dollinger G, Watson R (1993) Kinetic PCR analysis: real-time monitoring of DNA amplification reactions. Biotechnology 11:1026–1030

9. Heid CA, Stevens J, Livak KJ, Williams PM (1996) Real time quantitative PCR. Genome Res 6:986–994

10. Fleige S, Pfaffl MW (2006) RNA integrity and the effect on the real-time qRT-PCR performance. Mol Aspects Med 27:126–139

11. Guan H, Arany E, van Beek JP, Chamson-Reig A, Thyssen S, Hill DJ, Yang K (2005) Adipose tissue gene expression profiling reveals distinct molecular pathways that define visceral adiposity in offspring of maternal protein-restricted rats. Am J Physiol Endocrinol Metab 288: E663–673

12. Thellin O, Zorzi W, Lakaye B, De Borman B, Coumans B, Hennen G, Grisar T, Igout A, Heinen E (1999) Housekeeping genes as internal standards: use and limits. J Biotechnol 75:291–295

13. Vandesompele J, De Preter K, Pattyn F, Poppe B, Van Roy N, De Paepe A, Speleman F (2002) Accurate normalization of real-time quantitative RT-PCR data by geometric averaging of multiple internal control genes. Genome Biol, 3, RESEARCH0034

14. Suzuki T, Higgins PJ, Crawford DR (2000) Control selection for RNA quantitation. Biotechniques 29:332–337

15. Zhong H, Simons JW (1999) Direct comparison of GAPDH, beta-actin, cyclophilin, and 28 S rRNA as internal standards for quantifying RNA levels under hypoxia. Biochem Biophys Res Commun 259:523–526

16. Schmittgen TD, Zakrajsek BA (2000) Effect of experimental treatment on housekeeping gene expression: validation by real-time, quantitative RT-PCR. J Biochem Biophys Methods 46:69–81

17. Murphy RM, Watt KK, Cameron-Smith D, Gibbons CJ, Snow RJ (2003) Effects of creatine supplementation on housekeeping genes in human skeletal muscle using real-time RT-PCR. Physiol Genomics 12:163–174

Chapter 20
Study of Adipose Tissue Gene Expression by *In Situ* Hybridization

Michel Grino

Summary Adipose tissue synthesizes factors involved in the body's homeostasis. Thus, measurements of messenger ribonucleic acid (mRNA) concentrations are important to study the involvement of adipose tissue in various physiological and pathophysiological conditions, in particular in obesity. Because adipose tissue is highly heterogeneous, containing both a stromal and an adipocyte compartment, each one having different cellular composition and functional capacities, *in situ* hybridization is a powerful tool to analyze the discrete expression of the mRNAs coding for the various factors synthesized within this tissue. Presented here is a detailed protocol for *in situ* hybridization of mRNAs in adipose tissue using ^{35}S-labeled single-stranded probes with sufficient details for the readers unfamiliar with histologic techniques. Included are details of tissue sectioning and preparation, probe synthesis, hybridization reaction, and macro- and microscopic signal detection.

Key words Adipose tissue; *in situ* hybridization; mRNA; riboprobe; macroautoradiography; microautoradiography.

1 Introduction

Adipose tissue has long been considered as a passive depot for storing excess energy in the form of triglycerides. However, a new concept of a tissue that actively regulates the pathways responsible for energy balance and whose activity is controlled by a complex network of humoral and neuronal signals has recently emerged (*1*). Indeed, adipose tissue synthesizes factors involved in the regulation of arterial blood pressure (angiotensinogen, renin receptor, angiotensin type 1 and 2 receptors), fibrinolysis (plasminogen activator inhibitor type 1), coagulation (tissue factor), metabolism (leptin, adiponectin, resistin, retinol binding protein 4, visfatin), angiogenesis (adrenomedullin, vascular endothelial growth factor), adipose tissue development and organization (transforming growth factor β, insulin-like growth factor [IGF]-I and IGF binding proteins, factors that are involved in the modeling of the extracellular matrix), inflammatory reactions and

From: *Methods in Molecular Biology, Vol. 456:*
Adipose Tissue Protocols, Second Edition. Edited by: Kaiping Yang
© Humana Press, a part of Springer Science+Business Media, Totowa, NJ

immune response (interleukin [IL]-1β, IL-6, IL 8, tumor necrosis factor α [TNFα], TNFα receptors), and local glucocorticoid metabolism and action (11β-hydroxysteroid dehydrogenase type 1 [11β-HSD-1] and glucocorticoid receptors [GR]) *(2)*. As a consequence, any change in adipose tissue mass and/or distribution will have significant consequences on the body's homeostasis. Central obesity, which is characterized by increased adipose tissue deposition at the abdominal visceral level, around the viscera, is part of the metabolic syndrome, which associates insulin resistance, high blood pressure, and dyslipidemia, leading to increased prevalence of diabetes mellitus and coronary artery disease, which remain the leading cause of death in developed countries *(3)*. Part of the metabolic and cardiovascular complications associated with the metabolic syndrome are subsequent to an increased synthesis and secretion of deleterious factors and a decreased synthesis and secretion of beneficial factors from adipose tissue *(1,4)*. Thus, the aforementioned observations underscore the importance of studying adipose tissue metabolism, both under physiological and pathophysiological conditions. Measurements of mRNA concentrations are a means to study gene transcription levels and/or mRNA stability. This can be achieved with either solution hybridization *(5)*, Northern blotting *(6)*, quantitative RT-PCR *(7)*, or *in situ* hybridization *(8)*. Because adipose tissue is highly heterogeneous, containing both a stromal and an adipocyte compartment, each one having different cellular composition and functional capacities, *in situ* hybridization is a powerful tool to analyze the discrete expression of the mRNAs coding for the various factors synthesized within this tissue.

2 Materials

2.1 Slide Preparation

1. Subbing solution: weigh 2.5 g of gelatin (from porcine skin, Type A, ≈ 300 Bloom, Sigma, or other), add 1 L of water, heat to dissolve with constant stirring (do not exceed 45–50°C), then let cool to room temperature with stirring and add 250 mg of chromium potassium sulphate (*see* **Note 1**).
2. Subbing slides: place slides in slide racks and soak for at least 1 hour in soap in tap water. Rinse well in running tap water, then in water once. Dip slides (slowly to avoid generating foam) into subbing solution, drain on paper towel, cover with aluminium foil and let dry overnight. The day after, dip the slides into subbing solution again and let dry. Store in slide boxes at room temperature.

2.2 Probe Labeling

1. The cDNA of interest should be inserted in a plasmid containing the T3, T7, or SP6 polymerase promoter, thus allowing the synthesis of high specific activity single-stranded RNA probes. Plasmids should be linearized before labeling. To obtain the

antisense probe (that is complementary RNA) or the sense probe (that is mRNA, which will be used as a control), linearize the plasmid to allow the transcription from the 3' to the 5' end or from the 5' to 3' end of the insert, respectively. Digest 20 μg of plasmid overnight in a final volume of 400 μl with the appropriate enzyme and buffer (*see* **Note 2**). Load 2 μL of the digestion reaction on a 0.8% agarose gel (load one lane with a DNA ladder), electrophorese, and examine under ultraviolet light. The digested plasmid should appear as a single band with the expected molecular weight (vector + insert), indicating a complete linearization. If more than one band is visible, add more enzyme to the digestion reaction and incubate for another 12 h. When the digestion is completed, extract the reaction with 200 μL of phenol (equilibrated with TE)/chloroform/isoamylalcool [50/49/1, TE is 10 mM Tris-HCl, pH 7.4; 1 mM EDTA prepared with MilliQ water]), then with 200 μL of chloroform/isoamylalcool (98/2) and precipitate the plasmid by adding 40 μL of a solution of 3 M sodium acetate, pH 4.6, and 1 mL of absolute ethanol. Mix and place in a −80°C or −20°C freezer for at least 30 min or 12 h, respectively. Then spin for 20 min at 20 000 g at 4°C, and discard the supernatant. The pellet should be briefly dried, resuspended in 40 μL of TE, and stored at −20°C (*see* **Note 3**).

2. 10 mM ribonucleosides triphosphate (NTP): mix 10 μL each of 100 mM ATP, CTP, and GTP (Roche Applied Science, or other) with 70 μL of MilliQ water. 0.5 mM UTP: mix 0.5 μL of 100 mM UTP with 99.5 μL of MilliQ water. Store at −70°C or lower (*see* **Note 4**).

3. RNasin (Promega, or other). Store at −20°C.

4. ^{35}S-UTP (specific activity > 1250 Ci/mmole) is from Perkin Elmer (1 mCi/80 μl) or Amersham (1 mCi/50 μL). Store at −70°C or lower.

5. T3, T7, or SP6 RNA polymerase (Promega, or other). Store at −20°C; avoid exposure to frequent temperature changes.

6. Transcription buffer: 40 mM Tris-HCl, pH 7.9; 6 mM MgCl$_2$; 2 mM spermidine; 10 mM NaCl; 100 mM dithitreitol (DTT). (*see* **Note 5**)

7. Dnase (Promega, or other). Store at −20°C.

8. 5 M DTT is prepared in MilliQ water and made fresh as required. tRNA (from baker's yeast, Roche Applied Bioscience or other) is dissolved in TE at a concentration of 25 mg/ml and stored at −20°C. 4 M LiCl is prepared with MilliQ water, filtered or autoclaved, and stored at room temperature.

2.3   Prehybridization Solutions

1. Phosphate-buffered saline (PBS): 154 mM NaCl, 16 mM Na$_2$HPO$_4$, 4.4 mM NaH$_2$PO$_4$; pH should be around 7.2 and does not need to be adjusted. Either prepare in advance and autoclave before storage at room temperature or make fresh.

2. Fixative solution: prepare a 12% solution of a 37/41% or 36.5/38% formaldehyde solution (Fisher Scientific or Sigma, respectively) in PBS. Alternatively, prepare a 4% (w/v) paraformaldehyde solution in PBS (heat with constant stirring in a fume hood; do not exceed 60° C and then cool to room temperature before use). Make fresh as required.

3. Acetylation solution: 0.1 M triethanolamine (Sigma), 154 mM NaCl, adjust pH to
   8.0 with 30% NaOH solution (about 4.5 ml/L); transfer to a bottle. Either prepare
   in advance and autoclave before storage at room temperature or make fresh.
4. Ethanol: 100%, 95%, 80%, and 70%.

2.4 Hybridization Solutions and Device

1. Hybridization buffer: 10% dextran sulfate; 50% formamide; 600 mM NaCl;
   10 mM Tris-HCl, pH 7.4; 1 mM ethylene diamine tetra-acetic acid (EDTA); 1X
   Denhardt's solution; 0.5 mg/mL transfer RNA. Store at −20°C. (*see* **Note 6**). 5 M
   DTT is prepared in MilliQ water and made fresh as required.
2. Incubation trays are home made using cell culture dishes (245 × 245 × 25 mm,
   Nunc D4803) and plastic pipets, as illustrated in **Fig. 20.1**.
3. Tray buffer: 50% formamide, 600 mM NaCl, 10 mM Tris-HCl, pH 7.4, 1 mM
   EDTA. Store at 4°C.
4. 3 MM chromatography paper from Whatman. Saran wrap (*see* **Note 7**).
5. Glass coverslips (from 24 × 24 mm up to 24 × 50 mm, depending upon the size
   of the tissue section; *see* **Note 8**).

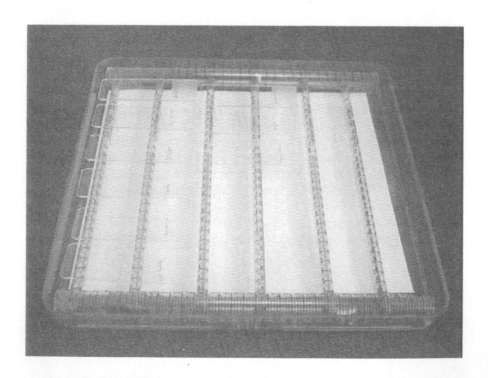

Fig. 20.1 Example of a home made hybridization tray

2.5 Washing Reagents

1. 20X SSC: $3M$ NaCl, $300\,\text{m}M$ sodium citrate, adjust to pH 7.2 with 3% NaOH solution; store at room temperature, does not need autoclaving.
2. β-mercapto-ethanol (β-ME, also 2-ME; Sigma, or other): obtained at a $14.4\,M$ solution. Store at 4°C.
3. RNase: dilute 100 mg of Rnase A (Roche Applied Science, or other) in 10 mL of 2X SSC, 50% glycerol, store at −20°C.

2.6 Signal Detection

1. Macroautoradiography: X-ray cassettes (18×24 or $24 \times 30\,\text{cm}$). BioMax MR films (Kodak). ^{14}C standards (General Electric Healthcare). GBX developer and fixer (Kodak), or an automatic film processor.
2. Microautoradiography: K5 (ILFORD) or NTB2 (Kodak) autoradiography emulsion, store at 4°C. Silica gel (Sigma, or other), or Drierite capsules (Hammond Drierite). Phenisol (ILFORD) or D19 (Kodak) developer. Hypam (ILFORD) or Kodak fixer (*see* **Note 9**).

3 Methods

3.1 Tissue Collection and Sections Preparation

1. Adipose tissue samples are frozen on dry ice as soon as possible after excision, wrapped into aluminium foil and kept at −70°C or lower. Do not use liquid nitrogen.
2. 20-μm thick frozen sections are cut in a cryostat, mounted on the gelatin-coated slides, placed on a piece of aluminium foil and allowed to dry at room temperature, and placed into slide boxes that are stored at −70°C or lower. The cryostat should be equipped with an independent specimen temperature control in order to maintain the adipose tissue sample at −35 to −40°C whereas the main refrigerating system is set at −25°C to −30°C. Depending upon the source of tissue (human subcutaneous adipose tissue obtained from obese patients consists of large adipocytes with little stroma and is difficult to cut) it could be useful to further refrigerate the knife with dry ice.

3.2 Probe Labeling

At room temperature, mix in an Eppendorf tube, 1 μL of linearized plasmid, 1 μL of 10 mM NTP solution, 0.5 μL of 0.5 mM UTP solution, 4 μL of 5X buffer, 2 μL of 100 mM DTT, 2.5 or 5.5 μL (when using ^{35}S-UTP from Perkin Elmer or Amersham, respectively) MilliQ water, 0.5 μL of RNasin, 8 or 5 μL of ^{35}S-UTP (when using ^{35}S-UTP from Perkin Elmer or Amersham, respectively), and 0.5 μL of RNA polymerase. Mix gently and incubate for 90 min at 37°C. Add 1 μL of Dnase, mix and incubate for 15 min at 37°C. Add 80 μL of MilliQ water, 10 μL of 4 M LiCl, and 300 μL of 100% ethanol, mix and place at −70°C or lower for at least 30 min.

Centrifuge at 20 000 g for 20 min at 4°C, remove supernatant (discard into radioactive waste), let stand at room temperature for 5–10 min to allow the remaining ethanol trapped in the pellet to evaporate, and resuspend in 100 μL of TE, 2 μL of 5 M DTT. Pipet 1 μL in a scintillation vial, add 10 mL of scintillation cocktail (Emulsifier-Safe, Perkin Elmer, or other) and count in a β-counter. Store the labeled probe at −70°C or lower. (*see* **Note 10**).

3.3 Prehybridization

1. The day before the experiment remove the slide boxes from the deep freezer and place at −20°C.
2. Place the slides on aluminium foil at room temperature for 10 min.
3. Place the slides in slide racks and incubate into fixative solution for 5 min if prepared with formaldehyde or for 15 min if prepared with paraformaldehyde (*see* **Note 11**).
4. Rinse twice in PBS.
5. Incubate into acetylating solution containing 0.25% acetic anhydride for 10 min (*see* **Note 12**).
6. Incubate into 70% ethanol 1 min, 80% 1 min, 95% 2 min, 100% 1 min, chloroform 5 min, 100% ethanol 1 min, 95% 1 min (*see* **Note 13**).
7. Stand upright to dry.

3.4 Hybridization

1. Hybridization solution: calculate the amount of solution needed (knowing that 0.05 μL/mm^2 is required, for example, 40 μL for a 24 × 32-mm coverslip) and add an extra 10%. Mix hybridization buffer with 5 M DTT (to obtain a final concentration of 100 mM), and probe (theoretical volume required to obtain 3 × 10^7 cpm/ml). Count 5 μL of hybridization buffer and adjust to 3 × 10^7 cpm/mL with additional probe (this is usually the case) or buffer and DTT (*see* **Note 14**).
2. Place a 39 × 24-cm piece of 3 MM paper into each hybridization tray and wet with 16 mL of tray buffer.
3. Pipet the required amount of hybridization solution to coverslips placed on a piece of BenchGuard in a fume hood. Avoid creating bubbles. Handle each slide upside down, touch it to the extremity of a coverslip and lower slowly, stopping when liquid contact is made (*see* **Fig. 20.2**). Flip the slide over when the cover-

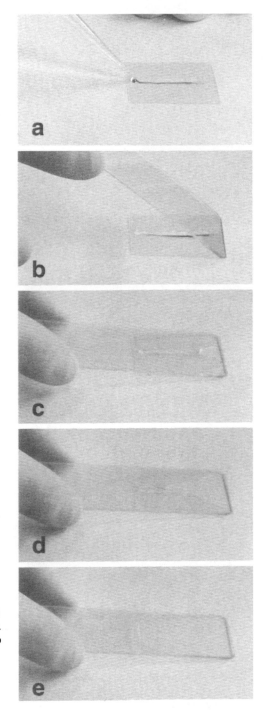

Fig. 20.2 Application of hybridization solution. (**A**) The hybridization solution is pipetted on the coverslip. (**B,C**) The slide (upside down) is slowly lowered on the coverslip until liquid contact is made. (**D,E**) Capillary action pulls the coverslip up to the slide, which is immediately flipped over and placed onto the hybridization tray. (Adapted from Simmons et al. [*11*])

slip is completely pulled up by capillary action, and rest it into the hybridization tray (*see* **Note 15**).
4. When a hybridization tray is filled up with slides, put the cover, wrap with Saran and incubate at 56°C for 20 h (*see* **Note 16**).

3.5   Washing

1. The hybridization trays should be opened in a fume hood. Place slides in slide racks and immerse into 500 mL of 2X SSC to allow the coverslips to slide off by capillarity (*see* **Note 17**).
2. Move the racks to a new container filled with 500 mL of 2X SSC and incubate for 30 min. Preheat 450 mL of 2X SSC to 30°C.
3. Pipet 500 μL of 10 mg/mL RNase into 50 mL of 2X SSC in a plastic tube, mix by vortexing, add to the preheated 2X SSC, mix and incubate the slides for 30 min at 37°C.
4. Incubate at room temperature into:

   1X SSC, 10 mM β-ME for 10 min twice
   0.5X SSC, 10 mM β-ME for 10 min
   0.1X SSC, 10 mM β-ME for 10 min (*see* **Note 18**)

5. Incubate at 65°C into 0.1X SSC, 10 mM β-ME for 30 min twice, let cool to room temperature (*see* **Note 19**).
6. Incubate into 0.1X SSC, 10 mM β-ME for 5 min at room temperature, rinse briefly in water and then in 70 % ethanol; blow dry with warm air.

3.6   Signal Detection

1. Macroautoradiographic detection: tape slides to cardboard or to a used X-ray film. Expose to BioMax MR together with ^{14}C standards. Exposure time is highly variable and should be determined for each gene studied. Develop in GBX developer for 3 min and fix in GBX fixer for 5 min, or in an automatic film processor.
2. Microautoradiographic detection: the dakroom should be equipped with a Kodak #2 Safelight or Wratten OB filter. Turn on water bath to 42°C. Scoop out approximately 7 mL of emulsion with a plastic spatula into a 15-mL plastic tube. Place the tube in the water bath for 10 min to melt, and then add an equivalent volume of water. Gently invert the tube several times to mix, then slowly pour the emulsion into a slide coating jar (*see* **Note 20**). Place the jar into the water bath and wait for one hour to allow air bubbles to rise. Dip the slides slowly into the emulsion, and then place the slides in an upright position. Let dry overnight, then place into black plastic slide boxes together with

10 g of silica gel (wrapped into a piece of gauze) or Drierite capsules, separated from the dipped slides by a clean slide. Tape the edges of the boxes with black tape and store at 4°C, protected from light and radioactivity. Develop slides for 4 min at 18°C with Phenisol (diluted 1 + 4) or with D19 (diluted 1+1) or at 20°C with full strengh ID19, rinse in water for 10 sec with slight agitation, fix for 5 min, wash in running tap water for 15 min, then in water for 5 min. Counterstain, if desired, for 3 min with 0.2% neutral red (*see* **Note 21**) or with other stain of choice.

Figure 20.3 shows an example of an *in situ* hybridization study of the mRNA coding for 11β-HSD-1 (the enzyme that converts inactive cortisone to active cortisol) in human abdominal adipose tissue. After hybridization of subcutaneous adipose tissue sections obtained from a lean subject with the antisense probe, an intense labeling was found in adipocytes and in the stromal compartment (stromal cell clusters, isolated stromal cells close to adipocytes, and walls of vessels; **Fig. 20.3A-C**, respectively). Hybridization with the sense probe did not give any detectable signal, demonstrating the specificity of the probe and of the hybridization procedure (**Fig. 20.3D**). 11β-HSD-1 mRNA expression was increased in the adipocyte compartment of subcutaneous adipose tissue obtained from an obese patient as compared with a nonobese patient (**Fig. 20.3, E and F**), and, in an obese patient, in the stromal compartment of visceral compared with subcutaneous tissue (**Fig. 20.3, G and H**). The above-mentioned observations demonstrate that 11β-HSD-1 mRNA is expressed in human adipose tissue, and suggest that the local reactivation of cortisol is increased in the adipose tissue of obese patients.

4   Notes

1. Unless stated otherwise, all solutions are prepared with Milli-RO 6 Plus water or equivalent, that is tap water with less 93–97% of ions, and less 1% of organic content, particles, and microorganisms. This standard is referred to as "water" in the text. Make fresh subbing solution each time.
2. Use the following formula to calculate the appropriate amount of enzyme: number of units = $1 \times 10^6 /$ (number of cleavage sites in λ DNA × size [in bp] of the plasmid [insert + vector]). Avoid using restriction enzymes that generates a protruding 3′ terminus because transcription will result in the synthesis of significant amounts of long molecules that are initiated at the terminus of the template (*9*).
3. All the solutions used for probe labeling or hybridization buffer should be prepared with water further purified with a MilliQ (or equivalent) system with a resistivity of 18.2 MΩ/cm, which is referred to as "MilliQ water" in the text. If desired, and in order to inactivate RNases, MilliQ water could be treated with diethylpyrocarbonate (DEPC: incubate with 0.1% DEPC for at least 12 h, then autoclave; be careful because native DEPC is a potent mutagen). However, I found it unnecessary to use DEPC-treated MilliQ water.

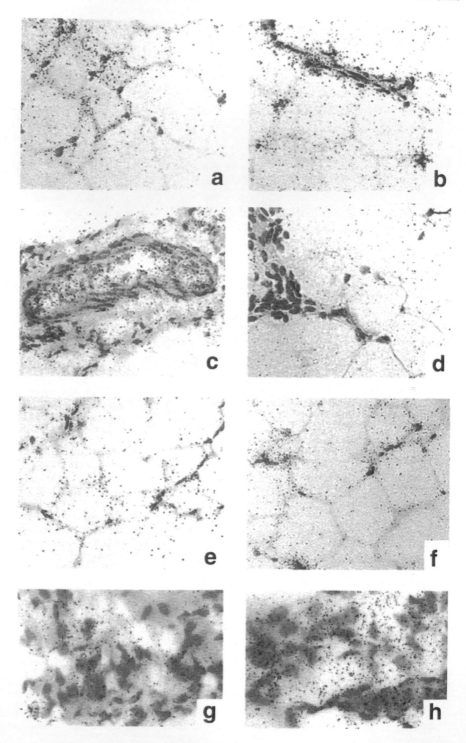

4. Even stored in a deep freezer, the shelf life of NTP is less than 1 yr.
5. The best results are obtained with the transcription buffer and the DTT provided with the enzyme.
6. To prepare 10 mL of hybridization buffer, weigh 1 g of dextran sulphate sodium salt (Amersham Bioscience, or other), add 2.6 mL of MilliQ water, and let stand overnight at 4°C. Add 5 mL of formamide (which should be of the highest possible quality, dispensed in aliquotes, and stored at −20°C), 1.2 mL of 5 *M* NaCl, 100 µL of 1 *M* Tris-HCl, pH 7.4; 50 µL of 0.2 M EDTA (adjusted to pH 8.0 with 30% NaOH); 200 µL of 50X Denhardt's solution (which is from Sigma [or other] or prepared as follows: dissolve 500 mg each of ficoll, polyvynilpyrrolidone, and bovine serum albumin [molecular biology grade] in a final volume of 50 ml of MilliQ water, filter, dispense in aliquotes and store at −20°C), and 200 µL of 25 mg/mL transfer RNA (from baker's yeast, Roche Applied Bioscience, or other, dissolved in TE).
7. Because high quality formamide is rather expensive, medium quality formamide (that is purity ≥98%) can be used to prepare the tray buffer.
8. The glass coverslips do not need any treatment before use. Simply avoid manipulating them with bare hands.
9. According to the manufacturer's instructions K5 emulsion is stable for at least two months. However, I routinely use the same batch of K5 up to 1 yr, without any appreciable increased background or decreased sensitivity. If Phenisol or D19 are not available, the developer (ID 19) can be prepared as follows: dissolve sequentially in 800 ml water 2.85 ml methanol, 72 g of sodium sulphite, 8.8 g of hydroquinone, 48 g of sodium carbonate, and 4 g of potassium bromure. Adjust to 1 L and store at 4°C in dark bottles.
10. Because ^{35}S emits β radiation, no shielding is necessary. However, it is an absolute requirement to work under a fume hood because ^{35}S degrades with time, even when stored in a deep freezer, and generates volatile compounds. Counting 1 µL of labeled probe should give between 1 and 2×10^6 cpm. Short (200–400 bp) probes give the best results. Longer probes can be hydrolyzed as described by Cox et al. *(10)*.

Fig. 20.3 Brightfield view of *in situ* hybridization for 11β-HSD-1 mRNA in adipose tissue from control or obese patients. Sections were counterstained with neutral red. Signal appears as silver grains. Expression of 11β-HSD-1 mRNA in subcutaneous adipose tissue from a lean patient **(A-D)**. Hybridization with the antisense probe: grain clusters are present over adipocytes **(A)**, stroma **(B)**, and walls of vessels **(C)**. **(D)** The result of a hybridization with the sense probe: note the lack of grain clusters, demonstrating the specificity of the probe and of the hybridization procedure. **(E)** and **(F)**, expression of 11β-HSD-1 mRNA in the adipocytes compartment of subcutaneous adipose tissue obtained from a lean **(E)** or an obese **(F)** patient. Expression of 11β-HSD-1 mRNA in the stromal compartment of subcutaneous **(G)** or visceral **(H)** adipose tissue obtained from an obese patient. Bar, 50 µm in **A-F** and 25 µm in **G** and **H**. Reproduced with permission from Paulmyer-Lacroix et al. *(12)*, Copyright 2002, The Endocrine Society

11. Although most, if not all, *in situ* hybridization protocols recommend to use baked or autoclaved dishes, I found that this precaution is not necessary and I use regular, hand-washed dishes. Similarly, I wear gloves to manipulate the dishes only during the radioactive steps.

12. Acetic anhydre should be stored protected from moisture (the bottle, whose cap is sealed with Parafilm, is kept in a closed recipient with some desiccant). Add acetic anhydre to the acetylation solution (placed into a bottle) at the last possible minute, shake vigorously to mix, and incubate immediately.

13. The same ethanol solutions can be used for the pre- and post-chloroform incubation. Chloroform should be stabilized with ethanol.

14. The labeled probe can be thawed and frozen several times. Make fresh hybridization solution each time.

15. Because hybridization solution is viscous and sticks to the wall of the tubes, be careful not to contaminate tip ejector and/or tip cone. In case of contamination, disassemble the pipette and wash the tip ejector and tip cone with diluted RBS detergent. Dispose properly of the tubes, tips, and BenchGuard into a radioactive waste box.

16. Although the hybridization temperature is a critical parameter of the hybridization reaction, I found not necessary to use a highly sophisticated oven. I use a regular, mechanically regulated oven.

17. The washing protocol is given for one, 50 slide rack. Please scale up or down as necessary. Use a rotator table for gentle agitation except for **steps 3** and **5**. Use specific slide racks and plastic containers throughout the washing procedure to limit RNase contamination. When removing the coverslips, it is convenient to maintain the racks 2–3 cm above the bottom of the container to allow the coverslips to slide off easily. Dispose of the first washing solution into radioactive waste. Check with local regulations to dispose of the radioactive waste.

18. Work under a fume hood and carefully cap the containers because β-ME is toxic and has a strong sulfurous odor.

19. Transfer slides into a container with washing solution at room temperature and place the container into a preheated 65°C water bath; preheat the second washing solution to 65°C to avoid heat shocking the slides.

20. If a slide coating jar (which has an approximate volume of 15 mL) is not available, a regular coplin jar with a block of clean slides to take up the excess space can be used. Exposure time is highly variable and should be determined for each gene studied. As a rule of thumb, the exposure time is at least three times those used for the BioMax MR film. It is convenient to hybridize and dip extra slides, which will be used to determine the appropriate exposure time.

21. Add 5 g of aluminium sulfate and 0.2 g of nuclear fast red to 100 mL of water, heat to dissolve and add a grain of thymol as a preservative. Store at room temperature protected from light. Filter before use.

Acknowledgments I thank Dr A Silaghi for her skillful help in photographic art.

References

1. Kershaw EE, Flier JS (2004) Adipose tissue as an endocrine organ. J Clin Endocrinol Metab 89:2548–2556
2. Hauner H (2005) Secretory factors from human adipose tissue and their functional role. Proc Nutr Soc 64:163–169
3. Park YW, Zhu S, Palaniappan L, Heshka S, Carnethon MR, Heymsfield SB (2003) The metabolic syndrome: prevalence and associated risk factors findings in the US population from the third National Health and Nutrition examination survey. Arch Intern Med 163:427–436
4. Matsuzawa Y (2006) The metabolic syndrome and adipocytokines. FEBS Lett 580:2917–2921
5. Walden PD, Ruan W, Feldman M, Kleinberg DL (1998) Evidence that the mammary fat pad mediates the action of growth hormone in mammary gland development. Endocrinology 139: 659–662
6. Wilson-Fritch L, Nicoloro S, Chouinard M, Lazar MA, Chui PC, Leszyk J, Straubhaar J, Czech MP, Corvera S (2004) Mitochondrial remodeling in adipose tissue associated with obesity and treatment with rosiglitazone. J Clin Invest 114:1281–1289
7. Desbriere R, Vuaroqueaux V, Achard V, Boullu-Ciocca S, Labuhn M, Dutour A, Grino M (2006) 11β-hydroxysteroid dehydrogenase type 1 mRNA is increased in both visceral and subcutaneous adipose tissue of obese patients. Obesity 14:794–798
8. Young WS (1992) Regulation of gene expression in the hypothalamus: hybridization histochemical studies. Ciba Found Symp 168:127–138
9. Schenborn ET, Mierendorf RC (1985) A novel transcription property of SP6 and T7 RNA polymerases: dependence on template structure. Nucl Acids Res 13:6223–6236
10. Cox, K.H., DeLeon DV, Angerer LM, Angerer RC (1984) Detection of mRNAs in sea urchin embryos by *in situ* hybridization using asymetric RNA probes. Dev Biol 101:485–502
11. Simmons DM, Arriza JL, Swanson SW (1989) A complete protocol for *in situ* hybridization of messenger RNAs in brain and other tissues with radiolabeled single-stranded RNA probes. J Histotechnol 12:169–181
12. Paulmyer-Lacroix O, Boullu S, Oliver C, Alessi M-C, Grino M (2002) Expression of the mRNA coding for 11β-hydroxysteroid dehydrogenase type 1 in adipose tissue from obese patients: an *in situ* hybridization study. J Clin Endocrinol Metab 87:2701–2705

Chapter 21
Application of Immunocytochemistry and Immunofluorescence Techniques to Adipose Tissue and Cell Cultures

Daniela Malide

Summary When isolated from tissue, white adipose cells are round, and their interior is filled with a large (80–120 µm) droplet of stored triglyceride, leaving a thin (1–2-µm) layer of cytoplasm between the lipid droplet and the plasma membrane. Their three-dimensional architecture, together with the fact that these cells ordinarily float in medium, have created major challenges when one attempts to perform microscopy techniques with these cells. Adipocytes serve as the principal energy reservoir in the body, and it is essential to overcome these difficulties to be able to study hormone-mediated responses in real adipose cells, which convey physiological significance that cannot be readily duplicated by the use of cultured model adipocytes. This chapter focuses on the use of confocal microscopy optical sectioning and computer-assisted image reconstruction in the whole adipose cell in the study of insulin-regulated protein trafficking. In addition, we illustrate the possibility to image whole-mount preparations of living adipose tissue, opening new ways to probe adipose cells in situ without disrupting their cellular interactions within living adipose tissue. Confocal microscopy constitutes an effective morphological approach to investigating adipose cell physiology and pathophysiology.

Key words Confocal microscopy; adipocytes; adipose tissue; immunofluorescence; glucose transport; lipid droplets; insulin action.

1 Introduction

The isolated rat adipose cell experimental system is the principal model for studies of the mechanism of insulin's stimulatory action on glucose transport, as reported *(1–5)*. Despite the successful use of this preparation in biochemical studies, the unique structure of the adipose cell, with its large central triglyceride storage droplet (80 µm) and thin (1–2 µm) rim of cytoplasm, has caused special problems for morphological approaches. Several methods have been used for studies of protein

From: *Methods in Molecular Biology, Vol. 456:*
Adipose Tissue Protocols, Second Edition. Edited by: Kaiping Yang
© Humana Press, a part of Springer Science + Business Media, Totowa, NJ

localization and trafficking by immunocytochemistry at both light microscopy and electron microscopy levels.

Studies that require sections of the adipose tissue and of the 3T3-L1 adipocyte cell line remain scarce because they are technically very difficult to prepare *(6–10)*. Although immunocytochemistry on ultrathin sections allows very precise subcellular localization, the sections give only a very limited view of the cell, and the sampling problem can become overwhelming. Thus, converting the information they provide into a three-dimensional (3-D) pattern may be difficult, especially for proteins that are unevenly distributed.

Other studies have focused on the use of plasma membrane sheets (lawns) as an assay for the translocation of the GLUT4 glucose transporter in response to insulin by light microscopy and electron microscopy immunocytochemistry *(11,12)*. This technique is limited to an examination of structures associated with the inner surface of the plasma membrane.

The specific aim of this chapter is to review an alternative morphological approach with the use of confocal microscopy optical sectioning and computer-assisted image reconstruction in the whole adipose cell, which we pioneered a decade ago and described in the first edition of this book *(13)*. As documented here, this approach allows the investigators to see the in situ localization of proteins and to trace the changes in response to different stimuli. Moreover, it overcomes some technical difficulties, particularly the sectioning, in studying the rat adipose cell and it opens a new, more accessible way to investigate the protein trafficking pathways in an insulin-responsive cell of physiological significance. In addition, we extend the use of confocal microscopy from isolated cells in suspension to intact adipose tissue. This facilitates visualization of responses in the adipose tissue in situ, without disturbing the interactions of adipose cells with those of the supportive stroma, containing vascular endothelial cells, macrophages, and poorly characterized stem cells. Given the importance of lipid and glucose handling by adipose cells for systemic insulin sensitivity, this offers an effective approach to investigating adipose cell physiology and pathophysiology that cannot be readily duplicated in model cell systems.

2   Materials

2.1   Adipose Cell Isolation (14)

1. Krebs-Ringer-bicarbonate-HEPES buffer (KRBH buffer) supplemented with 1% bovine serum albumin (BSA) and 200 n*M* adenosine.
2. Type I and type II collagenase; DNase I.
3. A shaking water bath and a low speed centrifuge.

2.2 Immunocytochemistry

1. Fixative: 4% paraformaldehyde (Electron Microscopy Sciences, Ft Washington, PA) in $0.15M$ phosphate-buffered saline (PBS), pH 7.4 (prepare fresh before each use).
2. Blocking and permeabilization buffer (B1) 1% BSA, 3% normal goat serum, 0.1% saponin in $0.15M$ PBS, pH 7.4.
3. Washing buffer (B2) 0.1% saponin in $0.15M$ PBS, pH 7.4.
4. Primary Antibodies: anti-GLUT4 affinity-purified rabbit IgG, 0.15 μg/mL, from Hoffman La Roche (Nutley, NJ); anti-beta tubulin mouse monoclonal ab JDR.3B8/ 1:100 from Sigma (St Louis, MO); anti-calnexin rabbit polyclonal ab SPA-860/ 1:100 from Stressgen Biotech (Victoria, British Columbia, Canada); anti-HA mouse monoclonal ab (HA-11),1 μg/mL, from Berkeley Antibody Co.(Richmond, CA); see also (13,15).
5. Secondary antibodies:

    Fluorescein isothiocyanate (FITC)-, Texas Red (TxR)-, and Cy5-conjugated antibodies specific for rabbit or mouse immunoglobulins (Ig), used at 15 μg/mL from Jackson ImmunoResearch.
    Rhodamine (Rhd)-conjugated lectin Lens culinaris agglutinin (50 μg/mL) from Vector Laboratories Inc.

6. Other materials: Vectashield mounting media from Vector Laboratories Inc.; glass microscopy slides, glass coverslips (# 1), plastic containers, a rocker platform.

3 Methods

The methods to obtain single cell suspensions of white and brown adipose cells from the rat adipose tissues are described in detail in other chapters of this text, and in scientific literature (14,16). Cells are incubated with insulin or other compounds at 37°C. Isolated rat adipose cells in suspension (floating cells) are used throughout the following staining protocols. Immunocytochemistry can be performed using direct or indirect techniques. Direct methods use a labeled specific (primary) antibody to bind directly (in one step) to the cellular epitope. Indirect methods comprise at least two steps: first an unlabeled primary antibody binds to the specific cellular epitope; in a second step the bound primary antibody is detected by a labeled-secondary antibody.

3.1 Indirect Immunofluorescence Methods Using Fixed and Permeabilized Cells (see Figs. 1 and 2; see Note 1)

1. Fix adipose cells in 4% PFA in $0.15M$ PBS, pH 7.4, for 20min at room temperature (RT). For ~ 1 mL packed cells, use 40mL of fixative solution with gentle shaking on a rocker platform (see Note 2). Use 15-mL or 50-mL polypropylene tubes (see Note 3).

2. Wash cells three times by centrifugation (210 g for 30 s) with 0.15 M PBS, pH 7.4 containing 50 mM glycine to quench aldehyde-induced nonspecific binding sites.

3. Permeabilize and block nonspecific binding sites with B1 for 45 min at RT. Use 1.5 ml polypropylene screw-cap tubes to incubate ~ 200 μL of packed cells in 1 mL of buffer B1.

4. Incubate cells with the primary antibody diluted in B1 for 2 h at RT with gentle shaking (see Note 4).

5. Aspirate the buffer beneath the floating cells using a long, fine pipet tip (gel loading tips) and add 100 μL of the antibody solution.

6. Wash cells three times, by centrifugation (210 g for 30 s) with buffer B2.

7. Incubate cells with the secondary antibody diluted in B1 for 45 min at RT with gentle shaking.

8. Wash cells three times by centrifugation (210 g for 30 s) with buffer B2.

9. Mount cells on glass slides using Vectashield mounting medium. Deposit a droplet of mounting media on a glass microscopy slide, add ~20 μL of packed cells, place a coverslip, and seal with nail polish.

Using this protocol, single, double, and triple labeling experiments can be successfully performed in isolated rat adipose cells (15,17–20).

3.2   Examination of Staining with a Fluorescence Microscope Equipped With a Confocal Laser Scanning System

We used a Nikon Optiphot 2 microscope attached to an MRC-1024 Bio-Rad confocal system controlled by Lasersharp image acquisition and analysis software from Bio-Rad Labs (see Notes 5–7). For each experimental condition, 10–15 cells were imaged separately by Kalman averaging 2–4 frames/image using a planapochromat 60X/1.4NA oil objective lens at optical zooms between 1 to 2.5. The excitation wavelengths used were 488-nm (FITC), 568-nm (Rhd), and 647-nm (Cy5), from a 15 mW krypton/argon laser. Images were acquired using 522DF32:605DF32, and 680DF32 bandpass emission filters. For three-dimensional (3-D) reconstruction, series of optical sections were collected at 0.5-μm intervals along the Z-axis. For presentation, digitized images were assembled, zoomed, and cropped using Imaris 5.0.3 software form Bitplane AG (Zurich, Switzerland) and montaged using Photoshop 7.0.2 software from Adobe Systems.

In our experience, careful optimization of fixation and permeabilization conditions allow satisfactory labeling of adipose cells for light microscopy. Confocal imaging allows precise visualization of fluorescent signals within a narrow plane of focus, with exclusion of out-of-focus blur, and the technique permits the reconstruction of 3-D structures from serial optical sections (21,22).

Several examples of the application of this technique for the localization of proteins used as compartment markers are illustrated: endoplasmic reticulum (see Fig. 21.1) and microtubules (see Fig. 21.2).

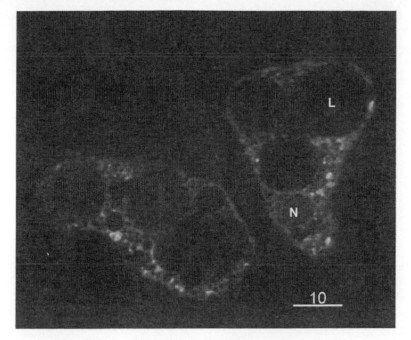

Fig. 21.1 Localization of the endoplasmic reticulum-marker calnexin in a single optical section of brown adipose cells. The staining appears intracellularly, displaying a reticular, honeycomb-like pattern throughout the cytoplasm and outlining the nuclear envelope. (N = nucleus; L= lipid droplet)

3.3 Cell Surface Immunofluorescence of Living Cells Stained in Nonpermeabilized State (see Fig. 21.3)

Living adipose cells can be immunostained using fluorescent-labeled lectins or antibodies in direct or indirect procedures.

3.3.1 Analysis of Protein Colocalization at the Cell Surface by Direct Staining of the Plasma Membrane Using Fluorescently Labeled Lectins (See Fig. 21.3A; Note 8)

1. Rinse adipose cells quickly with ice-cold $0.15\,M$ PBS, pH 7.4, and chill them to 4°C to stop protein trafficking.
2. Incubate cells with the rhodamine-labeled lectin Lens culinaris agglutinin for 30 min at 4°C in an ice/water bath with gentle shaking.
3. Wash cells three times with ice-cold $0.15\,M$ PBS, pH 7.4.
4. Fix cells with 4% PFA in $0.15\,M$ PBS, pH 7.4, for 20 min at RT.
5. Subsequently the cells can be permeabilized and stained using an antibody to a protein distributed in both intracellular and surface pools (e.g., GLUT4), following the steps described in **Subheading 3.1**.

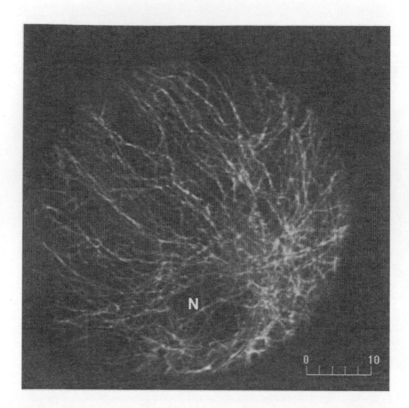

Fig. 21.2 Localization of the microtubule protein beta-tubulin in a whole white adipose cell. For 3-D reconstruction, series of optical sections are collected at 0.5-μm intervals along the z-axis. For visualization purposes a 3-D rendering is generated showing the extensive radial network of the microtubules (N = nucleus)

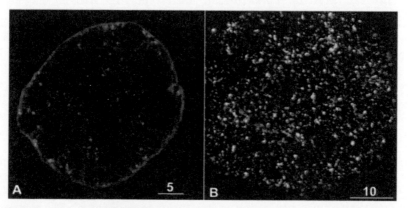

Fig. 21.3 Insulin induced GLUT4 localization to the cell surface. **(A)** Cross section. Sequential immunofluorescent-lectin staining of the plasma membrane followed by GLUT4 immunostaining on permeabilized brown adipose cells. In the absence of insulin, GLUT4 staining (green) shows a punctate pattern beneath and distinct from the plasma membrane identified by lectin staining (red). **(B)** Image tangential to cell surface. White adipose cells expressing HA-GLUT4-GFP incubated in the presence of insulin are immunostained for the cell surface HA-tag (red) using the protocol described in **Subheading 3.2, step 2** and GFP (total GLUT4) is visualized by direct fluorescence (green). In the merged image, the yellow color shows their high co-localization in most spots, indicating that GLUT4 is inserted in the plasma membrane

This allows direct examination (*see* **Fig. 21.3A**) and quantitative analysis of the protein localized at the cell surface *(15,20)*. Similarly, proteins present at the cell surface can be detected by indirect immunostaining using specific antibodies and performing the entire labeling procedure in an ice/water bath. Accurate quantitative analysis of the cell surface fluorescence can be performed and an example of MHC class I protein in brown adipose cells is illustrated in *(13,20)*.

3.3.2 Simultaneous Direct GFP (Green Fluorescent Protein) Detection and Indirect Antibody Immunostaining Using HA-GLUT4-GFP (see Fig. 21.3B)

GFP expression and localization can be visualized directly and can be used to "paint" particular cells or cellular processes. We have reported a chimeric GLUT4 molecule that has two tags: an HA tag in the extracellular loop of the GLUT4 molecule and a GFP at the cytoplasmic C-terminus *(23)*. Thus, GFP can be used a) in conjunction with the immunostaining of other proteins on fixed cells or b) for high-resolution imaging of subcellular events in living cells. Recent studies have successfully illustrated the latter possibility in rat adipocytes by time-lapse imaging of HA-GLUT4-GFP trafficking using confocal microscopy and total internal reflection fluorescence microscopy *(24,25)*. We illustrate the first possibility in **Fig. 21.3B**, showing insulin-induced GLUT4 translocation in cells expressing a HA-GLUT4-GFP-tagged protein. Transfected cells in the nonpermeabilized state are immunostained for the HA molecules using the protocol described in Section 3.2.1. Thus, only extracellulary exposed HA tag is detected and reveals the GLUT4 molecules inserted in the plasma membrane whereas GFP fluorescence reflects the total GLUT4 molecules.

3.4 Imaging Ex Vivo Intact Adipose Tissue by Confocal Microscopy (see Figs. 21.4 and 21.5)

Adipocyte cell size, shape, and number have a modulating effect on their metabolism and hormone actions and, therefore, morphological approaches that allow probing adipose cells in their native environment are highly desirable. As a step toward this goal, we chose to examine by fluorescence confocal microscopy ex vivo whole-mounts of epididymal adipose tissue. We selected regions that consist of only three to four layers of adipose cells, which allows imaging ~150 µm deep in the tissue using the confocal microscope. Deeper (250-µm) tissue penetration can be achieved by two-photon laser scanning microscopy using infrared pulses for excitation as recently reported *(26)*.

3.4.1   Autofluorescence of the Living Adipose
Tissue by Confocal Microscopy (see Fig. 21.4)

Although the precise origin of the autofluorescence of the unstained and unfixed adipose tissue remains to be elucidated, it depends on the excitation wavelength shown in **Fig. 21.4**. Images were obtained using a Zeiss 510 confocal system equipped with UV-Vis lasers (Carl Zeiss Inc, Jena, Germany) using a 20X/0.75NA Planapochromat objective, resolving up to ~ 120 μm deep into the adipose tissue. For higher details, a 40x/1.2NA C-Apochromat water immersion objective was used. Under 364-nm UV excitation, the signal comes mostly from structures in the cytoplasm of the adipose cells that appear brighter compared to the central lipid droplets (*see* **Fig. 21.4A**). In contrast, under 405-nm laser excitation, strong autofluorescence comes mostly from the lipid droplets and from the nuclei of the supporting stromal-vascular cells, whereas the cytoplasm appears as a dark rim (*see* **Fig. 21.4B**). The first row of adipose cells in contact with the coverslip is clearly visible and accurate measurements can be performed in this region. Cell borders can be resolved, and contour surfaces can be automatically segmented by software and thus surface areas, perimeters, diameters of individual cells can be obtained as it will be described (Wagner et al., manuscript in preparation).

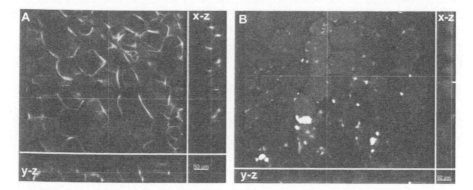

Fig. 21.4 Autofluorescence microscopy of live unstained whole-mount adipose tissue with confocal microscopy and UV illumination. Side-by-side autofluorescence series of images taken under 364-nm (**A**) and under 405-nm (**B**) laser illumination, respectively. Images are presented as single x-y-section; panels to the right-side and under **A** and **B** show x-z and y-z cross-sections taken at the crosshair lines, respectively. Note that, under 364-nm excitation (**A**), the outlines of the individual adipocytes can be easily visualized as bright rims compared to dark lipid droplets. In contrast, under 405-nm excitation (**B**), the autofluorescence comes from the lipid droplets of the individual cells whose cytoplasmic rim appears dark circles; very bright spots are consistent with the size and location of the nuclei of the stromal-vascular supporting cells

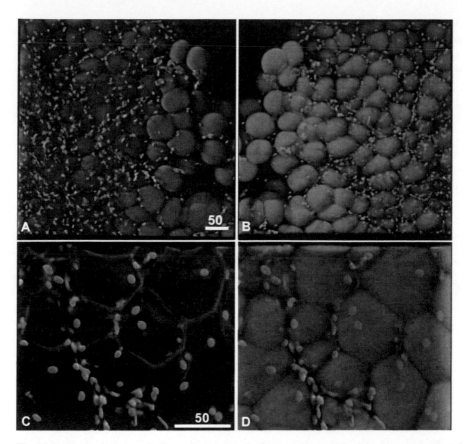

Fig. 21.5 Ex vivo imaging of adipose tissue stained with organelle specific dyes. Tissue was labeled with Hoechst (blue) to identify the nuclei, Bodipy 493/503 (green) for lipid droplets and MitoTracker Deep Red 633 (red) to identify mitochondria. Confocal microscopy images through ~150 μm thick region of the adipose tissue are reconstructed and 3-D "bottom-view"-closer to the coverglass (**A**) and "top-view" (**B**) is illustrated. Individual adipose cells can be easily identified from the stained unilocular/multilocular lipid droplet (green). Nuclei of the adipocytes and of the supportive stromal and vascular tissue are visible. MitoTracker reveals mitochondria but also labels collagen and reticular fiber network. Greater-magnification image (**C**) depicts the overlay of collagen fibers/mitochondria (red) and nuclei (blue) and image (**D**) the overlay with the adipose cells (green)

3.4.2 Confocal Microscopy of Fluorescently Labeled Cellular Compartments in Whole-Mounts of the Epididymal Adipose Tissue. (see Fig. 21.5)

Using fluorescent dyes improves depth penetration to ~150 μm into the tissue even using regular confocal microscope compared to autofluorescence imaging. Small fragments of the surgically excised epididymal adipose tissue were cut and incubated in culture media containing 5 μg/mL Hoechst 33342 (nuclei), 0.25 μg/ml

Bodipy® 493/503 (neutral lipids of the lipid droplets) *(27)* and/or 100 n*M* MitoTracker Deep Red 633 (mitochondria/collagen fibers) all from Molecular Probes (Eugene, OR) for 20 min at 37°C. The tissue fragments were washed with PBS and transferred to a coverglass-bottom culture chamber, incubated in culture media and the fluorescence was examined by confocal microscope. Images were acquired sequentially using a 364-nm (or 405-nm) laser line and emission between 420 and 480 nm for the autofluorescence signal and Hoechst 33342, a 488-nm laser line and emission between 505 and 550 nm for Bodipy® 493/503, and a 633-nm laser line with emission >650 nm for MitoTracker® Deep Red 633. The complex 3-D architecture of the adipose tissue can be visualized (*see* **Fig. 21.5**). Mature and small adipose cells are easily revealed by the staining of the lipid droplets (green) supported by stroma of collagen and reticular fibers (red; *see* **Note 9**), staining of the nuclei (blue) reveals clusters of small stromal vascular cells, endothelial cells occasionally in a pattern suggestive of blood capillaries.

4 Notes

1. We use confocal microscopy to investigate insulin action on isolated rat adipose cells. These techniques can be easily adapted for studies of the cultured 3T3-L1 adipocytes *(28)*. These methods are only guidelines for confocal microscopy of adipocytes, and alternative protocols remain to be designed for specific applications.
2. Other fixative and permeabilization procedures, such as −20°C methanol, can be used when working with isolated rat adipose cells or 3T3-L1 adipocyte cell line *(28–30)*.
3. In working with primary rat adipose cells, compared with cell lines, the major challenge derives from the fact that isolated adipose cells float in suspension throughout the staining and examination steps. Adipose cells in suspension are large, fragile cells, and should be manipulated gently, in order to avoid damage. These cells tend to lyse when contacting charged non-biological surfaces and particularly glass. Thus, uses of polypropylene and polyethylene containers, and the continued presence of BSA, are recommended throughout the procedures.
4. Adequate amounts of cells (~100–200 μL of packed cells) are required for these methods because the multiple washing steps will likely result in a significant loss of cells. In relation to the high number of cells in suspension, the antibodies should also be applied in excess.
5. Adipose cells float even when mounted on the microscopy slides; thus, working with an upright microscope is preferred. This is in contrast to cultured cells (attached to surfaces) when an inverted microscope is more appropriate.
6. A common problem with thick samples is light scattering interfering with accurate imaging beyond ~100 μm depth when using a conventional confocal microscope. This phenomenon often results in a progressive decrease in the fluorescent

signal in the optical sections taken further away from the coverslip. Thus, a multiphoton system may be considered to image the large size adipose cells and adipose tissue in the living animal and to perform accurate quantitative analyses (Wagner et al., manuscript in preparation).

7. For reasons that are unclear, long time storage of mounted cells on slides results in deterioration of the overall cell morphology and a loss of the lipid droplet(s). When possible, images should be acquired within a short time (days) after immunostaining.

8. When using the direct immunofluorescence of living cells it is important to ensure that the lectin-conjugate (or fluorescently-labeled antibodies) binds to carbohydrate moieties (or protein antigenic epitopes) exposed to the outer surface of the plasma membrane and is not allowed to enter the cytoplasm. If the cells to be stained are kept at 4°C to avoid endocytosis of the fluorochrome-conjugated lectin, the immunolabeling can be performed when cells are still alive. In our experience, this procedure results in much more reliable data, preventing possible fixative-induced redistribution of molecules within different domains of the plasma membrane. This method should be used for staining of aldehyde-sensitive epitopes exposed at the cell surface.

9. When staining live adipose tissue by incubating into MitoTracker Deep Red 633, in addition to labeling mitochondria there is also labeling of the collagen and fine reticular fibers, similar to previously reported labeling of reticular fibers by Cell Tracker Orange CMTMR (5-(and-6)-(4-chloromethyl) benzoyl) amino) tetramethylrhodamine *(31)*. This property can be useful to reveal the fiber network of the supporting stromal and vascular tissue.

Acknowledgments I thank Dr. Samuel W. Cushman for his generous support, helpful advice, and encouragement throughout this work and Drs. Ian A. Simpson and Evelyn Ralston for critical reading of the manuscripts and many helpful discussions.

References

1. Simpson IA and Cushman SW (1986) Hormonal regulation of mammalian glucose transport. Annu Rev Biochem 55:1059–1089
2. Birnbaum MJ (1992) The insulin-sensitive glucose transporter. Int Rev Cytol 137:239–297
3. Holman GD, Cushman SW (1994) Subcellular localization and trafficking of the GLUT4 glucose transporter isoform in insulin-responsive cells. Bioessays 16:753–759
4. Stephens JM, Pilch PF (1995) The metabolic regulation and vesicular transport of GLUT4, the major insulin-responsive glucose transporter. Endocr Rev 16:529–546
5. Bryant NJ, Govers R, James DE (2002) Regulated transport of the glucose transporter GLUT4. Nat Rev Mol Cell Biol 3:267–277
6. Blok J, Gibbs EM, Lienhard GE, Slot JW, Geuze HJ (1988) Insulin-induced translocation of glucose transporters from post-Golgi compartments to the plasma membrane of 3T3-L1 adipocytes. J Cell Biol 106:69–76
7. Martin S, Slot JW, James DE (1999) GLUT4 trafficking in insulin-sensitive cells. A morphological review. Cell Biochem Biophys 30:89–113

8. Slot JW, Geuze HJ, Gigengack S, Lienhard GE, James DE (1991) Immuno-localization of the insulin regulatable glucose transporter in brown adipose tissue of the rat. J Cell Biol 113:123–135
9. Smith RM, Charron MJ, Shah N, Lodish HF, Jarett L (1991) Immunoelectron microscopic demonstration of insulin-stimulated translocation of glucose transporters to the plasma membrane of isolated rat adipocytes and masking of the carboxyl-terminal epitope of intracellular GLUT4. Proc Natl Acad Sci U S A 88:6893–6897
10. Malide D, Ramm G, Cushman SW, Slot JW (2000) Immunoelectron microscopic evidence that GLUT4 translocation explains the stimulation of glucose transport in isolated rat white adipose cells. J Cell Sci 113:4203–4210
11. Robinson LJ, Pang S, Harris DS, Heuser J, James DE (1992) Translocation of the glucose transporter (GLUT4) to the cell surface in permeabilized 3T3-L1 adipocytes: effects of ATP insulin, and GTP gamma S and localization of GLUT4 to clathrin lattices. J Cell Biol 117:1181–1196
12. Voldstedlund M, Tranum-Jensen J, Vinten J (1993) Quantitation of Na+/K(+)-ATPase and glucose transporter isoforms in rat adipocyte plasma membrane by immunogold labeling. J Membr Biol 136:63–73
13. Malide D (2001) Confocal microscopy of adipocytes. Methods Mol Biol 155:53–64
14. Weber TM, Joost HG, Simpson IA, Cushman SW (1988) In: Kahn CR (ed) The insulin receptor. Alan R Liss Inc, New York, NY, Vol. part B, pp 171–187
15. Malide D, Dwyer NK, Blanchette-Mackie EJ, Cushman SW (1997) Immunocytochemical evidence that GLUT4 resides in a specialized translocation post-endosomal VAMP2-positive compartment in rat adipose cells in the absence of insulin. J Histochem Cytochem 45:1083–1096
16. Omatsu-Kanbe M, Zarnowski MJ, Cushman SW (1996) Hormonal regulation of glucose transport in a brown adipose cell preparation isolated from rats that shows a large response to insulin. Biochem J 315:25–31
17. Malide D, St-Denis JF, Keller SR, Cushman SW (1997) Vp165 and GLUT4 share similar vesicle pools along their trafficking pathways in rat adipose cells. FEBS Lett 409:461–468
18. Malide D, Cushman SW (1997) Morphological effects of wortmannin on the endosomal system and GLUT4-containing compartments in rat adipose cells. J Cell Sci 110: 2795–2806
19. Barr VA, Malide D, Zarnowski MJ, Taylor SI, Cushman SW (1997) Insulin stimulates both leptin secretion and production by rat white adipose tissue. Endocrinology 138:4463–4472
20. Malide D, Yewdell JW, Bennink JR, Cushman SW (2001) The export of major histocompatibility complex class I molecules from the endoplasmic reticulum of rat brown adipose cells is acutely stimulated by insulin. Mol Biol Cell 12:101–114
21. Matsumoto B (1993) Cell biological applications of confocal microscopy. Academic Press, San Diego, CA
22. Pawley J (1995) The handbook of biological confocal microscopy. IMR Press, Madison, WI
23. Dawson K, Aviles-Hernandez A, Cushman SW, Malide D (2001) Insulin-regulated trafficking of dual-labeled glucose transporter 4 in primary rat adipose cells. Biochem Biophys Res Commun 287:445–454
24. Oatey PB, Van Weering DH, Dobson SP, Gould GW, Tavare JM (1997) GLUT4 vesicle dynamics in living 3T3 L1 adipocytes visualized with green-fluorescent protein. Biochem J 327:637–642
25. Lizunov VA, Matsumoto H, Zimmerberg J, Cushman SW, Frolov VA (2005) Insulin stimulates the halting, tethering, and fusion of mobile GLUT4 vesicles in rat adipose cells. J Cell Biol 169:481–489
26. Larson DR, Zipfel WR, Williams RM, Clark SW, Bruchez MP, Wise FW, Webb WW (2003) Water-soluble quantum dots for multiphoton fluorescence imaging in vivo. Science 300:1434–1436
27. DiDonato D, Brasaemle DL (2003) Fixation methods for the study of lipid droplets by immunofluorescence microscopy. J Histochem Cytochem 51:773–780

28. Blanchette-Mackie EJ, Dwyer NK, Barber T, Coxey RA, Takeda T, Rondinone CM, Theodorakis JL, Greenberg AS, Londos C (1995) Perilipin is located on the surface layer of intracellular lipid droplets in adipocytes. J Lipid Res 36:1211–1226
29. Chakrabarti R, Buxton J, Joly M, Corvera S (1994) Insulin-sensitive association of GLUT-4 with endocytic clathrin-coated vesicles revealed with the use of brefeldin A. J Biol Chem 269:7926–7933
30. Martin S, Reaves B, Banting G, Gould GW (1994) Analysis of the co-localization of the insulin-responsive glucose transporter (GLUT4) and the trans Golgi network marker TGN38 within 3T3-L1 adipocytes. Biochem J 300:743–749
31. Miller MJ, Wei SH, Parker I, Cahalan MD (2002) Two-photon imaging of lymphocyte motility and antigen response in intact lymph node. Science 296:1869–1873

Chapter 22
Determination of Lipolysis in Isolated Primary Adipocytes

Srikant Viswanadha and Constantine Londos

Summary Lipolysis involves the sequential breakdown of triglycerides into free fatty acids and glycerol. The extent of lipolysis is therefore a key determinant of the energy status of an individual and also dictates insulin resistance. Here, we describe a protocol for estimating lipolysis in murine adipocytes. Glycerol released during the lipolytic reaction is estimated radiometrically to determine the extent of lipolysis within the cell and the data are normalized to cell number.

Key words Triglycerides; fatty acids; glycerol.

1 Introduction

The adipose tissue is a highly metabolic, multifunctional entity that regulates several endocrine and metabolic functions apart from energy storage in the form of triglycerides *(1)*. Energy status of the adipose depends on the net flux between triglyceride synthesis and adipocyte lipolysis. Under conditions of energy demand, such as during exercise or fasting, triglycerides are hydrolyzed to provide fatty acids as alternate sources of fuel. Aberrant lipolysis also results in increased fatty acids in circulation that, in turn, are predisposing factors for insulin resistance *(2)*.

Under basal conditions, triglyceride breakdown is minimal due to the protective effect of perilipin, which is a protein present on the surface of the lipid droplet *(3)*. However, lipolysis induced by beta adrenergic receptor agonists, such as epinephrine or isoproterenol, increases cAMP-dependent protein kinase activity via elevated adenylyl-cyclase and cAMP formation *(4)*. Phosphorylation of perilipin and hormone-sensitive lipase (HSL) caused by cAMP-dependent protein kinase causes a conformational change in the lipid droplet surface resulting in alleviation of the protective effect of perilipin and the migration of HSL to the droplet surface where lipolysis occurs *(5)*. More recently, it has been demonstrated that HSL is primarily involved in diglyceride breakdown while triglyceride hydrolysis is achieved by adipose triglyceride lipase (ATGL) *(6)*. Although the exact mechanisms involved in

From: *Methods in Molecular Biology, Vol. 456:*
Adipose Tissue Protocols, Second Edition. Edited by: Kaiping Yang
© Humana Press, a part of Springer Science + Business Media, Totowa, NJ

lipid hydrolysis are still debatable, the net effect of complete lipolytic breakdown of lipid is the release of free fatty acids and glycerol. Measurement of glycerol therefore provides an accurate estimate of adipocyte lipolysis.

2   Materials

2.1   Adipocyte Isolation and Lipolysis Assay

1. Adipocyte incubation solution (AIS): Krebs Ringer Bicarbonate HEPES buffer, containing $10\,mM$ sodium bicarbonate, and $30\,mM$ HEPES, pH 7.4, supplemented with 3% (w/v) fatty acid free bovine albumin fraction V (cat. no. 820025, ICN Biomedical Inc). Prepare fresh on the day of experiment (*See* **Note 1**).
2. Adipocyte wash buffer: $500\,nM$ adenosine (Sigma) in AIS. Prepare fresh on the day of experiment (*see* **Note 2**).
3. Collagenase buffer: $1–3\,mg/mL$ Type-1 collagenase (Worthington Biomedical Corp) in adipocyte wash buffer (*see* **Note 3**). Prepare fresh on the day of experiment.
4. Basal buffer for measuring basal lipolysis: $1\,U/mL$ adenosine deaminase (ADA; Calbiochem) and $100\,nM$ (-)-N^6-(2-phenyl-isopropyl)-adenosine (PIA; Sigma) in AIS (*see* **Note 4**). Prepare fresh on the day of experiment.
5. Stimulation buffer for measuring stimulated lipolysis: $1\,U/ml$ ADA and $10\,\mu M$ (-)-isoproterenol (ISO; Sigma) to AIS (*see* **Note 5**). Prepare fresh on the day of experiment.
6. Nylon mesh with 250-micron pore size.

2.2   Microscopy and Total Triglyceride Determination

1. Methylene blue: dissolve at $20\,mg/mL$ in water and stored at room temperature.
2. Precleaned micro slides: size 3×1"; thickness 0.93 to $1.05\,mm$ (Gold Seal®, Portsmouth, NH).
3. Microscopic cover slips: $22 \times 22\,mm$ No.1 (Fisher).
4. Spacers (Brinkmann Instruments; *see* **Note 6**).
5. $2N$ Dole's extraction mixture: isopropyl alcohol/heptane/$1\,N\ H_2SO_4$ (40:10:1).

2.3   Glycerol Assay

1. Assay Buffer: $2\,mM\ MgCl_2$ (Sigma), $100\,nM$ triethanolamine-HCl (Sigma), and $2\,mg/mL$ bovine serum albumin. Store in aliquots at $-20°C$.
2. Glycerol (Invitrogen) standards: prepare in AIS: 6000:3000:1500:750, and $375\,pmol$ glycerol/$50\,\mu L$.

3. 500 U/ml glycerokinase (Roche). Store at 4°C.
4. 50 mM ATP (Sigma). Store aliquots at −20°C.
5. [(-^{32P}] ATP (Perkin-Elmer).
6. Acid mix: 2 N HClO$_4$ (Sigma) and 0.2 mM H$_3$PO$_4$ (Mallinckrodt). Store at room temperature.
7. 100 mM ammonium molybdate (Sigma). Store at room temperature.
8. 200 mM triethylamine (Sigma). Store at room temperature.
9. Costar Plate Sealers (Corning Inc.).

3 Methods

Adipocyte lipolysis involves the concerted actions of several key enzymes; namely, ATGL, HSL, and monoglyceride lipase (MGL). ATGL catalyzes the removal a fatty acid from triglycerides resulting in diglyceride formation (6). However, HSL is chiefly instrumental in the breakdown of diglycerides to monoglycerides, which are further hydrolyzed into free fatty acids and glycerol by MGL. Because fatty acids are released in sequential steps of triglyceride breakdown, their estimation alone does not indicate complete hydrolysis of lipid. Therefore, in addition to free fatty acid estimation, analyses of residual diglyceride or monoglyceride concentrations within the cell are necessary to truly evaluate the extent of lipid breakdown. Determination of glycerol release therefore serves as a convenient and better approach to measure complete lipolytic breakdown of triglycerides within adipocytes.

3.1 Adipocyte Isolation

1. Harvest fresh fat pads from animals or subjects (See **Note 7**). Weigh and mince thoroughly (~2- to 3-mm pieces in diameter) in collagenase solution (3 mL per gram of tissue).
2. Transfer the mixture to a 30-mL narrow-mouthed polypropylene bottle (Nalgene) and incubate at 37°C with shaking at ~220 rpm for 1 h.
3. After digestion, filter the mixture through a 250-μm gauze mesh into a 50-mL conical polypropylene tube (Becton Dickinson; see **Note 8**) and allow to stand for 2–3 min. Adipocytes float on the top.
4. Remove the infranatant containing the collagenase solution using a long needle and syringe.
5. Add 10 mL of adipocyte wash buffer to the adipocytes and allow to stand for 2–3 min. Remove infranatant once again and add 10 mL of adipocyte wash buffer. Repeat the wash procedure three times.
6. After the final wash, centrifuge tubes containing the adipocytes in wash buffer for 30 s at 800 rpm.
7. Remove and discard the infranatant. Resuspend 25–50 μL of packed adipocytes in 5 mL of AIS for subsequent distribution into tubes for lipolysis assay (see **Note 9**).

3.2 *Microscopy*

1. Add 10 microliters of re-suspended adipocytes plus one drop of Methylene blue to a microfuge tube.
2. Insert spacer over cover slip. place 6 microliters of the above adipocyte suspension at the center of the cover slip.
3. View adipocytes under a confocal microscope.
4. Measure the diameters of approx. 100 cells.

3.3 *Determination of Total Triglyceride Content*

1. Add 1 mL of adipocyte suspension to 5 mL of 2 N Dole extraction mixture in glass tubes.
2. Add 3 mL of heptane and 2 mL of 0.9% NaCl to the tubes.
3. Incubate overnight at 4°C.
4. Centrifuge tubes at 1000 g for 30 min at room temperature.
5. Transfer upper phase to preweighed scintillation vials.
6. Dry under nitrogen at 37°C. After drying, weigh the vials to calculate the amount of lipid extracted from 1 ml of adipocyte suspension.

3.4 *Lipolysis*

1. Add 200 µL of adipocyte suspension (*See* **Note 10**) to 5 mL of polypropylene, 12 × 75-mm incubation tubes (*see* **Note 11**) containing 400 µL of basal or stimulation buffer and incubate for 1 h at 37°C. After incubation, pipet 200 µL of infranatant into microfuge tubes and store at −20°C until assayed for glycerol.
2. While the aforementioned tubes are incubating, add 200 µL of adipocyte suspension to tubes containing 400 µL of AIS and let stand at room temperature for 1–2 min.
3. Remove 200 µL of infranatant into microfuge tubes and store at −20°C. These samples serve as blanks for the glycerol assay.

3.5 Glycerol Assay *(see* Note 12*)*

1. Add 12 µL of 50 mM ATP to 5 mL of thawed assay buffer.
2. Add enough [(-^{32}P] ATP to reach 30000 counts per min (CPM) in 50 µL (*see* **Note 13**).
3. Add 20 µL of glycerokinase to the assay buffer.
4. Pipet 50 µL of blanks (AIS), standards, and unknowns into each well of a 96-well plate.

5. Add 50 μL of assay buffer to each sample. Seal plate with Costar plate sealer and incubate at 37°C for 30 min.
6. Add 100 μL of acid mix and seal plate.
7. Incubate at 90°C in a heating block or floating water bath (*see* **Note 14**) for 40 min.
8. Cool plate to room temperature. Add 50 μL of 100 mM ammonium molybdate and 50 μL of 200 mM triethylamine.
9. Centrifuge plate at 100 g for 20 min.
10. Count 100 μL of supernatant in a scintillation counter (*see* **Note 15**).

3.6 Calculations

1. Adipocyte radius (r) = diameter / 2.
2. Mean adipocyte mass = Mean adipocyte volume ($4/3 \, \pi r^3$) × density (0.915) (assuming that the fat cell is spherical and is composed of mainly TG).
3. Total TG in 1 mL of adipocyte suspension = wt. of extracted lipid based on Dole's extraction procedure.
4. Number of adipocytes per milliliter = Total TG in 1 mL of suspension / mean adipocyte mass.
5. Number of adipocytes per 200 μL (volume of adipocyte suspension added to the microfuge tubes for lipolysis) = Number of adipocytes per milliliter / 5 = Number of adipocytes per 600 μL of mix (adipocyte suspension + buffer for lipolysis; *see* **Note 16**).
6. Number of adipocytes per 50 μL (volume used for glycerol assay) = number of adipocytes per 600 μL of mix / 12.
7. Factor for 10^6 cells = 10^6 / number of adipocytes per 50 μL.
8. Convert CPM values obtained from glycerol samples to concentrations (pmol / 50 μL) based on the standard curve. Subtract the sample blanks and plate blanks accordingly. Glycerol (pmol / 10^6 adipocytes; *see* **Note 17**) = Glycerol concentrations (pmol / 50 μL) × factor for 10^6 cells.

4 Notes

1. All solutions should be prepared using deionized water. All apparatus used for this procedure must be thoroughly cleaned and glycerol-free. The Krebs Ringer Bicarbonate Buffer can be prepared several days in advance and stored at room temperature. However, BSA must be added only on the day of assay to make up AIS.
2. Optimal concentration of adenosine used in this assay was based on the finding that mouse adipocytes were approximately 10-fold less sensitive compared with rat adipocytes (unpublished observation, John Tansey and Constantine

Londos). The protocol is suitable for isolated mouse and rat primary adipocytes and will most likely work for human fat cells.

3. Because of the lot-to-lot variation in collagenase activity, the optimal concentration of collagenase may range from 1 to 3 mg/mL. Preliminary analysis of the optimum concentration of collagenase required to digest adipose tissue samples is therefore desirable before carrying out the actual experiment.

4. It is important to use ADA supplied in ammonium sulfate and not as glycerol suspensions. Aliquots of PIA can be stored at −20°C for several months.

5. Solutions of ISO are light sensitive. Tubes must therefore be wrapped in aluminum foil and stored in a 4°C refrigerator. Although slight browning may occur after several weeks of storage, the potency does not appear to be affected.

6. Use of spacers prevents the disruption of adipocytes without altering their morphology.

7. Fresh adipose tissue is absolutely necessary for the assay. Tissue samples may be placed in 1X PBS (pH 7.4) prior to weight determination and collagenase digestion.

8. Glassware should not be used anytime during adipocyte isolation and washes as it may result in adipocyte lysis.

9. The volume of packed adipocytes used for resuspension was based on our previous finding *(7)* that maximal lipolytic response is achieved using 10,000 to 15,000 cells per milliter of the final incubation medium (**Fig. 22.1**).

10. Adipocytes rapidly rise to the top on standing. Therefore, agitation of the tube is essential prior to the addition of adipocyte suspension to each microfuge tube containing lipolytic buffer to ensure homogeneity of the suspension.

11. Shaking speed depends on the geometry of the vessel and the volume of the incubation medium. Vigorous shaking might cause disintegration of cells while slow shaking could prevent adequate mixing of the adipocytes with the incubation medium. Therefore, a visual inspection of the suspension is essential to determine the optimum shaking speed. A uniform milky appearance of the adipose suspension assumes adequate mixing.

12. The radiometric glycerol assay *(8)* is used to measure glycerol levels in the low nmol range that cannot be detected by standard spectrophotometric procedures. Alternatively, bioluminescence *(9)* or fluorometric *(10)* methods can also be used to measure glycerol release from adipocyte lipolysis.

13. Optimum amount of radioactivity ranges from 30,000 to 40,000 CPM. Lower CPM can result in undetectable glycerol levels while CPM greater than 40,000 could result in greater background values.

14. If using a water bath for incubation, the plate has to be wrapped securely in an aluminum foil to prevent evaporation.

15. Supernatant (100 µL) has to be manually transferred from each well of the 96-well plate to scintillation vials unless a plate reader capable of measuring β-emission is used.

16. Pipet tips should be changed between each sample. Care should be taken to avoid contact with the precipitate as it contains high counts of excess $\gamma^{32}P$ that has not be transferred from $\gamma^{32}ATP$ to the glycerol-3-phosphate.

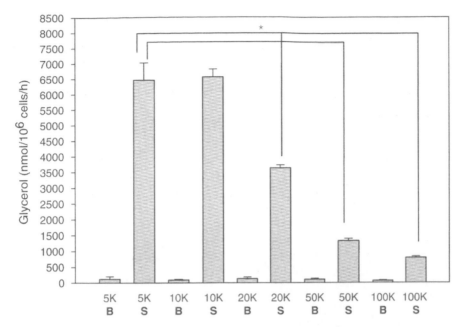

Fig. 22.1 Change in basal (**B**) and stimulated (**S**) lipolysis using different adipocyte concentrations. Suspensions of adipocytes were incubated at 37°C with shaking at 150 rpm. Values on the vertical axis represent glycerol produced (: ± SD) of triplicate incubations (from a single adipocyte isolation procedure). SD bars not visible were smaller than the thickness of the line describing the data bar. *Significant differences between treatments (stimulated lipolysis) joined by lines; $p < 0.05$. Overall treatment effect was also significant for basal lipolysis ($p < 0.05$). Basal activities were measured with ADA plus PIA, and stimulated with Isoproterenol plus ADA. The values under the bars indicate the total number of cells per incubation tube × 10³; thus, 10 K = 10,000 cells per 600 :L = ~16,000 cell per milliliter. (Reproduced from Viswanadha and Londos (7), with permission from J Lipid Res)

17. Because of the low concentration of glycerol in the incubation medium, it is likely that it is present as a homogenous solution rather than being present entirely in the infranatant. Therefore, the entire volume of 600 μL is used for calculating the concentration of adipocytes rather than just the 200 μL of infranatant extracted after incubation.

Data normalization is a critical factor in interpreting results. Cell protein concentrations are not suitable for normalization, because the vast majority of protein in a typical adipocyte suspension is contributed by the BSA added to protect the cells, whereas the protein contributed by the cells is but a miniscule fraction of the total protein present. Similarly, DNA present may be contributed by smaller cells that cling to the adipocytes (11). Thus, the most efficacious manner to normalize lipolysis data is by cell number.

Acknowledgments This project was supported by the NIDDK/NIH intramural research program.

References

1. Kershaw EE, Flier JS (2004) Adipose tissue as an endocrine organ. J Clin Endocrinol Metab 89:2548–2566
2. Boden G, Lebed B, Schatz M, Homko C, Lemieux S (2001) Effects of acute changes of plasma free fatty acids on intramyocellular fat content and insulin resistance in healthy subjects. Diabetes 50:1612–1617
3. Brasaemle DL, Rubin B, Harten IA, Gruia-Gray J, Kimmel AR, Londos C (2000) Perilipin A increases triacylglycerol storage by decreasing the rate of triacylglycerol hydrolysis. J Biol Chem 275:38486–38493
4. Steinberg D, Mayer SE, Khoo JC, Miller EA, Miller RE, Fredholm B, Eichner R (1975) Hormonal regulation of lipase, phosphorylase, and glycogen synthase in adipose tissue. Adv. Cyclic Nucleotide Res 5:549–568
5. Clifford GM, Londos C, Kraemer FB, Vernon RG, Yeaman GJ (2000) Translocation of hormone-sensitive lipase and perilipin upon lipolytic stimulation of rat adipocytes. J Biol Chem 275:5011–5015
6. Haemmerle G, Lass A, Zimmermann R, Gorkiewicz G, Meyer C, Rozman J, Heldmaier G, Maier R, Theussl C, Eder S, Kratky D, Wagner EF, Klingenspor M, Hoefler G, Zechner R (2006) Defective lipolysis and altered energy metabolism in mice lacking adipose triglyceride lipase. Science 312:734–737
7. Viswanadha S, Londos C (2006) Optimized conditions for measuring lipolysis in murine primary adipocytes. J Lipid Res 47:1859–1864
8. Brasaemle DL, Levin DM, Adler-Wailes DC, Londos C (1999) The lipolytic stimulation of 3T3-L1 adipocytes promotes the translocation of cytosolic hormone-sensitive lipase to the surfaces of lipid storage droplets. Biochim Biophys Acta 1483:251–262
9. Mauriege P, Imbeault P, Prud'homme D, Tremblay A, Nadeau A, Despres JP (2000) Subcutaneous adipose tissue metabolism at menopause: Importance of body fatness and regional fat distribution. J Clin Endocrinol Metab 85:2446–2454
10. You T, Berman DM, Ryan AS, Nicklas BJ (2006) Effects of hypocaloric diet and exercise training on inflammation and adipocyte lipolysis in obese postmenopausal women. J Clin Endocrinol Metab 89:1739–1746
11. Klyde BJ, Hirsh J (1979) Isotopic labeling of DNA in rat adipose tissue: evidence for proliferating cells associated with mature adipocytes. J Lipid Res 20:691–704

Chapter 23
Study of Glucose Uptake in Adipose Cells

Jun Shi and Konstantin V. Kandror

Summary Glucose is the main metabolic fuel in mammalian cells. Glucose entry into cells is facilitated by a family of ubiquitously expressed glucose transporter proteins. Typically, glucose transporters are localized on the plasma membrane. One notable exception is the glucose transporter isoform 4 (Glut4), which is specifically expressed in insulin sensitive tissues, i.e., skeletal muscle, heart muscle, and fat, and is responsible for the insulin effect on blood glucose clearance *(1)*. Under basal conditions, Glut4 is compartmentalized in intracellular membrane vesicles and thus has no access to the extracellular space. Upon insulin administration, Glut4-containing vesicles fuse with the plasma membrane and deliver the transporter to its site of action. As a result, Glut4 content on the plasma membrane is increased, and glucose uptake in the cell is significantly elevated. Here, we describe two complementary techniques. The first one uses tritiated 2-deoxyglucose and is designed to measure insulin-stimulated glucose transport into cultured adipose cells. The second allows one to quantify the degree of Glut4 translocation from an intracellular compartment to the plasma membrane.

Key words Insulin; Glut4; glucose; adipocytes; flow cytometry; cell differentiation.

1 Introduction

Insulin-dependent translocation of Glut4 to the plasma membrane causes a significant increase in the rate of glucose transport inside the cell. However, it is not yet totally clear whether the increase in glucose transport is directly proportional to the amount of Glut4 at the plasma membrane or whether there are additional mechanisms that regulate "intrinsic activity" of the transporter *(2)*. Therefore, it is essential to carefully compare the effect of insulin on glucose transport and on Glut4 translocation under different experimental conditions. It is possible that such studies may reveal new regulatory mechanisms and lead to the discovery of novel drug targets.

From: *Methods in Molecular Biology, Vol. 456:*
Adipose Tissue Protocols, Second Edition. Edited by: Kaiping Yang
© Humana Press, a part of Springer Science+Business Media, Totowa, NJ

From the point of view of accuracy and convenience, the best way to analyze Glut4 translocation from the intracellular pool to the cell surface may be to follow an epitope on the extracellular portion of the Glut4 molecule, which becomes accessible to specific antibodies added to the cell medium only after the Glut4 molecule is incorporated into the plasma membrane. Unfortunately, useful antibodies against endogenous exofacial epitopes of Glut4 are not yet available, so such an epitope has to be engineered into the molecule. Several different tags (mostly, Myc and HA) have been successfully used for this purpose. Of course, it is absolutely essential that the intracellular compartmentalization and insulin responsiveness of the exogenously expressed reporter molecule mimic that of endogenous Glut4. Luckily, available evidence suggests that this, indeed, is the case; at least when the expression levels of the exofacially tagged Glut4 do not exceed the levels of the endogenous transporter *(3)*. This, however, may not be true when the amounts of exogenous Glut4 significantly (approx. 10-fold) exceed the levels of endogenous Glut4 *(4)*. Therefore, care should be taken in each particular case to make sure that translocation of the exogenously expressed epitope-tagged Glut4 reflects the behavior of the endogenous transporter.

2 Materials

2.1 Insulin-Stimulated Glucose Uptake

1. Dulbecco's Modified Eagle's Medium (DMEM) without sodium pyruvate: supplement with 4.5 g/L D-glucose, 4.5 g/L L-glutamine, and 10% bovine serum or fetal bovine serum (*see* **Note 1**). Store at 4°C for up to 1 mo, protecting it from light.
2. Krebs Ringer HEPES buffer without glucose (KRH (−) glucose): 121 mM NaCl, 4.9 mM KCl, 1.2 mM $MgSO_4$, 0.33 mM $CaCl_2$, 12 mM HEPES, pH 7.4. Sterilize by filtration through a 0.2-μm filter and store at 4°C up to 1 yr.
3. Krebs Ringer HEPES buffer with glucose (KRH (+) glucose): 121 mM NaCl, 4.9 mM KCl, 1.2 mM $MgSO_4$, 0.33 mM $CaCl_2$, 12 mM HEPES, 25 mM D- (+)-Glucose, pH 7.4. Sterilize through a 0.2-μm filter and store at 4°C.
4. 0.1% SDS in KRH (−) glucose: 0.04 g SDS in 40 mL of KRH (−) glucose. Sterilize by filtration through a 0.2-μm filter and store at room temperature for 1–2 mo.
5. Cytochalasin B: 5 mM cytochalasin B (Sigma) in 100% EtOH. Store at −20°C (*see* **Note 2**).
6. Cold 2-DOG (100X stock solution): 100 mM 2-deoxy-D-glucose in KRH (−) glucose. Store in 50-μL aliquots at −80°C.
7. 100 μM Insulin (1000X stock solution): Dissolve 6 mg of insulin in 10 mL of 5 mM HCl and pass through a 0.2-μm filter. Store in 500-μL aliquots at −80°C for up to 1 mo. The aliquots are intended for single use only. Avoid repeated freeze/thaw cycles.
8. Radioactive glucose cocktail: Place 10 μL of ^{3}H-2-deoxy-D-glucose in ethanol: water solution (specific activity, 5–10 Ci (185–370 GBq) / mmol) into a 2.0-mL

tube, remove the cap, and leave open for 5 min to evaporate ethanol (Note: this is radioactive material. Handle with care, avoid spills, and put up signs around your work place to warn other people in the lab). Add 1.6 mL of KRH (−) glucose to the tube and 16 μL of cold 2-DOG (100X) stock solution (*see* **Note 3**).

2.2 Measurements of Glut4 Translocation by Flow Cytometry

1. Dulbecco's Modified Eagle's Medium (DMEM): same as in **Subheading 2.1., step 1.**
2. 100 μ*M* Insulin (1000X stock solution): same as in **Subheading 2.1., step 7.**
3. Phosphate-buffered saline without calcium and magnesium (PBS): 137 m*M* NaCl, 2.67 m*M* KCl, 8 m*M* Na_2HPO_4 1.47 m*M* KH_2PO_4. Dissolve 8 g of NaCl, 0.2 g of KCl, 0.2 g of KH_2PO_4 and 2.16 g of Na_2HPO_4 $7H_2O$ in 800 mL of distilled H_2O. Adjust pH to 7.4 with HCl, and add distilled H_2O to 1 liter. Sterilize the buffer by autoclaving for 20 min at 15 psi (1.05 kg/cm²) on liquid cycle or by filtration through a 0.2-μ*M* filter. Store at room temperature.
4. Phosphate-buffered saline with calcium and magnesium (PBS⁺⁺): 137 m*M* NaCl, 2.67 m*M* KCl, 8 m*M* Na_2HPO_4, 1.47 m*M* KH_2PO_4, 0.9 m*M* $CaCl_2$, 0.5 m*M* $MgCl_2$. Dissolve 8 g of NaCl, 0.2 g of KCl, 0.2 g of KH_2PO_4, 2.16 g of Na_2HPO_4 $7H_2O$, 0.1 g of $CaCl_2$ and 0.1 g of $MgCl_2/6H_2O$ in 800 mL of distilled H_2O. Adjust pH to 7.4 with HCl, and add distilled H_2O to 1 liter. Sterilize the buffer by filtration through a 0.2-μ*M* filter (Avoid autoclaving in order to prevent precipitation of calcium salts.) Store at room temperature.
5. PBS⁺⁺ with 5% bovine serum albumin (BSA) and 5% donkey serum: Add 5% (w/v) of BSA and 5% (v/v) of donkey serum to PBS⁺⁺ (*see* **Note 4**).
6. PBS with 5% fetal bovine serum: Add 5% (v/v) of fetal bovine serum to PBS (*see* **Note 5**).
7. Primary antibodies: Anti-myc monoclonal antibody (Cell Signaling Technology), non-specific mouse IgG (ZYMED Laboratories), phycoerythrin (PE)-conjugated monoclonal antibody against the transferrin receptor (BD Pharmingen).
8. Secondary antibody: PE-conjugated donkey F(ab)₂ anti-mouse IgG (Jackson Immunoresearch Laboratories).

3 Methods

3.1 Insulin-Stimulated Glucose Uptake

Generally, these experiments are performed with 3T3-L1 adipocytes cultured in 6-well plates between day 5 and day 8 of differentiation. If cells are growing in a different type of plate, change the volumes of the reagents using **Table 23.1**. Any experiment should include the following four conditions: (1) basal glucose uptake measured in the absence of insulin, (2) insulin-stimulated glucose uptake, (3) basal

Table 23.1 Volumes of solutions for measurements of insulin-stimulated glucose transport

Culture vessel	Surface area per well (cm²)	Relative surface area (vs. 6-well)	Volume of plating medium	Volume of KRH (−) glucose added after starvation	Volume of radioactive glucose cocktail
96-well	0.3	0.03	100 μL	45 μL	5 μL
24-well	2	0.2	500 μL	225 μL	25 μL
12-well	4	0.4	1 mL	450 μL	50 μL
35-mm	10	1	2 mL	900 μL	100 μL
6-well	10	1	2 mL	900 μL	100 μL
60-mm	20	2	4 mL	1.8 mL	200 μL
10-cm	60	6	10 mL	5.4 mL	600 μL

glucose uptake measured in the presence of cytochalasin B, and (4) insulin-stimulated glucose uptake measured in the presence of cytochalasin B. Each experimental condition should be analyzed in triplicates.

1. Rinse cells in each well twice with 2 mL of DMEM (without serum) warmed to 37°C. Add DMEM to cells slowly and carefully by the side of the well in order to avoid detachment of cells.
2. Incubate cells with 2 mL of serum-free DMEM per well for at least 2 h at 37°C.
3. Wash cells in each well twice with 2 mL of KRH (−) glucose buffer warmed to 37°C.
4. Prepare working solution of insulin or carrier (5 mM HCl) in KRH (−) glucose, warm it to 37°C and add 900 μL into each well for 13 min.
5. At the end of the 13 min incubation, add 5 μL of Cytochalasin B stock solution into wells defined as cytochalasin B-treated cells. Add 5 μL of 100% EtOH to other wells and shake gently. This step should take approx. 2 min.
6. Immediately after step 5, add 100 μL of the radioactive glucose cocktail to all wells.
7. Incubate samples at 37°C for 4 min and then transfer them on ice (*see* **Note 6**).
8. Aspirate the radioactive glucose cocktail, and add ice-cold KRH (+) glucose (2 mL per well) to terminate the reaction (*see* **Note 7**).
9. Repeat the washing step with ice-cold KRH (+) glucose for three more times.
10. Transfer to room temperature, add 400 μL of 0.1% SDS in KRH (−) glucose to each well, incubate at room temperature for 10 min, and thoroughly resuspend to homogeneity (*see* **Note 8**).
11. Transfer cell lysates to 1.5-mL Eppendorf tubes.
12. Vortex vigorously to break up clumps.
13. Put 300 μL of lysates in scintillation vials containing 3 mL of Ecolite(+)™ Liquid Scintillation Fluid (MP Biochemicals) and count in a scintillation counter for 1 min per vial.

14. These numbers represent "Counts in the samples" (*see* **Subheading 3.1., step 15**). In parallel, mix 10 μL of the radioactive glucose cocktail with 290 μL of 0.1% SDS in KRH (−) glucose and count this mixture under the same conditions. This number represents "Counts in the cocktail" (*see* **Subheading 3.1., step 15**).

15. Perform protein assay of every sample using BCA reagent (Pierce Biotechnology).

16. Use 0.1% SDS in KRH (−) glucose as blank.

17. Calculate the amount of intracellular 2-deoxyglucose using the following formula:

$$\left[\frac{[Counts\,in\,the\,sample]\times1000}{[Counts\,in\,the\,cocktail]\times0.03\times[C]\times t}\right]\text{pmol/mg}\times\text{min}$$

where [C] is protein concentration in mg/ml and t is the total time of incubation with radioactive glucose in min. Typically, insulin should increase glucose uptake in 3T3-L1 adipocytes 5- to 15-fold.

3.2 Measurements of Glut4 Translocation by Flow Cytometry

Because no specific antibodies against extracellular epitopes of Glut4 are currently commercially available; this protocol requires cells stably transfected with epitope-tagged Glut4. Several such cell lines have been described in the literature (*3,5–10*). Here, we describe the protocol for 3T3-L1 adipocytes stably expressing Glut4 with seven myc epitopes in the first extracellular loop and EGFP fused in frame with the cytoplasmic tail (myc$_7$-Glut4-EGFP). EGFP (green) fluorescence shows the total amount of the reporter Glut4 molecule expressed in the cell. This information may become useful when comparing different cell lines and conditions that may affect levels of myc$_7$-Glut4-EGFP expression. Staining of nonpermeabilized cells with the anti-myc antibody followed by PE-conjugated secondary antibody (i.e., red fluorescence) shows the amount of the reporter protein incorporated on the plasma membrane.

Because commercial antibodies generally are expensive, use plates with the smallest surface area possible. The following protocol was developed for cells grown in 6-well plates as a 10 cm² well usually provides sufficient number of 3T3-L1 adipocytes. For other cell types, such as undifferentiated 3T3-L1 pre-adipocytes, 12-well plates may be quite appropriate. In this case, volumes of all solutions should be adjusted accordingly (*see* **Table 23.2**). Each sample should contain at least 100,000 cells. In some cases, the technical quality of the data may be improved by taking more cells for the analysis.

Remember that depending on the type of the FACS machine, you may need to adjust the green and the red fluorescent channels. Therefore, in addition to cells analyzed in your experiment, you have to prepare additional plates of cells for

Table 23.2 Volumes of solutions for flow cytometry

Culture vessel	Surface area per well (cm^2)	Relative surface area (vs. 6-well)	Volume of plating medium	Volume of primary- or secondary antibody-containing solution	Volume of PBS with 5% fetal bovine serum for harvesting of cells
96-well	0.3	0.03	100 µL	50 µL	50 µL
24-well	2	0.2	500 µL	250 µL	250 µL
12-well	4	0.4	1 mL	500 µL	500 µL
35-mm	10	1	2 mL	1 mL	1 mL
6-well	10	1	2 mL	1 mL	1 mL
60-mm	20	2	4 mL	2 mL	2 mL
10-cm	60	6	10 mL	6 mL	6 mL

controls described below. In order to save the machine time, which can be expensive, do all the preliminary culturing work and cell staining (*see* **Subheading 3.2., steps 1-15**) before final measurements. Note that staining of "experimental" and "control" cells may be performed simultaneously as each individual step does not require much time. Bring an ice bath to the FACS facility with harvested stained cells in suspension and protected from light and perform cell sorting (*see* **Subheading 3.2., steps 16-24**).

1. Wash myc$_7$-Glut4-EGFP-expressing cells twice with PBS (2 ml/well) prewarmed to 37°C.
2. Transfer myc$_7$-Glut4-EGFP-expressing cells to DMEM without serum for at least 3 h.
3. Stimulate cells with 100 n*M* insulin for 15 min (*see* **Note 9**).
4. Quickly transfer the plates to 4°C and rinse three times quickly with ice cold PBS^{++}.
5. Incubate the cells with 1:500 dilution of anti-myc monoclonal antibody or nonspecific mIgG (with the same final concentration as the anti-myc antibody) in PBS^{++} with 5% BSA and 5% donkey serum for 1.5 h at 4°C.
6. Rinse the cells three times with ice-cold PBS^{++} (2 mL/well). Then wash twice with 2 ml/well of PBS^{++} for 5 min each time at 4°C.
7. Incubate the cells for 1 h at 4°C with 1:200 dilution of PE-conjugated donkey F(ab)$_2$ anti-mouse IgG secondary antibody (the reconstituted stock solution from Jackson Immunoresearch is 0.5 mg/mL) in PBS^{++} with 5% BSA and 5% donkey serum.
8. Rinse the cells three times with ice-cold PBS^{++} (2 mL/well). Then wash twice with PBS^{++} (2 mL/well) for 10 min each at 4°C. Rinse once quickly with PBS.
9. Gently scrape the cells in 1 mL of PBS with 5% fetal bovine serum. Place the tube with cell suspension on ice. These are "experimental cells".
10. While staining "experimental cells," transfer the plates with wild type 3T3-L1 adipocytes to 4°C and wash with ice cold PBS^{++} for three times.
11. Dilute PE-conjugated monoclonal antibody against the transferrin receptor (BD Pharmingen) 1:500 using PBS^{++} with 5% BSA and 5% donkey serum. Incubate cells with this solution for 1.5 h at 4°C.

12. Rinse the cells three times with ice-cold PBS++ (2 mL/well). Then wash them twice with 2 mL/well of PBS++ for 5 min each time at 4°C. Quickly rinse once with PBS at 4°C.

13. Gently scrape the cells in 1 mL of PBS with 5% fetal bovine serum. Place the tube with cell suspension on ice. These cells represent "red control".

14. Simultaneously with **steps 8–12**, wash myc$_7$-Glut4-EGFP-expressing 3T3-L1 adipocytes with PBS three times. Gently scrape the cells in 1 mL of PBS with 5% fetal bovine serum. Place the tube with cell suspension on ice. These cells represent the "green control".

15. Simultaneously with **steps 12–14**, wash wild type 3T3-L1 adipocytes with PBS three times. Gently scrape the cells in 1 mL of PBS with 5% fetal bovine serum. Place the tube with cell suspension on ice. These cells represent "background control".

16. Filter "background control" cells (**step 15**) through a 40-μM cell strainer (BD Falcon; *see* **Note 10**).

17. Following the manufacturer's instructions, adjust the forward scatter and the side scatter settings so that the whole population of single cells fits into the window. A minor fraction of dead cell and cell clumps at the upper left corner of the chart is usually acceptable.

18. Adjust the settings of both green fluorescent and red fluorescent channels in order to have the peak of the background fluorescence at approx. 10.

19. Filter the "green control" cells (**step 14**) through a 40-μM cell strainer (*see* **Note 10**).

20. Perform the flow cytometry analysis. If required, readjust color compensation in the red channel to the background level, as any signal in the red channel is caused by "leaking" green fluorescence.

21. Filter the "red control" cells (**step 13**) through a 40-μM cell strainer (*see* **Note 10**).

22. Perform the flow cytometry analysis. If required, readjust color compensation in the green channel to the background level, as any signal in the green channel is caused by "leaking" red fluorescence.

23. Filter the suspension of "experimental cells" through a 40 μM cell strainer (*see* **Note 10**).

24. Perform the flow cytometry analysis; measure red and green fluorescence.

25. For basal cells, subtract the median value (10) of red fluorescence obtained with nonspecific mIgG from red fluorescence obtained with anti-myc antibodies. Repeat the same operation for insulin-treated cells. Calculate the insulin-induced increase in red fluorescence.

4 Notes

1. 3T3-L1 pre-adipocytes are grown in DMEM supplemented with 10% bovine serum. Differentiating cells and 3T3-L1 adipocytes should be cultured in DMEM supplemented with 10% fetal bovine serum. Importantly, fetal bovine serum obtained from different vendors and even different lots of serum from the same vendor may have different biological activity in respect to differentiating 3T3-L1 cells. It is highly recommended to obtain small (50–100 mL) samples

from different vendors, and to test their ability to maintain differentiation of 3T3-L1 preadipocytes simultaneously split from one plate. The most potent and cost effective serum should be purchased in large quantities for future use. Fetal bovine serum should be stored at −20°C protected from light. After thawing a bottle at 4°C in the dark overnight, 50-mL aliquots should be prepared and stored at −20°C in the dark. Each 50-mL aliquot is intended for single use only and should not be re-frozen. It is not necessary to inactivate fetal bovine serum by heat for use with 3T3-L1 cells.

2. Cytochalasin B is a cell-permeable mycotoxin. It potently inhibits glucose transport by direct binding to glucose transporters. Therefore, glucose uptake measured in cytochalasin B-treated cells corresponds mainly to passive diffusion. Treatment of cells with insulin should not significantly change this parameter.

3. The total volume of the radioactive glucose cocktail required for the experiment will depend on the number and size of your samples. Each well of a 6-well plate requires 100 μl of the cocktail. Adjust the total volume of the cocktail according to **Table 23.1** (*see* **Note 6**), but keep the ratio between radioactive and non-radioactive 2-deoxyglucose.

4. This solution should be prepared on the day of the experiment. Calculate the total volume of this solution considering 2 mL for each well of a 6-well plate. If longer storage is required, add NaN_3 to the final concentration of 0.02% and store at 4°C for a few days. NaN_3 is highly toxic, avoid inhalation and skin contacts. NaN_3 should be prepared in distilled water as a 2% stock solution. Keep the stock solution tightly closed and store in a cool dry place. Note that PBS with calcium and magnesium is used to facilitate cell adhesion to plates.

5. This solution should also be prepared on the day of the experiment. Calculate the total volume of this solution considering 1 mL for each well of a 6-well plate. Note that PBS without calcium and magnesium is used in order to reduce adhesion between cells, and to avoid formation of cell clumps.

6. Incubation of 3T3-L1 adipocytes with the radioactive glucose cocktail at 37°C for 3–4 min is sufficient to obtain significant readings of ^3H. Other cells, however, may require longer incubations. For example, when glucose transport is measured in 3T3-L1 pre-adipocytes, the incubation should be extended to 15 min.

7. At this step, it is extremely important to be as fast as possible. Hold a Pasteur pipette connected to the vacuum pump in one hand and a pipet aid in the other. Immediately after the radioactive glucose cocktail is aspirated with the help of the vacuum pump, add ice-cold KRH (+) glucose using the pipet aid. Keep plates on ice and use ice-cold buffer for this will considerably diminish residual glucose uptake during the washing steps. Remember that, at this point, it is more important to act fast than to apply accurate volumes of KRH (+) glucose. If the resulting numbers show significant experimental error, try to add 10 μM of cytochalasin B into ice-cold KRH (+) glucose during the first wash.

8. Use 400 μL of 0.1% SDS in KRH (−) glucose regardless of the plate format. Note that the lysate may become viscous, especially at low temperatures. Use a 1-mL pipet set at 300 μL to carefully re-suspend the lysate. Be very cautious because the samples are radioactive. Try not to produce foam as it will decrease recovery.

9. There are two ways to add insulin to cultured cells. You can remove the starvation media by aspiration and add fresh DMEM into each well. Then, add the required amount of the $100\,\mu M$ insulin stock solution or carrier ($5\,mM$ HCl) directly into the wells with the plates tilted to 45 degrees. Avoid touching the cell monolayer with the tip of the pipet. Alternatively, prepare the required volume of DMEM with $100\,\mu M$ insulin or carrier in a 15-mL or a 50-mL tube. Remove the starvation media by aspiration and add insulin-containing DMEM to each well.

10. Filtration through the cell strainer helps to break cell clumps and avoid clogging of the FACScan machine. Filter the cell suspension immediately before the analysis to avoid reaggregation.

References

1. Saltiel AR and Kahn CR (2001) Insulin signalling and the regulation of glucose and lipid metabolism. Nature 414:799–806
2. Kandror KV (2003) A long search for Glut4 activation. Sci STKE 2003:PE5
3. Shi J, Kandror KV (2005) Sortilin is essential and sufficient for the formation of Glut4-storage vesicles in 3T3-L1 adipocytes. Dev Cell 9:99–108
4. Carvalho E, Schellhorn SE, Zabolotny JM, Martin S, Tozzo E, Peroni OD, Houseknecht KL, Mundt A, James DE, Kahn BB (2004) GLUT4 overexpression or deficiency in adipocytes of transgenic mice alters the composition of GLUT4 vesicles and the subcellular localization of GLUT4 and insulin-responsive aminopeptidase. J Biol Chem 279:21598–21605
5. Kanai F, Nishioka Y, Hayashi H, Kamohara S, Todaka M, Ebina Y (1993) Direct demonstration of insulin-induced GLUT4 translocation to the surface of intact cells by insertion of a c-myc epitope into an exofacial GLUT4 domain. J Biol Chem 268:14523–14526
6. Kishi K, Muromoto N, Nakaya Y, Miyata I, Hagi A, Hayashi H, Ebina Y (1998) Bradykinin directly triggers GLUT4 translocation via an insulin-independent pathway. Diabetes 47:550–558
7. Ueyama A, Yaworsky KL, Wang Q, Ebina Y, Klip A (1999) GLUT-4myc ectopic expression in L6 myoblasts generates a GLUT-4-specific pool conferring insulin sensitivity. Am J Physiol 277:E572–E578
8. Karylowski O, Zeigerer A, Cohen A, McGraw TE (2004) GLUT4 is retained by an intracellular cycle of vesicle formation and fusion with endosomes. Mol Biol Cell 15:870–882
9. Govers R, Coster AC, James DE (2004) Insulin increases cell surface GLUT4 levels by dose dependently discharging GLUT4 into a cell surface recycling pathway. Mol Cell Biol 24:6456–6466
10. Bogan JS, McKee AE and Lodish HF (2001) Insulin-responsive compartments containing Glut4 in 3T3-L1 and CHO cells: regulation by amino acid concentrations. Mol Cell Biol 21:4785–4806

Chapter 24
Measurement of Phosphoinositide 3-Kinase and Its Products to Study Adipogenic Signal Transduction

Alexander Sorisky and AnneMarie Gagnon

Summary Adipogenesis is an important component of adipose tissue development and growth. Alterations in adipogenesis may promote adipose tissue insulin resistance and inflammation. The ability of preadipocytes to differentiate into mature adipocytes depends on the activation of phosphoinositide 3-kinase (PI3K). This chapter describes the methodology used to measure the cellular accumulation of phosphoinositide products of PI3K. This approach involves labeling the cells with myo-[2–^{3}H]-inositol, extraction and deacylation of the phosphoinositides, and HPLC separation of the deacylated derivatives. The assay of PI3K activity itself is also described in detail. The ability to analyze PI3K and its phosphoinositide products is a useful tool for ongoing endeavours to understand adipogenesis and adipose tissue dysfunction.

Key words Preadipocyte; adipocyte; adipogenesis; phosphoinositide 3-kinase; inositol phospholipid; HPLC; insulin.

1 Introduction

The study of signal transduction pathways that regulate adipogenesis owes much to the development of the mouse 3T3-L1 cell line, originally established from day 17 disaggregated Swiss 3T3 mouse embryos (1,2). These cells have provided a reproducible and responsive model system to study adipocyte differentiation. At the same time, ongoing efforts by many laboratories have generated protocols that have made primary culture of adipose stromal preadipocytes more accessible for studies on adipogenic signal transduction (3).

When caloric balance is chronically positive, which is a situation that is occurring on a global scale, adipose tissue accumulates as it stores excess energy as triacylglycerol. This expansion occurs via enlargement of existing adipocytes as well as through the recruitment of preadipocytes and their differentiation into adipocytes. Recent studies suggest that significant obesity-associated insulin resistance and

From: *Methods in Molecular Biology, Vol. 456:*
Adipose Tissue Protocols, Second Edition. Edited by: Kaiping Yang
© Humana Press, a part of Springer Science + Business Media, Totowa, NJ

inflammation occurs when adipogenesis wanes and compensatory adipocyte hypertrophy develops. Hypertrophied adipocytes are insulin-resistant and adipokine production is shifted towards a proinflammatory profile *(4)*. Therefore, it is of interest to study the signaling pathways that regulate adipogenesis.

Insulin and IGF-1 were identified early on as proadipogenic agonists. Over the years, many cytosolic downstream signaling molecules activated by these agonists have been implicated in the induction of adipocyte differentiation, including Ras, Raf, ERK1/2, p38 MAPK, and, the focus of this chapter, type 1 phosphoinositide 3-kinase (PI3K).

PI3K acts on phosphatidylinositol-4,5-bisphosphate, abbreviated as PI(4,5)P2, to form PI(3,4,5)P3. PI(3,4,5)P3 can be converted to PI(3,4)P2 through dephosphorylation at the 5 position. The measurement of PI(3,4,5)P3 and PI(3,4)P2 is challenging because their levels in the membrane are two to three orders of magnitude lower than PI(4,5)P2, from which they must be distinguished. These two 3-phosphorylated phosphoinositides recruit a variety of effector proteins, such as protein kinase B (PKB, also known as Akt), but the array of targets for each of the lipids appears to only partially overlap, suggesting unique cellular effects *(5)*. Therefore, it is important to measure both of these lipids to study PI3K signal transduction, and this can be accomplished by the assay for radiolabeled phosphoinositides described here.

For applications in which only the measurement of PI(3,4,5)P3 is required, two approaches can be considered. Thin-layer chromatography methods have been described for cellular extracts of [^{32}P]-labeled phosphoinositides; they will resolve PI(3,4,5)P3, but will not distinguish PI(3,4)P2 from PI(4,5)P2 *(6)*. There is also an enzyme-linked immunosorbent assay-based commercial PI(3,4,5)P3 kit from Echelon. Forced cellular expression of various tagged-PH domains with relative (but not complete) specificity for different phospohinositides have also been used to visualize subcellular location of these lipids by confocal microscopy and for use in mass assays *(7)*.

2   Materials

2.1   *Phosphoinositide Analysis*

1. Equipment: Laminar flow hood; cell culture incubator with 10% CO_2 atmosphere; Speedvac, that can be fitted interchangeably with an organic or aqueous trap; nitrogen gas tank; scintillation counter.
2. Cell culture reagents: 3T3-L1 preadipocytes (ATCC) or human stromal preadipocytes isolated according to Hauner *(3)* or commercially available (Zen-Bio or Cambrex); Dulbecco's modified Eagle's medium (DMEM) supplemented with serum (percentage and type indicated in Methods), antibiotics; inositol-free DMEM (Specialty Media, cat. no. SLM-100).
3. Inositol derivatives: myo-[2–^{3}H]-inositol in sterile H_2O (cat. no. ART0116A), [^{3}H]-PI(4,5)P2 and [^{3}H]-PI4P (cat. no. ART0183–5 and ART0185–5), all from American Radiolabeled Chemicals; [^{3}H]-inositol 1,3,4 trisphosphate (cat. no. NET990), [^{3}H]-

inositol 1,3,4,5 tetrakisphosphate (cat. no. NET941), [³H]-inositol 1,4,5 trisphosphate (cat. no. NET911), all from PerkinElmer; carrier phosphoinositides (Sigma, cat. no. P-6023) dissolved at 8.3 mg/mL in $CHCl_3$:MeOH:H_2O, 20:9:1; v:v:v.

4. Plastic ware: polypropylene tubes: 11.5-mL (100 × 15.7-mm) and 8-mL (100 × 13-mm) tubes, 2 mL Spin-X tubes fitted with 0.45 micron nylon filters, 2-mL tubes, and cell culture dishes (20 × 100 mm).

5. HPLC reagents: HPLC-grade solvents (chloroform, methanol), HPLC solvent filtering membranes (Pall Corporation, cat. no. 6655), HPLC guard cartridge anion exchanger (Whatman, cat. no. 4641–0005); 1.25 M $(NH_4)_2HPO_4$ pH 3.2 (pH adjusted with phosphoric acid, then filtered), HPLC column (Whatman Partisphere SAX WVS 4.6 × 250 mm, cat. no. 4621–1505).

6. Other materials: Agonists (e.g., insulin, PDGF), Sigmacote siliconizing solution (Sigma, cat. no. SL2), deacylation reagent (25% methylamine in H_2O:MeOH: n-butanol; 4:4:1; v:v:v), wash solution (n-butanol:petroleum ether:formic acid ethyl ester, 20:4:1; v:v:v), UltimaFlow scintillation cocktail for high-salt samples (PerkinElmer, cat. no. 6013599); 2.4 M HCl.

7. KRH buffer: 25 mM HEPES, pH 7.4, 125 mM NaCl, 4.8 mM KCl, 2.6 mM $CaCl_2$, 1.2 mM $MgSO_4$, 5.6 mM glucose; can be stored at −20°C in single-use aliquots.

2.2 PI 3-Kinase Assay

1. Standards: 1 mg/mL PI3P and PI4P, dissolved in $CHCl_3$:MeOH:H_2O, 1:2:0.8; v:v:v (Echelon, cat. no. P3016 and P4016).

2. TLC supplies: TLC plates (K5 silica gel 150A, 20×20 cm; 250 μm layer thickness; Whatman, cat. no. 4850–820), Hamilton syringe.

3. Lysis buffer: 1X PBS, 1% Triton X-100, 200 μM Na-orthovanadate, 0.1 mg/mL phenylmethylsulphonylfluoride, 10 μg/mL aprotinin, 10 μg/mL leupeptin, 4 μg/ mL benzamidine, 50 mM NaF, 1 mM β-glycerophosphate.

4. PI 3-kinase buffer: 20 mM Tris-HCl, pH 7.5, 0.1 M NaCl, 0.5 mM EGTA, 0.2 mg/mL L-α-phosphatidylinositol (PI; from a 20 mg/mL stock in DMSO, Sigma cat.#P-0639).

5. ATP-$MgCl_2$ cocktail: 10 mM $MgCl_2$, 10 μM cold ATP, 20 μCi/tube [³²P]-γATP; 0.5 M LiCl/0.1 M Tris-HCl, pH 7.5; 0.1 M NaCl/1 mM EDTA/20 mM Tris-HCl, pH 7.5.

3 Methods

3.1 Cell Culture and Radiotracer Labeling

The protocol described here assumes the preadipocytes (3T3-L1 or human) have been seeded in 100-mm cell culture dishes, and grown to confluence in the appropriate serum-supplemented DMEM with antibiotics.

1. Gently rinse the cells one time with inositol-free DMEM supplemented with 10% calf serum and antibiotics. Keep the cells in inositol-free DMEM supplemented with 10% calf serum and antibiotics for 24 h.
2. Place the cells overnight (16 h) in inositol-free DMEM supplemented with 0.5% calf serum, antibiotics, and 100 μCi/mL of myo-[2-^{3}H]-inositol. To save on consumption of myo-[2-^{3}H]-inositol, the volume of the labeling medium can be reduced overnight to 2 mL/dish (*see* **Notes 1** and **2**).

3.2 Cell Stimulation and Lipid Extraction

1. Following the overnight period, remove labeling medium and replace with 5 mL of KRH buffer at 37°C. Allow 15 min for equilibration, and then add agonist (e.g., insulin or PDGF) at desired concentration.
2. At the appropriate time point (e.g., 0–60 min), aspirate KRH buffer and immediately lyse the cells in 2 mL of ice-cold 2.4 M HCl. Scrape cells thoroughly and transfer the lysate to a chilled large 11.5-mL polypropylene tube (*see* **Note 3**). Add 0.5 mL of ice-cold KRH buffer to each plate, scrape, and pool with the preceding HCl extract.
3. Add 3 mL of ice-cold CHCl$_3$:MeOH, (1:2, v:v) and vortex. There should be a single phase.
4. Add 2 mL of ice-cold CHCl$_3$, add 18 μL of a 8.3 mg/mL stock solution of carrier phosphoinositides (*see* **Note 4**; final concentration 20 μg/mL) and vortex. There should now be 2 phases.
5. Centrifuge at 250 g, 4°C, for 5 min to separate the phases. The lower organic phase must now be removed. Use a siliconized glass pipette (*see* **Note 5**), and aspirate the lower phase (*see* **Note 6**) and transfer to a corresponding new chilled 8-mL polypropylene tube, each prerinsed with CHCl$_3$ for each sample (*see* **Note 7**).
6. Wash the remaining aqueous phase by adding 3 mL of CHCl$_3$, vortex, and centrifuge at 250 g, 4°C, for 5 min. Remove the lower organic phase, as described previously, and pool with the corresponding samples from the first transfer in the same 8-mL polypropylene tube.
7. Completely and carefully dry the pooled organic phase samples under a nitrogen stream (~ 1 to 1.5 h; use a gentle flow to avoid splashing).

3.3 Deacylation of Lipids

This process converts the phosphoinositide lipids into water-soluble glycerophosphoinositol species that can then be resolved based on the number and position of negatively-charged phosphate groups on the inositol ring.

1. Add 3 mL of deacylation reagent (25% methylamine in H$_2$O:MeOH:n-butanol; 4:4:1, v:v:v) to the dried lipids, and cap the tubes. Vortex well initially and incubate at 53°C for 40 min, continuing to vortex every 10 min.

2. After 40 min, immediately uncap tubes and directly dry the reaction mixture in a Speedvac using an organic trap (*see* **Note 8**).
3. Resuspend the dried samples in 1 mL of H_2O. Vortex well. Add 1.2 mL of n-butanol:petroleum ether:formic acid ethyl ester, 20:4:1, v:v:v. Vortex and centrifuge for 5 min at 250 g, room temperature. There should be 2 phases.
4. Aspirate and discard the upper organic layer. Wash aqueous phase by adding 1.2 ml of n-butanol:petroleum ether:formic acid ethyl ester (20:4:1; v:v:v). Vortex well. Centrifuge at 250 g, room temperature, for 5 min. Discard upper organic phase. Dry the samples in the Speedvac fitted with an aqueous trap (*see* **Note 8**).
5. Resuspend the dried samples in 0.5 mL of H_2O. Transfer to Spin-X tube filters, and centrifuge at 16,000 g, room temperature, for 5 min. Rinse the polypro-plylene tubes with another 0.5 mL of H_2O, transfer to Spin-X tube filters, and centrifuge again as explained previously. The pooled 1-mL samples are then ready for HPLC analysis. Samples at this stage may also be maintained for approximately 1 wk at 4°C. If required, samples can be capped and stored for several weeks at −20°C. There are two earlier stop points in the sample prepara-tion: 1) once the $CHCl_3$-extracted lipids have been dried, prior to addition of the deacylation reagent, and 2) after the evaporation step in the Speedvac following the deacylation reaction.

3.4 HPLC Separation

The negatively charged glycerophosphoinositols can be resolved using an anion-exchange HPLC column with ammonium sulfate buffer. The separation column used is a Whatman Partisphere SAX HPLC column. A Whatman anion guard column is placed proximal to the separation column. In this configuration, by changing the guard column approximately every 20 runs, the separation column will often last for 300 runs or more (*see* **Note 9 (6)**).

A gradient flow rate of 1 mL/min of increasing salt concentration is used, achieved by combining H_2O with 1.25 M $(NH_4)_2HPO_4$ pH 3.2 (adjust pH with phosphoric acid; *see* **Note 10**). We use the gradient shown in **Table 24.1**. Eluted fractions are collected, either every 30 (0.5 mL) or 60 s (1 mL), in scin-tillation cocktail vials. The scintillation cocktail added to each vial (4 mL) must be able to dissolve the high salt fractions; we use UltimaFlow for this reason (*see* **Note 11**).

The performance of the column can be assessed by intermittently running a 1-mL sample of 0.1 mM ATP (100 nmol) through the HPLC system, monitoring its elution time with an ultraviolet lamp detector. Samples can also be spiked with 100 nmol ATP to be used as an internal control. There are no commercially availa-ble radiolabeled 3-phosphorylated phosphoinositides, but radiolabeled PI(4,5)P2 and PI4P can be purchased, deacylated, and run as external standards. In addition, radiolabeled inositol phosphate species can be purchased for further assessment of column performance. See **Table 24.2** for expected elution times.

Table 24.1 Gradient protocol

	Time (min)	Composition of gradignt (%)		Duration of gradient (min)
		H₂O	Salt	
Separation	0	100	0	10
	10	80	20	50
	60	55	45	45
	105	0	100	1
Wash	106	0	100	10
	116	100	0	1
	117	100	0	18
	135	0	0	1

Table 24.2 Expected elution times

	Expected elution time (min)
GroPIns4P	34
GroPIns(3,4)P2	55
GroPIns(4,5)P2	60
GroPIns(3,4,5)P3	82
ATP	64
Ins(1,3,4)P3	72
Ins(1,4,5)P3	76
Ins(1,3,4,5)P4	97

3.5 In Vitro *PI 3-Kinase Assay*

To measure the agonist-stimulated association of PI3K activity with a docking protein of interest (e.g., tyrosine phosphorylated proteins, or IRS proteins), incubate cell lysates with an appropriate immunoprecipitating antibody that has been attached to agarose beads, using standard immunoprecipitation protocols. The next step is to test for the presence of PI3K activity in the immunoprecipitate (1.5-mL tubes).

1. Wash immunoprecipitates as follows (500 μL per tube per wash; *see* **Note 12**):

   3X with immunoprecipitation lysis buffer
   1X with PBS
   1X with 0.5 *M* LiCl, 0.1 M Tris-HCl, pH 7.5
   1X with H₂O
   1X with 0.1 *M* NaCl, 1 mM EDTA, 20 m*M* Tris-HCl, pH 7.5

2. Resuspend final pellet in 100 μL of PI 3-kinase buffer (at room temperature) and incubate at room temperature for 10 min (*see* **Note 13**).
3. Prepare ATP-MgCl₂ cocktail.
4. Add required volume of cocktail to each tube and incubate for 3 min.

5. Stop reaction with 300 µl of CHCl₃:MeOH:11.6N HCl, 100:200:2, v:v:v. Vortex. Add 200 µl of CHCl₃, vortex, and spin for 5 min at 16,000 g. Remove and discard aqueous upper phase.
6. Wash organic phase twice with 300 µL of MeOH:1 N HCl; 1:1, v:v. Vortex and spin at 16,000 g for 5 min. Discard aqueous upper phase after each wash.
7. Dry organic phase under a nitrogen stream.

3.6 TLC of Labeled PI3P

1. Run K5 silica plates in 200 mL of CHCl₃:MeOH; 1:1, v:v. Air dry. Dip in 1% potassium oxalate/2 mM EDTA in EtOH:H₂O; 1:1, v:v. Air dry and keep in a dry cool place.
2. Bake TLC plate at 110°C for 45–60 min just before use; allow to cool down for 5–10 min before spotting samples.
3. With a soft pencil, draw a line one inch from the bottom of the plate (loading line). Score plate with a sharp metal edge to create 1- to 2-inch lanes.
4. Spot 10 µL of PI3P and PI4P standard (from Echelon), in outside lane (*see* **Note 14**).
5. Resuspend dried samples in 30 µL of CHCl₃ and then apply to the TLC plate with a Hamilton syringe. Rinse sample tube with 30 µL of CHCl₃ and re-apply.
6. Run plate in developing chamber in MeOH:CHCl₃:NH₄OH:H₂O; 100:70:15:25, v:v:v:v.
7. When solvent has reached the top of the TLC plate, remove plates, air dry in fume hood, and then stain them with iodine in a glass chamber to localize lipid standards (stain yellow).
8. Wrap plates in plastic wrap, and expose to radiographic film for about 1 h, or use a phosphoimager to detect bands.

4 Notes

1. Take appropriate precautions for the use of radioactivity. When using organic solvents (e.g., CHCl₃, methylamine) during the lipid extraction and deacylation, a fume hood should be used.
2. Placing the cells in inositol-free DMEM with 10% calf serum for 24 h reduces the inositol concentration to ~4 µM inositol in the medium, since calf serum has ~40 µM inositol. This serves to maximize the effectiveness of subsequent overnight labeling of the cellular phosphoinositides with the myo-[2-³H]-inositol, during which the cells are in a medium with a lower concentration of 0.5% calf serum, further reducing the inositol concentration to ~0.2 µM.

3. It is very important to use polypropylene tubes following the lysis of the cells, as this material is resilient to the solvents used for the lipid extraction. The use of other plastics tubes will create a gooey and radioactive mess!

4. Addition of the carrier phosphoinositides encourages the movement of the relatively polar phosphoinositides from the aqueous phase into the organic phase.

5. To siliconize the pipettes, pipet Sigmacote up and down several times. Allow empty coated pipets to dry overnight in a beaker. Make sure to perform these steps in a ventilated chemical hood because the siliconizing solution is hazardous. Once dried, rinse pipets with H_2O to wash away byproducts. Allow pipets to dry again.

6. Proper removal of the lower organic phase is a crucial step to the success of the phosphoinositide extraction. At this point, the tube will have an upper aqueous phase, a semi-solid whitish "disk" of proteins at the interface, and the lower organic phase. To accomplish this, a siliconized glass pipet is best used. It must be siliconized (see **Note 5**); otherwise, the charged phosphoinositides may stick to the glass. The pipet must travel through the aqueous supranatant without taking up any of this phase, must bypass and not disturb the "disk," and then should aspirate the infranatant (leaving behind a few microliters is acceptable, because there will be an upcoming wash of the aqueous phase and a second transfer of the resulting organic phase at that time). Furthermore, the pipette itself must be primed, otherwise the aspirated organic phase will not be held securely during the transfer, and will drip out. To prime the pipet, aspirate and eject ~1 mL of $CHCl_3$ several times before each transfer of the lower phase for each sample (separate pipets should be used for each sample).

7. Prepare tubes pre-rinsed with $CHCl_3$ as follows: fill first tube with $CHCl_3$, transfer $CHCl_3$ to next tube, and so on, continuing to transfer to other tubes until all required tubes have been rinsed.

8. The time for complete drying varies, but usually occurs within 1 day. The $CHCl_3$ rinse treatment of these tubes reduces the drying time at this step, as well as after the deacylation step.

9. An indication of the column deteriorating will be a gradual compression of the times of elution.

10. Both the ammonium phosphate solution (salt) and the H_2O are filtered through a 0.2-micron filter. Note also that it is critical to avoid any interaction of the salt buffer with methanol, which is often the solvent used for the HPLC instrumentation when not in use. Therefore, completely wash all lines with H_2O before running any salt buffer.

11. If precipitation occurs when scintillation cocktail is added to high salt eluates, this will substantially quench the number of ^3H counts detected, and lead to falsely low results.

12. When aspirating, be careful not to disturb the beads and use a gel loading tip to reduce the chances of inadvertently losing the immunoprecipitated material.

13. The immunoprecipitated PI3K can act on PI, PI4P, or PI(4,5)P2 in vitro, although PI(4,5)P2 is the in vivo target. PI is generally used as a substrate in the in vitro assay as it is the least expensive.
14. Using PI3P and PI4P as standards will differentiate between PI 3-kinase and PI 4-kinase activity acting on the PI substrate.

Acknowledgments This work was supported by a grant from the Canadian Institutes of Health Research (to A. Sorisky).

References

1. Todaro GJ, Green H (1963) Quantitative studies of the growth of mouse embryo cells in culture and their development into established lines. J Cell Biol 17:299–313
2. Green H, Kehinde O (1975) An established preadipose cell line and its differentiation in culture. II. Factors affecting the adipose conversion. Cell 5:19–27
3. Hauner H, Skurk T, Wabitsch M (2001) Cultures of human adipose precursor cells. Methods Mol Biol 155:239–247
4. Heilbronn L, Smith SR, Ravussin E (2004) Failure of fat cell proliferation, mitochondrial function and fat oxidation results in ectopic fat storage, insulin resistance and type II diabetes mellitus. Int J Obes 28(Suppl 4):S12–S21
5. Di Paola G, De Camilli P (2006) Phosphoinositides in cell regulation and membrane dynamics. Nature 443:651–657
6. Carter AN, Downes CP (1993) Signaling by neurotrophic factors: activation of phosphoinositide 3-kinase by nerve growth factor Neuroprotocols 3:107–118
7. Downes CP, Gray A, Watt SA, Lucocq JM (2003) Advances in procedures for the detection and localization of inositol phospholipid signals in cells, tissues, and enzyme assays. Methods Enzymol 366:64–84

Index

Printed in the United States of America